# Understanding Structural Equation Models

The field of structural equation models (SEMs) is rapidly expanding. A researcher who wants to select and apply SEMs to their data faces several challenges: (1) They can often become extremely complex, with many parameters to estimate. Small samples or those with relatively few variables often cannot support this complexity reliably, leading to under-identified models, poor power, or unstable estimates; (2) Researchers must choose an appropriate measurement model, and these choices are not often well understood in advance; (3) No single "correct" SEM exists, although "better" ones do, and the existence of competing plausible alternatives is often overlooked; and (4) Critical examination of model assumptions involving the linearity of parameters and the existence of influential or outlying observations is often overlooked. This book provides an overview of SEMs as a flexible, skeptical, and iterative scientific process.

**Key Features:**

- Emphasis on multiverse analysis, right-sizing statistical models to data, and the generation of plausible skeptical alternatives
- Robust assumption checking (LOESS regression, regression and SEM diagnostics)
- Detailed, visual coverage of a variety of path diagrams, their links to matrix-based specifications, and data exploration using heat-map visualization and tests of dimensionality
- A variety of SEMs including mediational models, psychometrics (e.g., parallel, tau-equivalent, congeneric measurement), growth curve models, exploratory factor analysis, multigroup, categorical, and exploratory structural equation modeling

This text is designed for graduate students, early-career researchers, and advanced undergraduates who wish to move beyond plug-and-play SEMs to a deeper, more philosophical and data-conscious understanding. Its careful balance of theory, worked examples, and emphasis on skepticism will help its audience build confidence in using SEMs flexibly and responsibly for a broad range of social and behavioral science research.

## Chapman & Hall/CRC
## Statistics in the Social and Behavioral Sciences Series

**Series Editors**
*Jeff Gill, Steven Heeringa, Wim J. van der Linden, and Tom Snijders*

**Recently Published Titles**

**Crime Mapping and Spatial Data Analysis using R**
*Juanjo Medina and Reka Solymosi*

**Computational Aspects of Psychometric Methods: With R**
*Patricia Martinková and Adéla Hladká*

**Principles of Psychological Assessment**
With Applied Examples in R
*Isaac T. Petersen*

**Multilevel Modeling Using R, Third Edition**
*W. Holmes Finch, Jocelyn E. Bolin, and Ken Kelley*

**Polling, Prediction, and Testing, Second Edition**
*Ole J. Forsberg*

**Generalized Kernel Equating with Applications in R**
*Marie Wiberg, Jorge Gonzalez, and Alina A. von Davier*

**Applied Survey Data Analysis, Third Edition**
*Brady T. West, Steven G. Heeringa, and Patricia A. Berglund*

**Visualization for Social Data Science**
*Roger Beecham*

**Introduction to Bayesian Data Analysis for Cognitive Science**
*Bruno Nicenboim, Daniel J. Schad, and Shravan Vasishth*

**Understanding Structural Equation Models**
Models of Relationships Between Variables
*Phillip K. Wood*

For more information about this series, please visit: https://www.routledge.com/Chapman--HallCRC-Statistics-in-the-Social-and-Behavioral-Sciences/book-series/CHSTSOBESCI

# Understanding Structural Equation Models

## Models of Relationships Between Variables

Phillip K. Wood

CRC Press
Taylor & Francis Group
Boca Raton London New York

CRC Press is an imprint of the
Taylor & Francis Group, an **informa** business
A CHAPMAN & HALL BOOK

First edition published 2026
by CRC Press
2385 NW Executive Center Drive, Suite 320, Boca Raton FL 33431

and by CRC Press
4 Park Square, Milton Park, Abingdon, Oxon, OX14 4RN

*CRC Press is an imprint of Taylor & Francis Group, LLC*



ISBN: 978-1-032-97756-0 (hbk)
ISBN: 978-1-032-96242-9 (pbk)
ISBN: 978-1-003-59526-7 (ebk)

DOI: 10.1201/9781003595267

Typeset in Palatino
by SPi Technologies India Pvt Ltd (Straive)

# *Contents*

# *Preface*

The sophistication and ability of structural equation modeling (SEM) to model complex behaviors has increased dramatically during my years of using SEM in teaching, scholarship, and consulting. Using SEM can also be challenging and sometimes frustrating. The interpretation of conflicting evidence and problems with the replicability of findings has increased rather than improved with the advent of arguably better SEM models, however. I hope that better use of SEM would result if researchers understood how different types of SEM models were related, if the conceptual advantages of better psychometric models were communicated, and if a more skeptical approach to analytic strategy was used. This includes looking at the data, considering the plausible alternatives that a reasonable skeptic might offer to a favored model, and rigorously assumption checking both the proposed model and the data used to support it.

I am mindful that readers vary in terms of their exposure to statistics when taking up a new topic. The most common fear my students express is that their statistics class will begin at a level they don't follow and which has no connection to what they already know. For this reason, the pace of the introductory chapters of the book may be too slow for some, and for those I encourage skimming the material and focusing more directly on subsequent chapters. For others, some topics like the mechanics of exactly how SEM software arrives at solutions may be daunting, and for those I also advocate skimming, but they should also consider the practical takeaway messages at the ends of the chapters. In coursework, depending on the mix of discussion of varieties of SEM models and the desire to include practical experience with SEM software, choices about the most relevant chapters may have to be made, but I hope that the book will continue to be useful as a reference for continued use of SEM.

**Phillip K. Wood**

# *The Author*

**Phillip K. Wood** is Professor of Psychological Sciences at the University of Missouri–Columbia, where he has taught graduate seminars in quantitative methods, including beginning and advanced structural equation modeling (SEM), for over 30 years. He earned his Ph.D. in educational psychology and measurement from the University of Minnesota, and earlier degrees from the University of Iowa and Wartburg College.

Dr. Wood's research spans advanced latent variable modeling techniques—particularly SEM, latent growth, growth-mixture models, state-trait modeling, longitudinal data analysis, and models for longitudinally intensive data as applied to developmental processes, substance abuse within young adult populations, and lifespan development.

A strong advocate of methodological transparency and reproducibility, Dr. Wood maintains open-access resources, including SAS, Mplus, lavaan, and Onyx code, accessible through his university-hosted repositories. He regularly moderates the Transcontinental Karl Popper Conference, which explores philosophy of science in psychological research, highlighting his commitment to the interplay between methodological rigor and theoretical skepticism.

Combining decades of classroom instruction with cutting-edge research, Dr. Wood brings a practical, data-conscious perspective fueled by a belief that SEM should be inquisitive, skeptical, and disciplined—a perfect guide for readers navigating the complexities of latent variable modeling.

# 1

# *Introduction*

Use of structural equation models (SEMs) in research is increasing exponentially. For example, Google Scholar citations of the phrase "structural equation model" increased from 477 in 1990 to approximately 75,000 in 2020, with an annual median increase of 18%. This is likely an understatement of the increasing influence of SEM, though, as it does not include use of specific types of SEMs such as growth curves, factor analysis, or the rapidly increasing number of acronyms for other particular SEMs. Rapid development of new structural models as well as advances in our ability to easily collect large amounts of data make use of more sophisticated models feasible. This book gives guidance to the reader in light of the "quiet methodological revolution" underway in which many SEMs can be entertained, The "best" SEM model for the data at hand involves a balance model parsimony and complexity (Rodgers, 2010).

## Using SEMs is a Process

In my experiences using and assisting other researchers with SEMs, I have come to realize that using SEMs, or any statistical model for that matter, is a process of inquiry. To be sure, lifelong literacy in the SEM requires an understanding underlying mathematical models and theory (as covered, for example, by Bollen, 1989; Kaplan, 2008; Mulaik, 2009a, 2009b). A given degree of computer programming is also necessary to execute SEMs (as covered by Byrne, 2012; Geiser, 2023; Loehlin & Beaujean, 2017). To be honest, however, the final structural model published in a research article was almost never the initially conceived model, nor was it derived from first principles. Recent articles have outlined remedies for the most common pitfalls involved in conceptualizing SEMs, arriving at a well-fitting model, trouble-shooting software, and interpreting the results (e.g., Newsom, Hachem, Granger, & Denning, 2023). This chapter offers an expanded description of this process, with specific ties to the chapters that follow.

The decisions involved in choosing an SEM often began with a research design that considers the context of the questions involved and (hopefully) conflicting research perspectives. A general "proto-SEM" may guide initial

DOI: 10.1201/9781003595267-1

specification of a model. After data collection, the researchers should carefully examine the data at hand with a skeptical eye as to whether that SEM is appropriate. This skeptical evaluation of the SEM may involve questions such as: Do the SEM's measurement assumptions appear to have been met? Is it reasonable to believe in the effects of interest or overall fit of the model, or could these results be an artifact of a small number of influential observations or some misspecification of relationships between variables? Unanticipated model modifications may be necessary for technical reasons described later in this book, but the final published model may exclude some sets of variables due to lack of effect or poor operationalization. New groups of variables may be included if they appear to constitute third variable explanations for initial effects found. The final model is also not the only model considered by the researchers. To the extent possible, several side analyses have been conducted to assess the robustness of important effects. Optimally, some of these alternative models represent counterarguments that a reasonable skeptic could advance in opposition to the favored model.

Use of several SEMs, rather than a single initial model for the data, may raise concerns in some about "hypothesizing after results are known" (HARKing), selective reporting of results, or changing the hypothesis after the fact ("p-hacking"). So long as the process itself is openly pre-registered and additional analyses are in the service of refuting, not favoring, a model, the effect of this process should be more conservative statements of effects. Possible limitations of a proposed model can be often concretely evaluated in interpretation of findings. This wider exploration mirrors concerns about whether pre-registration of a single analysis promotes best practices (e.g., Morey, 2019), and the possibility of a "multiverse" analysis of families of statistical analyses has been explored as a better approach (Steegen, Tuerlinckx, Gelman, & Vanpaemel, 2016). The process of inquiry presented in this book aims to outline how a researcher might navigate such a multiverse in a principled fashion.

## A Hitchhiker's Guide to the Statistical Multiverse

This hidden universe of alternative models has been noted by articles observing that different researchers in a given area often deploy a wide variety of models (e.g., Bastiaansen et al., 2020; Breznau et al., 2022). Candidate SEMs discussed in the following chapters are legion, but they are not all equally informative, nor are many appropriate for a given research situation. Given that some researchers believe that one must commit to a particular model as "true," it is necessary to briefly discuss what we mean, philosophically, when we advance a given model for the data.

## "Right-Sizing" the Model

Any statistical model that we use is, technically speaking, "wrong" in that it is a systematic oversimplification of the data at hand (Popper, 2002). The researcher's task, therefore, is to select an appropriately oversimplified model for the data at hand. More complicated statistical models are appropriate when data consist of larger numbers of observations, more reliable assessments, or more variables/assessments.

Simpler models are more appropriate when fewer observations, less internally consistent assessments, and fewer variables are under consideration. Application of complicated models to more limited data may not only result in failures of estimation or improper solutions (Savalei & Kolenikov, 2008) but also have poorer statistical power to test effects of interest (see, for example the comparison of the merits of repeated measures versus MANOVA discussed by Maxwell, Delaney, & Kelley, 2017, pp. 773–777). I and others have stressed the importance of this aspect of statistical modeling given the plethora of available models and the increasing complexity of data ascertainment (Rodgers, 2010; Wood, 2023; Wood, Steinley, & Jackson, 2015; Wood, Wiedermann, & Wood, 2024).

## Reality, Models, and Theory

### Inductivism

Karl Popper used the termed "inductivist" to describe the view that theories arise from elementary observations and are expanded by accretion of new observations to produce "true" theories. More directly relevant to the discussion of multivariate structural models, Cattell (1988, p. 10) criticized the scientifically precise classical bivariate "brass instrument" approach to developing psychological models and stressed the need for multivariate probabilistic evaluations of theories that undergo extension, refinement, and revolution in the course of scientific inquiry. Theories do not arise like Venus from the oceans of volumes of data; instead they are the creative conjecture of a researcher who explores an initial conjecture in service of a possible solution to a problem (Popper, 2002). Modern technology allows us to gather, store, and analyze large amounts of data, which can lead to rather elaborate statistical models. Such an ornate theory is no guarantee of "truth," however. If having a subtle yet sophisticated theory based on centuries of observational data were a guarantor of scientific quality, astrologers would receive Nobel Prizes. Theories, and the models based on those theories, are useful to the extent that they generate new testable hypotheses.

## Models Come Not From Data, But Rather From Our Responses to Problems

### *Genetic Epistemology*

Theories and their accompanying statistical models are not descriptions of "nature as it is." The researcher chooses to observe some things and not others because of their prior beliefs about the problem at hand and what is relevant. Evaluation of theories involves an evolutionary epistemology – a survival of the fittest of ideas, if you will – in which new theories are constantly creatively generated, extended, refined, elaborated, or discarded. They do this in order to compete for survival in the face of critical tests and the competing explanations of other theories (Campbell, 1974).

From this perspective, a good model cannot confirm a particular theory but is advanced as the "least worst" explanation in the face of the counter-explanations offered by reasonable skeptics. Although a study involving a single SEM can provide some external validity by instantiating parameter estimates in a single sample, a critical examination of a model involves checking whether the model's assumptions are realistic and, if possible, is a better alternative relative to competing models.

### *Some Models Are Better Than Others*

All this is not to say that any SEM is just as good as any other. In addition to evaluating whether a given model is appropriate to the data at hand, it is important to understand that some SEMs have conceptual advantages over others. As discussed in the following chapters, some SEMs deal with the problem of measurement error in predictor variables or the fact that, in some situations, the researcher is interested in knowing a shared effect of several predictors on a criterion rather than estimating whether a unique effect exists for each individual predictor variable. Some SEMs, for conceptual reasons, are used to estimate patterns of individual-level growth and decline rather than simple overall changes in mean level. Multigroup SEMs allow the researcher to examine the external validity of an SEM across known or unknown subgroups. Yet other models allow the researcher to test for the effects of method of assessment, or how subgroups of questions could be thought of as sharing a common shared cause. A researcher, then, can choose a model for any of these reasons and decline others. Even so, interpretation of the final model reported must acknowledge that other models may provide a better explanation but cannot be explored given the limitations of the data at hand. These limitations can include not having a large or diverse enough sample, enough variables representing a construct, or use of only a single method of assessment, for example.

## Ways to Think About SEMs

The broad goal of justifying a model and defending it against competing alternatives may seem intimidating given, at this point, that one is interested in learning how to model a single SEM. This book presents several individual SEMs as a way to identify central issues involved in choosing and evaluating SEMs. The examples are often covered in translational articles for a given research area, but they are also a jumping-off point for the more sophisticated models that will be proposed in coming years. The ways that SEMs have been represented provide a framework for highlighting some of the central issues involved in choosing and evaluating an SEM.

There is no single "orthodox" way to think about structural models. Rather, five ways of thinking about SEMs are: scatterplots, path diagrams, single equations, matrix algebra, and vector diagrams. Each of these can be informative in the development and critical evaluation of an SEM, and for that reason a chapter is presented on each of these topics.

### Scatterplots

Students often prefer a simple bivariate scatterplot for summarizing relationships. Such plots are a useful tool for determining the functional form between variables (e.g., linear or quadratic), assessing whether ceiling or floor effects are present in the data and, for smaller sample studies, whether influential or outlying observations may affect conclusions about the fit or estimated parameters of an SEM.

### Equations

Writing a single equation as a part of a model (e.g., $X = \lambda_1 F + e$) is a way to refer to individual parts of an SEM. Programs such as lavaan and Mplus require the programmer to specify the model using such equations.

### Matrix Algebra

Matrix algebra is a compact way of summarizing several individual equations and expresses the general form of a model. Many discussions involving SEMs express their models in matrix form, and in some situations, such as behavior genetic or family-level data, it is often more convenient to express the SEM in matrix form. Programs such as OpenMx or Proc Calis in SAS allow the programmer to specify the model in matrix form.

### Path Diagrams

Path diagrams, perhaps the most ubiquitous figures used to report SEMs, are useful because they summarize relationships between latent variables and, with a bit of practice, allow the researcher to understand possible third-variable explanations in the data as well as unique or indirect effects.

### *Vector Diagrams*

Vector diagrams are a convenient way to graphically show the strength of associations between variables as well as their variation. Vector diagrams can be a useful way to re-express relationships in SEMs such as exploratory factor models.

## Points to Consider in Evaluation/Choosing SEMs

The process of inquiry is not a series of steps analogous to baking a loaf of bread, and the navigation between the ways of thinking about SEMs described in the previous section vary as a function of the research context and the data at hand. Critical examination of SEMs often involves the following major themes.

### Measurement and Psychometric Assumptions

Researchers are generally familiar with the fact that choice of an appropriate statistical model involves the assumption of whether the variables under consideration represent ratio, interval, or ordered categorical assessments. This applies to SEM models as well. Parsimonious models for smaller data sets often assume the interval or ratio measurement, while larger data sets may assume ordered categorical measurement, for example. The researcher must also decide whether the primary research interest concerns patterns of covariation in the data (a primary concern in confirmatory factor models), whether the SEM's purpose is to examine mean levels (as is done in SEM analogues of linear models such as analysis of variance), or whether both patterns of covariation and mean level should be considered (as discussed later in this section for models of growth or decline).

SEMs also differ from more familiar models in their ability to directly estimate measurement error within the SEM rather than relying on ancillary analysis such as an item analysis. As will be discussed, the presence of measurement error in predictor variables causes interpretational difficulties in regression, ANOVA, or other general linear models and, in some situations, latent variable models are an attractive alternative. Measurement assumptions play a critical role in the specification and assessment of models of data (Borsboom, 2006).

### Assumption Checking

As mentioned earlier, fitting an SEM and reporting its overall fit and patterns of statistical significance is a risky enterprise because we do not know whether it is reasonable to believe that the model's assumptions have been met and whether it would seem reasonable to believe that a replication of the study would yield similar results. Several diagnostics for structural models have been proposed which allow the researcher to assess whether the fit of the model and its estimated parameters are affected by influential or outlying observations. Graphical techniques such as heat maps and LOESS regressions can help the researcher identify the shape of functional relationships in the data and patterns of multivariate covariation.

### Growth Versus Change in Longitudinal Data

When the data under consideration are longitudinal, the researcher must also decide whether differential growth or decline is present, or whether only mean differences exist across time (i.e., change). In models in which growth or decline is present, elevation in scores is accompanied by a systematic change in variation in scores. In models which document only change, however, the primary interest is only in mean differences across measurement occasions. Models in which growth/decline are thought present are referred to an integration of "variable centered" models, which examine only relationships between variables, with "person-centered" approaches, which seek to identify developmental trajectories (Muthén & Muthén, 2000). SEMs vary in the form chosen for growth/decline across measurement occasions or whether the form is specific to given measurement occasions (Wood et al., 2015).

### Group Differences in SEMs

Questions of external validity speak to the generalizability of a model to different groups. Different patterns of covariation, change, and growth/decline may vary as a function of known variables (such as biological sex or ethnicity) or, under finite mixture models, as a function of groups not known to the researcher beforehand (Wood et al., 2023; Wood, Wiedermann, & Steinley, 2025).

## Outline

The book first discusses three ways to represent data for an SEM and introduces some useful basic matrix algebra. The next two chapters introduce path notation. Common statistical models are introduced as examples. In addition to providing practice in using the tracing rules for path diagrams, these examples serve as examples of basic types of relationships that SEMs estimate and how more common statistical techniques such as regression can be modeled as SEMs. After this, the problem of measurement error in predictor variables is introduced, as is the use of confirmatory factor analysis (CFA) as a remedy. After discussing the model fit of CFA models, various ways to check the assumptions of the model are discussed.

# 2

## *Data Representation and Basic Matrix Algebra*

As noted in the last chapter, researchers' models are simplifications of what they research. All data analysis is an act of journalism. Data are re-expressed to bring forward some aspects of the data for ease of analysis and discussion, but doing so also involves hiding other aspects of the data. The emphasized characteristics of the data determine the kind of questions we can answer. A given re-expression of our data means that some questions cannot be addressed which can be addressed under other re-expressions. In this chapter we consider a simple data set to illustrate these re-expressions for different types of SEMs. Using a small data set also provides the context for the first two ways of modeling data: the simple (or scalar) algebraic equation, and linear algebra. Matrix algebra allows us to see that the same formulae are involved across the transformations discussed here.

### Prereader

When analyzing data, we often think about structural models based on three transformations of the data. The data under this are transformed to standardized form (in which mean level and differences in variability are removed), deviation form (in which only the mean level of the variables is removed), and the adjoined transformations in which a single variable with value one is added to the data.

The *standardized* (or z-score) matrix is the basis for such statistics as the correlation coefficient, standardized regression weights, and standardized factor loadings. This matrix is used to consider variables in a common metric and can sometimes be used to express measures of strength of association or effect sizes in the model.

The *deviation* score matrix, in which the raw data are transformed so that variable means are zero, is useful when the focus is on estimating differences in variability across variables and characterizing the functional relationship between predictor variables and a criterion. Whereas the standardized score matrix answers questions of association involving an increase in standardized units and can help the researcher determine the relative effect size of

DOI: 10.1201/9781003595267-2

different predictor variables, the deviation score matrix answers in the metric of the variables and can answer such questions as "If we increase the dosage of a drug by one gram, how many points decrease in cholesterol may we expect?" – a functional relationship between variables, which may be linear, polynomial, or some other nonlinear form. Note, though, that neither the standardized nor deviation score metrics can be used to generate models that explain level of performance (e.g., we cannot easily forecast a specific cholesterol level for an individual, for example).

Such questions can be addressed with the *adjoined* score matrix, however. The adjoined matrix consists of the raw data with an additional column of ones. Models based on the adjoined matrix deal with both level or elevation information as well as patterns of variation and covariation between variables. The adjoined matrix is useful when testing models of growth over time, actuarial prediction of a criterion, and in models involving missing data.

Finally, working with these data sets and calculating summary statistics gives the reader some practice in working with the data using matrix operations. Many software packages use this approach to specifying the model or report the analyses in matrix form, so working with elementary options on matrices provides a groundwork for later work.

## Summary Statistics

Using a simple raw data set also shows how common statistics can be generated. Summary statistics can often be used to fit a structural model from published data. The *Publication Manual of the American Psychological Association* (American Psychological Association, 2001, pp. 23–24) refers to these summary statistics as "informationally adequate statistics," and statisticians often refer to the "sufficient statistics" necessary to replicate a distribution without actually manipulating the raw data, but the ideas behind these terms share many similarities. For structural equation models, Raykov, Tomer, and Nesselroade (1991) and, more recently, McDonald and Ho (2002) contain a discussion of the summary statistics necessary to replicate a variety of structural model analyses.

## Multiple Regression as SEM

This chapter presents material you have already encountered in multiple regression or analysis of variance. My objective here is to highlight terms used using already familiar material. I also introduce a few terms from matrix algebra, which, again, may be already part of some students' vocabulary; but doing so ensures a common vocabulary for a variety of readers. Fortunately, a relatively small number of ideas in matrix algebra are sufficient for expressing SEMs. Knowing how to specify SEMs in their matrix form has a lifelong literacy function as new models are introduced in this form. For those interested in the computer programming aspect of SEMs, many SEM programs such as OpenMx and SAS's Proc Calis allow the user to specify the model in matrix form. This is useful for some classes of SEMs such as those dealing with family data, behavior genetics, or neuroscience applications. Thus, although the chapters that follow will use path diagrams or equation forms to introduce models, the relevance of matrix algebra will not go away any time soon.

## A Small Data Set Example

### "Let's Do Lunch"

As an example, consider my oldest child, Jules, who takes their lunch to school every day for a week.[1] This lunch has three parts: something to drink, a sandwich, and a desert. Each day they assign a rating of how much they like each part of the lunch, using a five-point scale from "1" = Strongly Dislike to "10" = Really Like. Data from this small study are shown under "Raw Data" in Table 2.1.

If we want to talk about the data in matrix form, we might write it out as:

$$X = \begin{bmatrix} 1 & 4 & 4 \\ 2 & 7 & 2 \\ 3 & 5 & 6 \\ 4 & 8 & 10 \\ 5 & 6 & 8 \end{bmatrix}.$$

Notice that the labels for the rows and columns aren't present in the equation. You need to remember what the columns and rows of your data mean as you specify matrices.

Suppose for a moment you wanted to know something about whether a given day received good or bad ratings on the variables. It's hard to do this, however, because you have to mentally look across each column of the data. You also don't know anything about how variable scores on the three variables are, nor can you inspect the rows for an idea of what patterns co-occur

**TABLE 2.1**

Food Preferences for One Week's Lunches

| | | Adjoined Matrix | | | Deviation Matrix | | | Standardized Matrix | | |
|---|---|---|---|---|---|---|---|---|---|---|
| | | Raw Data | | | | | | | | |
| Day of Week | | Sandwich | Beverage | Dessert | Sandwich | Beverage | Dessert | Sandwich | Beverage | Dessert |
| Monday | 1 | 1 | 4 | 4 | –2 | –2 | –2 | $-\sqrt{2}$ | $-\sqrt{2}$ | $\frac{-2}{\sqrt{8}}$ |
| Tuesday | 1 | 2 | 7 | 2 | –1 | 1 | –4 | $\frac{-1}{\sqrt{2}}$ | $\frac{1}{\sqrt{2}}$ | $\frac{-4}{\sqrt{8}}$ |
| Wednesday | 1 | 3 | 5 | 6 | 0 | –1 | 0 | 0 | $\frac{-1}{\sqrt{2}}$ | 0 |
| Thursday | 1 | 4 | 8 | 10 | 1 | 2 | 4 | $\frac{1}{\sqrt{2}}$ | $\sqrt{2}$ | $\frac{4}{\sqrt{8}}$ |
| Friday | 1 | 5 | 6 | 8 | 2 | 0 | 2 | $\sqrt{2}$ | 0 | $\frac{2}{\sqrt{8}}$ |
| Mean | | 3 | 6 | 6 | 0 | 0 | 0 | 0 | 0 | 0 |
| Standard Deviation (Sample) | | $\sqrt{2}$ | $\sqrt{2}$ | $\sqrt{8}$ | $\sqrt{2}$ | $\sqrt{2}$ | $\sqrt{8}$ | 1 | 1 | 1 |

across the variables. Another problem with analyzing the raw data is that it is inconvenient to use matrix algebra to calculate desired statistics such as standard deviations or correlations from the raw data. One can, though, calculate the means of the three variables by adding the rows together and dividing by the sample size or, in matrix form: $m = \begin{bmatrix} 3 & 6 & 6 \end{bmatrix}$.

## Matrix Algebra Terminology

Let's build a bit of vocabulary: A *matrix* (such as $X$ or $m$) is any arrangement of numbers (or expressions that represent numbers) organized in a square or rectangular fashion. Within a particular matrix, we might make mention of a particular vertical *column* (as when talking about one variable in a data set) or horizontal *row* (as when talking about the data from one individual in our study). A matrix can consist of any number of rows and columns, but we have special names for three types of matrices: A *scalar* is any matrix with only one row and one column (i.e., represents a single number). A *row vector*, such as $m$, is a matrix with only one (horizontal) row and more than one column. A *column vector* is a matrix with only one (vertical) column and more than one row. Often we use uppercase letters to refer to matrices with more than one column and row, while lowercase letters refer to scalars and vectors. The uppercase letter $X$ is often used to refer to the raw data of a study.

Sometimes we also might want to talk about a particular number in the matrix. We call this a cell (or entry) of the matrix. When we want to refer to it, we will refer to it by its letter and then two subscripts to indicate the row and column, respectively. For example, in the data set above, $R$ is a matrix of correlations:

$$R = \begin{bmatrix} 1 & 0.5 & 0.8 \\ 0.5 & 1 & 0.4 \\ 0.8 & 0.4 & 1 \end{bmatrix}; R_{2,3} = 0.4.$$

Notice that $R$ is a *symmetric* matrix, meaning that every number above the *diagonal* (going from upper left to lower right) is the same as its corresponding value below the diagonal.

## Getting to the Deviation Score Matrix

In order to arrive at a deviation score matrix, we'll need to subtract the mean of each variable from each of the observations from the raw data. To do this, we're going to subtract the means from the raw data using matrix algebra. Just as in "regular" algebra, matrices can be added and subtracted, but there is one requirement: the matrices have to have the same number of rows and columns, so we're not going to be able to subtract vector $m$ from our matrix $X$ in order for us to subtract the corresponding elements of the matrices. We'll need to expand our vector matrix into a large enough matrix $M$.

### Matrix Multiplication

In order for us to do matrix multiplication, the matrices need to have the proper size. The number of rows and columns in a matrix is called its *order*. The raw score matrix $X$ has order 5 × 3 and the vector $m$ has order $1\times3$, for example. We often use subscripts to show the order of a matrix, such as $X_{5\times3}$ and $m_{1\times3}$. In order for us to multiply two matrices together, the number of columns of the first matrix needs to equal the number of rows of the second matrix. The resulting matrix will have as many rows as the first matrix and as many columns as the second matrix. Notice that when multiplying matrices, it is important to specify the order of multiplication. Except in special circumstances, AB is not equal to BA.

Each entry in the final matrix is calculated by multiplying the numbers in the $r$th row of the first matrix with the numbers in the $c$th column of the second matrix and adding up these products. For example, in order for us to make a matrix of column means the same size as $X$, we will need to multiply $m$ by a column vector of ones. (The element in the first row, first column of $M\left(M_{1,1}\right)$ is 3. $M_{1,2}$ is 6, $M_{1,3}=6$, $M_{2,1}=3$, etc.) That is:

$$M=\begin{bmatrix}1\\1\\1\\1\\1\end{bmatrix}\begin{bmatrix}3 & 6 & 6\end{bmatrix}=\begin{bmatrix}3 & 6 & 6\\3 & 6 & 6\\3 & 6 & 6\\3 & 6 & 6\\3 & 6 & 6\end{bmatrix} \tag{2.1}$$

From this we can then calculate the deviation score matrix shown in the middle of Table 2.1 by subtraction:

$$x=\begin{bmatrix}1 & 4 & 4\\2 & 7 & 2\\3 & 5 & 6\\4 & 8 & 10\\5 & 6 & 7\end{bmatrix}-\begin{bmatrix}3 & 6 & 6\\3 & 6 & 6\\3 & 6 & 6\\3 & 6 & 6\\3 & 6 & 6\end{bmatrix}=\begin{bmatrix}1-3 & 4-6 & 4-6\\2-3 & 7-6 & 2-6\\3-3 & 5-6 & 6-6\\4-3 & 8-6 & 10-6\\5-3 & 6-6 & 7-6\end{bmatrix}=\begin{bmatrix}-2 & -2 & -2\\-1 & 1 & -4\\0 & -1 & 0\\1 & 2 & 4\\-2 & 0 & 2\end{bmatrix} \tag{2.2}$$

Often the lowercase $x$ is used to refer to the data in deviation form despite the comment earlier that lowercase is usually used for vectors.

Examination of any particular observation (row) in the data now tells you which variables scored above or below the average in the data. You can also get a rough idea of variability given that, for example, scores for Dessert seem more variable than for the other two foods. A rough idea of positive and negative correlations in the data can be made by observing the number of times that negative and positive values agree across variables across occasions.

## Calculating Variances and Covariances

The deviation score matrix is convenient for statistical questions in which questions of variability and covariation are of interest. It is also easy to calculate variances and covariances with it, which is a useful summary matrix. To calculate a variance/covariance matrix we'll need to introduce another simple idea, that of the matrix *transpose*. The transpose of a matrix can be formed by flipping the rows and columns of a matrix. In our example it looks like this:

$$x'x = \begin{bmatrix} -2 & -1 & 0 & 1 & -2 \\ -2 & 1 & -1 & 2 & 0 \\ -2 & -4 & 0 & 4 & 2 \end{bmatrix} \begin{bmatrix} -2 & -2 & -2 \\ -1 & 1 & -4 \\ 0 & -1 & 0 \\ 1 & 2 & 4 \\ -2 & 0 & 2 \end{bmatrix} = \begin{bmatrix} 10 & 5 & 16 \\ 5 & 10 & 8 \\ 16 & 8 & 40 \end{bmatrix} \tag{2.3}$$

You may have encountered $x'x$ in your prior coursework where it is called the *uncorrected sums of squares* (USS). To get the sample variance/covariance matrix $S$, we divide each of the elements of the USS by the sample size. We can do this by multiplying $x'x$ by a diagonal matrix with $1/N$'s on the diagonal like this:

$$S = \begin{bmatrix} \frac{1}{5} & 0 & 0 \\ 0 & \frac{1}{5} & 0 \\ 0 & 0 & \frac{1}{5} \end{bmatrix} \begin{bmatrix} 10 & 5 & 16 \\ 5 & 10 & 8 \\ 16 & 8 & 40 \end{bmatrix} = \begin{bmatrix} 2 & 1 & \frac{16}{5} \\ 1 & 2 & \frac{8}{5} \\ \frac{16}{5} & \frac{8}{5} & 8 \end{bmatrix} \tag{2.4}$$

In matrix form, we might write this as: $S = N^{-1}x'x$. Variances of the variables can be found on the *diagonal* of $S$, going from the upper left-hand corner to the lower right-hand corner. In our example, the variances of the sandwich and drink ratings were both equal to 2, while the variance of the dessert variable was 8. The matrix $S$ looks like a correlation in that it is symmetric, meaning that it is a mirror image along the diagonal. These numbers represent covariances rather than correlations, however. You can tell the correlation between variables is positive or negative by looking at the sign of the covariance.

One example analysis you may be familiar with is the calculation of multiple regression weights. This will allow us to introduce one final matrix idea, that of matrix division or the inverse. In order to do this, we first need to separate the criterion variable, $y$, from the predictor variables in our study. This merely means that we are taking the original data matrix of our study and are breaking it down into two other matrices. In the case of a regression, the matrix containing the criterion information, $y$, is a column vector with as many rows as there are observations in the data. The matrix of predictor variables, usually again denoted as $X$, is a rectangular matrix having as many columns as there are predictors and as many rows as there are observations.

### Calculating and Using Unstandardized Regression Weights

As we will see in subsequent chapters, regression models are different from many SEMs in that a formula can be used to arrive at the regression weights. Regression models are said to have a "closed form" for this reason, to distinguish them from many structural models where an iterative search procedure has to be used to recover SEM parameters. In order to write a formula which will produce a vector of all the regression weights for a regression model, we calculate: $b = (x'x)^{-1} x'y$.

### Matrix Inverses and Identity Matrices

Notice that this formula has a superscript$^{-1}$ in it. This is used to indicate division in matrix algebra and is called the *inverse*. The concept of the matrix inverse is the matrix algebra analogue to division, but this requires a bit of explanation. To explain, just as we can mathematically rewrite the fraction $\frac{1}{8}$ as $8^{-1}$, the inverse of any matrix $A$ is denoted as $A^{-1}$. This notation is needed rather than using fractions because, as we noted earlier, matrix multiplication is not commutative. Having a denominator in a matrix expression does not inform the order in which to do the math. Only square matrices have inverses, but not all square matrices have inverses. Multiplying any matrix $X$ by its inverse $X^{-1}$ results in an identity matrix. An identity matrix is the analogue to the number 1 in regular scalar algebra and consists of a diagonal matrix of ones with zeros elsewhere. The identity matrix has the same number of rows and columns as $X$. To say this in matrix algebra form, we can write $AA^{-1} = A^{-1}A = \mathrm{I}$.

Let's use this formula to calculate the regression weights for the example. The column vector of the dessert data is:

$$y = \begin{bmatrix} -2 \\ -4 \\ 0 \\ 4 \\ 2 \end{bmatrix}.$$

and the predictor variables of the sandwich and beverage ratings are in another matrix:

$$x = \begin{bmatrix} -2 & -2 \\ -1 & 1 \\ 0 & -1 \\ 1 & 2 \\ 2 & 0 \end{bmatrix}.$$

Calculating regression weights is given as:

$$b = (x'x)^{-1} x'y = \begin{bmatrix} 10 & 5 \\ 5 & 10 \end{bmatrix}^{-1} \begin{bmatrix} -2 & -1 & 0 & 1 & 2 \\ -2 & 1 & -1 & 2 & 0 \end{bmatrix} \begin{bmatrix} -2 \\ -4 \\ 0 \\ 4 \\ 2 \end{bmatrix}.$$

Now, the inverse of $x'x$ is:

$$(x'x)^{-1} = \begin{bmatrix} \frac{10}{75} & \frac{-5}{75} \\ \frac{-5}{75} & \frac{10}{75} \end{bmatrix}.$$

$b$ can then be calculated as:

$$b = \begin{bmatrix} \frac{10}{75} & \frac{-5}{75} \\ \frac{-5}{75} & \frac{10}{75} \end{bmatrix} \begin{bmatrix} -2 & -1 & 0 & 1 & 2 \\ -2 & 1 & -1 & 2 & 0 \end{bmatrix} \begin{bmatrix} -2 \\ -4 \\ 0 \\ 4 \\ 2 \end{bmatrix}$$

$$= \begin{bmatrix} \frac{-2}{15} & \frac{-1}{5} & \frac{1}{15} & 0 & \frac{4}{15} \\ \frac{-2}{15} & \frac{1}{5} & \frac{-2}{15} & \frac{1}{5} & \frac{-2}{15} \end{bmatrix} \begin{bmatrix} -2 \\ -4 \\ 0 \\ 4 \\ 2 \end{bmatrix} = \begin{bmatrix} \frac{8}{5} \\ 0 \end{bmatrix} = \begin{bmatrix} 1.6 \\ 0 \end{bmatrix}. \quad (2.5)$$

The vector $b$ has as many rows as there are predictor variables in the study and is just a convenient way to write the two regression weights, $\begin{bmatrix} 1.6 \\ 0 \end{bmatrix} = \begin{bmatrix} b_{\text{Sandwich}} \\ b_{\text{Beverage}} \end{bmatrix}$. The first row corresponds to the first predictor (i.e., that predictor which was in the first column of the $x$), the second row corresponds to the second predictor and so on if we had more than two predictors. For this example, then, the unstandardized regression weight predicting dessert from the rating assigned to Sandwich is 1.6, while the unstandardized regression weight predicting Dessert from the Drink is zero, meaning that the observed covariance between the Drink and Dessert variables can be accounted for by the fact that both of these variables are related to the Sandwich variable. In conclusion, Juls can look at the Sandwich on a particular day and use that to predict how good the Dessert will be.

The unstandardized regression weight is interpreted as the amount of gain in the criterion which would result if we were to observe a one-unit change in the predictor variable and held all other predictor variables constant. Although some regression texts describe this "holding constant" process as constraining all other predictors to the mean, this is unnecessarily restrictive. In keeping with the interpretation of a regression model as a statement about statistics that we might expect in a subsequent study, all that is required is that we agree to observe only individuals who have some constant value on the other predictors.

### Unstandardized Regression Weights via Covariance Matrices

Having the raw data is not completely necessary. It is possible to calculate unstandardized regression weights using parts of the variance/covariance matrix. You might want to do this, for example, when you are re-analyzing the data of another study and do not have the raw data but only summary statistics. To do this, we take the original variance/covariance matrix and partition it into the variance/covariance matrix of the predictors (call it $C_x$) and the covariances of the predictors with the criterion (call it $C_{xy}$). The unstandardized regression weights $\beta$ are then calculated (using our data example) as:

$$b = C_x^{-1}C_{xy} = \begin{bmatrix} 2 & 1 \\ 1 & 2 \end{bmatrix}^{-1} \begin{bmatrix} 3.2 \\ 1.6 \end{bmatrix} = \begin{bmatrix} \frac{2}{3} & \frac{-1}{3} \\ \frac{-1}{3} & \frac{2}{3} \end{bmatrix} \begin{bmatrix} 3.2 \\ 1.6 \end{bmatrix} = \begin{bmatrix} 1.6 \\ 0 \end{bmatrix}.$$

## The Standardized Score Matrix, *z*

Although a deviation score matrix quickly gets us to unstandardized regression weights, it has limitations. For example, when looking at a deviation score matrix, it is difficult to tell whether a given individual scored well above the mean across all variables or only on some. Consider the deviation score data example presented earlier. If we examine the data from one day and observe that both values are positive (which tells us that both values are above average ratings), we don't know whether either or both values represent an extremely high rating unless both variables have the same variability. In this particular artificial case, they do, but this is not usually the case. In the same fashion, it is difficult for us to compare the regression weights for different predictors because they are usually in different metrics (e.g., some variables may be Likert-format questions and others grade point averages). If we wish to compare the relative effect sizes of associations between variables, it is necessary to standardize the variables to a common metric, usually means of zero and standard deviations of one.

In order to do this, we divide each deviation score by the standard deviation associated with that variable.[2] The matrix algebra formula for doing this is $Z = xD^{-1}$, where $D$ is a diagonal matrix of standard deviations of the variables in the study (and, accordingly, $D^{-1}$ is a diagonal matrix with each diagonal element equal to the reciprocal of these standard deviations). For the data from our example, this would look like this (I've included a decimal representation for those of you doing these analyses on your calculator):

$$\begin{bmatrix} -2 & -2 & -2 \\ -1 & 1 & -4 \\ 0 & -1 & 0 \\ 1 & 2 & 4 \\ -1 & 0 & 2 \end{bmatrix} \begin{bmatrix} \frac{1}{\sqrt{2}} & 0 & 0 \\ 0 & \frac{1}{\sqrt{2}} & 0 \\ 0 & 0 & \frac{1}{\sqrt{8}} \end{bmatrix} = \begin{bmatrix} -\sqrt{2} & -\sqrt{2} & \frac{-2}{\sqrt{8}} \\ \frac{-1}{\sqrt{2}} & \frac{1}{\sqrt{2}} & \frac{-4}{\sqrt{8}} \\ 0 & \frac{-1}{\sqrt{2}} & 0 \\ \frac{1}{\sqrt{2}} & \sqrt{2} & \frac{4}{\sqrt{8}} \\ \sqrt{2} & 0 & \frac{2}{\sqrt{8}} \end{bmatrix}$$

$$= \begin{bmatrix} -1.41 & -1.41 & -0.71 \\ -0.71 & 0.71 & -1.41 \\ 0 & -0.71 & 0 \\ 0.71 & 1.41 & 1.41 \\ 1.41 & 0 & 0.71 \end{bmatrix}$$

The point of looking at these standardized values is that we can now explore whether a particularly high or low value on one variable for a given observation is accompanied by a similarly high value on some other variable. It's slightly easier to see, for example, that the values for the sandwich variable (column 1) and the dessert variable (column 3) are arrived at simply by switching some of the numbers across rows. With this score matrix we can calculate a correlation matrix, using the same formula as for the variance/covariance matrix.

The standardized score matrix $z$ is pre-multiplied by its transpose and divided by its sample size to produce a matrix $R$. In matrix form we can write $R = N^{-1}z'z$. For our example, this would look like:

$$R = \begin{bmatrix} \frac{1}{5} & 0 & 0 \\ 0 & \frac{1}{5} & 0 \\ 0 & 0 & \frac{1}{5} \end{bmatrix} \begin{bmatrix} -\sqrt{2} & \frac{-1}{\sqrt{2}} & 0 & \frac{1}{\sqrt{2}} & \sqrt{2} \\ -\sqrt{2} & \frac{1}{\sqrt{2}} & \frac{-1}{\sqrt{2}} & \sqrt{2} & 0 \\ \frac{-2}{\sqrt{8}} & \frac{-4}{\sqrt{8}} & 0 & \frac{4}{\sqrt{8}} & \frac{2}{\sqrt{8}} \end{bmatrix} \begin{bmatrix} -\sqrt{2} & -\sqrt{2} & \frac{-2}{\sqrt{8}} \\ \frac{-1}{\sqrt{2}} & \frac{1}{\sqrt{2}} & \frac{-4}{\sqrt{8}} \\ 0 & \frac{-1}{\sqrt{2}} & 0 \\ \frac{1}{\sqrt{2}} & \sqrt{2} & \frac{4}{\sqrt{8}} \\ \sqrt{2} & 0 & \frac{2}{\sqrt{8}} \end{bmatrix}$$

$$
= \begin{bmatrix} \frac{1}{5} & 0 & 0 \\ 0 & \frac{1}{5} & 0 \\ 0 & 0 & \frac{1}{5} \end{bmatrix} \begin{bmatrix} 5 & 2.5 & 4 \\ 2.5 & 5 & 2 \\ 4 & 2 & 5 \end{bmatrix} = \begin{bmatrix} 1 & 0.5 & 0.8 \\ 0.5 & 1 & 0.4 \\ 0.8 & 0.4 & 1 \end{bmatrix}. \tag{2.6}
$$

Just as with the variance/covariance matrix, the principal diagonal of the correlation matrix is composed of 1's, reflecting the fact that each variable in the study now has a variance (and standard deviation) of one. When we apply the regression formula we used before, but now with standardized values, $b = (z'z)^{-1} z'z_y$, this also results in the two standardized regression coefficients.

## The Adjoined Matrix, *X*

Sometimes a researcher is interested in predicting the actual score on the criterion variable. To do that, they'll need not just the unstandardized regression weights in the model, but also the regression intercept. Fortunately, there is a useful modification of the raw score matrix which we can use to make a vector of both the unstandardized regression weights and the intercept, the adjoined matrix. The adjoined matrix (also described as $X$) is identical to the raw score matrix, except that it also contains an additional column which has the value "1" for every row, as you can see on the left-hand side of Table 2.1. The placement of ones can be in any column, and one has to remember that the resulting vector will have the intercept in that column.

### Average Sum of Squares and Cross-Products Matrix

A symmetric summary matrix, the average sums of squares and cross-products, is associated with the adjoined matrix: $C_{\text{adj}} = \frac{1}{N} X'X$. This matrix will again be a symmetric matrix, but the individual elements will be a little different from what you may be used to. In our example, for the column of ones and three variables, $S$, $B$ and $D$, this matrix will have four rows and columns with the following amounts:

$$C = \begin{bmatrix} 1 & \frac{\Sigma S}{N} & \frac{\Sigma B}{N} & \frac{\Sigma D}{N} \\ \frac{\Sigma S}{N} & \frac{\Sigma S^2}{N} & \frac{\Sigma SB}{N} & \frac{\Sigma SD}{N} \\ \frac{\Sigma S}{N} & \frac{\Sigma SB}{N} & \frac{\Sigma B^2}{N} & \frac{\Sigma SD}{N} \\ \frac{\Sigma D}{N} & \frac{\Sigma SD}{N} & \frac{\Sigma BD}{N} & \frac{\Sigma D^2}{N} \end{bmatrix}. \tag{2.7}$$

The element in the row and column corresponding to the column of ones will always be the number 1. Notice that the row and column contain the formulae for the means of variables *S*, *B*, and *D*. (Summation operators indicate summing over all rows [observations] in the data set.) The diagonal elements of this matrix are now the average sum of the squares of the raw scores of the data (i.e., the square of each raw number for that variable added across all observations and divided by the sample size). The remaining off-diagonal elements of the matrix are the average sums of the raw-score cross product terms of the variables (i.e., the product of pairs of variables summed across all observations and divided by *N*).

For the example data set, the Average Sum of Squares and Cross-Products Matrix looks like this:

$$\text{SSCP} = \begin{bmatrix} \frac{1}{5} & 0 & 0 & 0 \\ 0 & \frac{1}{5} & 0 & 0 \\ 0 & 0 & \frac{1}{5} & 0 \\ 0 & 0 & 0 & \frac{1}{5} \end{bmatrix} \begin{bmatrix} 1 & 1 & 1 & 1 & 1 \\ 1 & 2 & 3 & 4 & 5 \\ 4 & 7 & 5 & 8 & 6 \\ 4 & 2 & 6 & 10 & 8 \end{bmatrix} \begin{bmatrix} 1 & 1 & 4 & 4 \\ 1 & 2 & 7 & 2 \\ 1 & 3 & 5 & 6 \\ 1 & 4 & 8 & 10 \\ 1 & 5 & 6 & 8 \end{bmatrix}$$

$$= \begin{bmatrix} \frac{1}{5} & 0 & 0 & 0 \\ 0 & \frac{1}{5} & 0 & 0 \\ 0 & 0 & \frac{1}{5} & 0 \\ 0 & 0 & 0 & \frac{1}{5} \end{bmatrix} \begin{bmatrix} 5 & 15 & 30 & 30 \\ 15 & 55 & 95 & 106 \\ 30 & 95 & 190 & 188 \\ 30 & 106 & 188 & 220 \end{bmatrix}$$

$$= \begin{bmatrix} 1 & 3 & 6 & 6 \\ 3 & 11 & 19 & 21.2 \\ 6 & 19 & 38 & 37.6 \\ 6 & 21.2 & 37.6 & 44 \end{bmatrix} \tag{2.8}$$

which gives you the means of the Sandwich, Beverage, and Dessert variables in the first row and column and average cross-products elsewhere. It is important to remember that these average cross-products are not equal to the covariances between variables as you can verify by comparing these values with those found in Eq. (2.4).

### Unstandardized Regression Weights and Intercept

Dividing the data into a vector of raw $Y$ scores and a matrix of the column of ones and the predictor scores allows us to again use the standard formula for calculating regression weights $b = (X'X)^{-1} X'Y$ to now provide the regression intercept (often called $b_0$, $k$, or $c$) and the unstandardized regression weights. For the example data set, this looks like the following:

$$X = \begin{bmatrix} 1 & 1 & 4 \\ 1 & 2 & 7 \\ 1 & 3 & 5 \\ 1 & 4 & 8 \\ 1 & 5 & 6 \end{bmatrix} \text{and}\, Y = \begin{bmatrix} 4 \\ 2 \\ 6 \\ 10 \\ 8 \end{bmatrix}.$$

$$b = \left( \begin{bmatrix} 1 & 1 & 1 & 1 & 1 \\ 1 & 2 & 3 & 4 & 5 \\ 4 & 7 & 5 & 8 & 6 \end{bmatrix} \begin{bmatrix} 1 & 1 & 4 \\ 1 & 2 & 7 \\ 1 & 3 & 5 \\ 1 & 4 & 8 \\ 1 & 5 & 6 \end{bmatrix} \right)^{-1} \begin{bmatrix} 1 & 1 & 1 & 1 & 1 \\ 1 & 2 & 3 & 4 & 5 \\ 4 & 7 & 5 & 8 & 6 \end{bmatrix} \begin{bmatrix} 4 \\ 2 \\ 6 \\ 10 \\ 8 \end{bmatrix}$$

$$= \begin{bmatrix} 3.8 & 0 & -0.6 \\ 0 & \frac{1}{7.5} & \frac{-1}{15} \\ -0.6 & \frac{-1}{15} & \frac{1}{157.5} \end{bmatrix} \begin{bmatrix} 30 \\ 106 \\ 188 \end{bmatrix} = \begin{bmatrix} 1.2 \\ 1.6 \\ 0 \end{bmatrix} \tag{2.9}$$

which gives us the same unstandardized regression weights for the Sandwich and Beverage variables as before in Eq. (2.5), but with the regression intercept of 1.2.

## Concluding Remarks

As we will see in the next chapter, a structural equation model is a statement about the average cross-products of a given score matrix. If standardized data are used, this cross-products matrix is a correlation coefficient. If a deviation score matrix is used, the cross-products matrix is a variance/covariance matrix, and if the adjoined score matrix is used, the model is built on the average sum of squares and cross-products of the raw data.

Before concluding this introduction, it's helpful to comment about some common situations when calculating an inverse is not possible and to introduce a small number of equations that we sometimes encounter in structural models.

### When Operations Involving About Matrix Inverses are Not Possible

Sometimes in regression models the inverse of cross-products of predictor variables, $(x'x)^{-1}$, can't be calculated. Basically, if one of the predictor variables in your model can be calculated perfectly from some combination of other predictor variables, the inverse of the matrix cannot be calculated. In practice, this can happen in four situations:

1. When the researcher attempts to enter subtests and total or composite scores as predictors, or if, as in some instruments, a score on one scale is the reverse of some other scale (such as personality scales, which define introversion scales as the opposite of extroversion). Such scales are said to be "ipsatively" scored.
2. If fewer observations are present than variables.
3. If no variability exists on one or more variables.
4. If some variables considered alone or in combination with other variables perfectly predict some other variables in the data set.

In practice a number of error messages may be produced by statistical software when this happens. Although sometimes software will inform you that an inverse cannot be calculated, other messages include "Some estimated parameters are linearly dependent on other parameters," "Determinant is zero," "cross-products matrix is not full rank," or "matrix is rank deficient," which are all ways of saying that steps involved in calculating the inverse could not be performed.

### Some Other Useful Matrix Algebra from Regression

The algebra here uses $Y$ and $X$ as the vector and matrix of criterion and predictor scores, with $b$ as a vector of regression weights. If you are new to matrix algebra, it may be helpful for you to verify the results of doing the calculations on the data of Table 2.1 either by hand or by using a spreadsheet's matrix functions.

### *Predicted Criterion Scores*

Predicted $Y$ scores can be generated with $\hat{Y} = Xb$. These will be the predicted raw, deviation, and standardized criterion scores depending on whether the adjoined, deviation, or standardized score metrics are used.

### *Errors of Prediction*

A column vector of errors of prediction can be generated by $= Y - \hat{Y}$. These errors of prediction will be the same for the adjoined and deviation score matrices and will represent the errors of a standardized criterion variable.

### *Proportion of Variance Accounted for $(R^2)$*

The proportion of variance accounted for by the predictor variables in a regression can be calculated in the standardized or deviation score metric (but not the adjoined matrix) by $R^2_{Y.X} = \hat{Y}'\hat{Y}(Y'Y)^{-1}$.

## Notes

1 If this example is not "serious enough," replace these example variables with any other three variables in your area, say ratings of three types of behaviors exhibited by an inpatient, scores of five children on three vocabulary tests, or three student evaluations of five different professors.

2 Unfortunately, some writers refer to this process as "normalization," leading some students to conclude that the variables are now "normally distributed" or are "multivariate normal." We use the term *standardized score matrix* here to avoid this confusion. The resulting version of the data is not now composed of "normal" variables. A distribution of scores that is skewed or bimodal will continue to be so after standardization.

# 3

# *Path Diagrams*

> Philosophy is written in this grand book, which stands continually open before our eyes (I say the "Universe") but cannot be understood without first learning to comprehend the language and know the characters as it is written. It is written in mathematical language, and its characters are triangles, circles, and other geometric figures, without which it is impossible to humanly understand a word.
>
> **– Galileo Galilei, *Il Saggiatore* (1623)**

The quote from Galileo is amusing because path diagrams are indeed a system of triangles, circles, and squares which represent structural relationships. These relationships are often linear (meaning that they involve only the mathematical properties of addition and multiplication). Path models summarize several relationships in the data simultaneously and are, for example, useful in understanding how correlations between predictor variables affect regression estimates. They are also useful in understanding how relationships between variables may be explained by some unmeasured variable. In all but more complicated SEMs, the path diagram is a frequently reported figure. Understanding the basics of a path diagram also helps communicate the idea of model parsimony, which is discussed in subsequent chapters.

Some statistical software packages allow the user to fit the model based on a path diagram (e.g., Amos, Mplus, lavaangui in R, and Onyx), while others require the user to write each part of a path diagram as a separate element (e.g., lavaan, SAS Proc Calis, Mplus). Most SEM packages now come with the ability to draw the path diagram corresponding to the fitted model. Although well-suited for some SEMs, path diagrams can be cumbersome when many variables are being analyzed or when the data are structurally complex (such as family-level data or more complicated behavior genetic models). In these cases, it is often more helpful to think of the SEM in matrix form. Regardless of the software platform you choose, understanding the logic of path notation generally will provide a set of long-term literacy skills.

Path diagrams are used on data from one of the three score matrices introduced in the previous chapter. Some questions of the data cannot be answered that are more appropriate for other score matrices. For example, if the researcher is interested in growth and decline in a longitudinal study, using the deviation or standardized score matrices cannot model these processes. Conversely, if the goal is to assess effect sizes among multiple variables


DOI: 10.1201/9781003595267-3

in different scales a standardized score matrix can be more convenient. For technical reasons, statistical software does not often directly analyze the standardized version of the data but uses either the adjoined or deviation score forms of the data, which are then transformed into standardized coefficients.

## What Is a Path Diagram?

A path diagram is analogous to an electronic circuit. As we will see in our discussion of feedback loops, estimation of feedback and reciprocal relationships actually borrows directly from Mason's (1956) feedback models for circuits. In an electronic circuit a small number of elements such as a power source and "and" and "or" switches must be connected properly to carry the current and perform the intended function. Path diagrams, though, are designed to assess variation (expressed as a variance), covariation (a covariance), and elevation (mean effects modeled by the column of ones mentioned in the adjoined matrix in the last chapter. Similar to an electric circuit, some simple rules must be followed in path analysis in order for the diagram to "work." These rules are referred to as the "tracing rules," and this book uses a particular explication of these called the Reticular Action Model (RAM) model (cf. McArdle & McDonald, 1984). This interface is used in some software (such as Onyx or lavaangui) but other software (e.g., Amos) uses slightly different graphical conventions. These choices are sometimes made to reduce clutter or because of differences in the matrix algebra underlying the software.

### Path Diagram Elements

A manifest variable is represented in a path diagram by enclosing it within a rectangle. An unobserved or latent variable is represented by an oval. Variables' names are recorded inside the rectangle or oval.

Estimated parameters in a SEM are used to convey variances, covariances, or regression weights. Regression weights are sometimes termed as "path coefficients." Variances are expressed in a path diagram by means of a double-headed arrow termed a "sling." Covariances are expressed by means of a double-headed arrow, which spans two variables, and regression weights are expressed by means of a single-headed arrow. Each variable in a diagram must either contain a sling or, alternatively, be explained completely by reference to other variables in the diagram.

To see these elements in the context of a path diagram, Figure 3.1 shows the parameters for a two-predictor regression equation based on a deviation score matrix. Parameters of the model are shown with text names. Note that

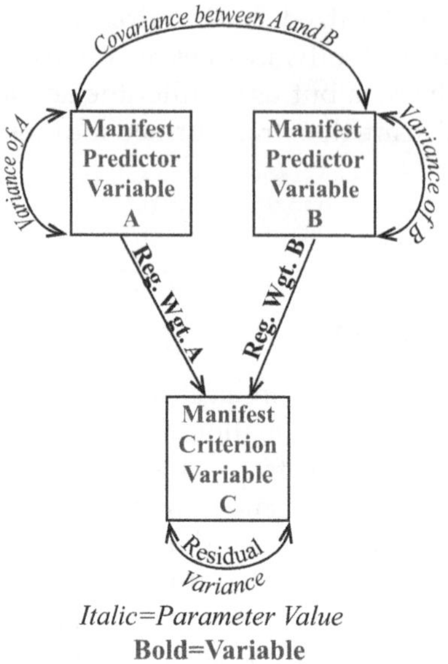


**FIGURE 3.1**
Two-predictor regression model showing variables and parameters.

although the sling value associated with predictor variables corresponds to their respective variances, the sling associated with an endogenous variable C corresponds to the variance of errors of prediction – i.e., the variance of the criterion for any fixed values of all predictors, also known as a conditional variance. If, instead of a deviation score matrix (or variance/covariance matrix), the standardized score matrix (or correlation matrix) was used for the diagram, the variances of $A$ and $B$ would be one, the covariance between $A$ and $B$ would be the correlation between them, standardized regression weights would be shown in the diagram, and the error variance of $C$ would correspond to the proportion of variability in C due to error, $1 - R^2_{C.AB}$.

## Path Diagrams Based on the Adjoined Matrix

As noted in the last chapter, the adjoined matrix must be used when a SEM addresses questions of growth, decline, or level of performance in addition to variation and covariation. The adjoined score matrix contains a "variable" consisting of ones across all observations and a triangle (usually with the number 1 inside) is used to represent it, as shown in Figure 3.2. In this diagram, the arrows from the column of ones to the predictor variables represent

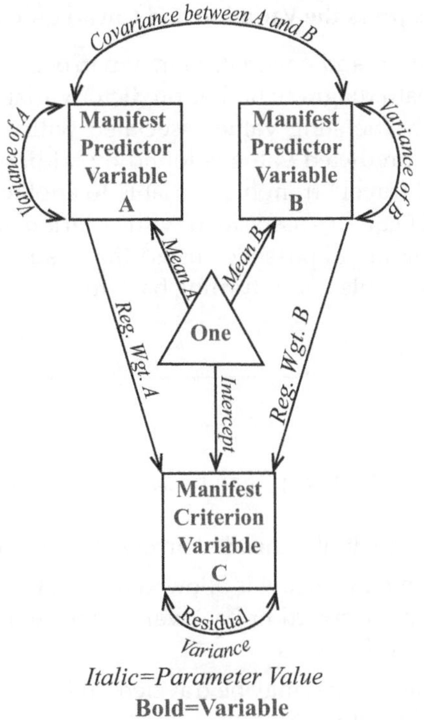


**FIGURE 3.2**
Unstandardized two-predictor regression model with intercept.

their respective means, the paths from the variables to the dependent variable represent the unstandardized regression weights, and the path from the column of ones to the dependent variable represents the regression intercept. In order to see how these parameters numbers were determined, it's necessary to learn how to use the tracing rules for path diagrams.

## The Tracing Rules for Path Diagrams

As mentioned in Chapter 2, a path diagram is an expression of a model relative to one of the three average cross-products matrices calculated from the data. The tracing rules described in the next section show how parameters in a path diagram are an attempt to reproduce these cross-product matrices. We discuss how structural equation modeling programs actually estimate parameter estimates at a later point, but to begin, we show here how a path diagram reproduces the three types of average cross-product matrices for the simple case of a single-predictor regression model.

### Path Diagrams Re-Express the Variation and Covariation in the Data

The observed variation and covariation in a matrix can be expressed in one of three ways in a path diagram: (a) For predictor variables, the average sum of squares is simply the sling value associated with that variable. (b) For cross-products, the predicted value is found by adding up all of the different ways one may "travel" from one variable to another in the diagram. (c) For average sums of squares associated with criterion variables, the value is calculated by adding up all possible "trips" that result from "traveling" out from the criterion variable and returning back to it.

## Tracing Rules

The rules for an individual "trip" can be summarized as follows:

1. One and only one double-headed arrow may be collected.
2. Movement along any arrow is allowed during the trip, regardless of whether it points forward or backward, with two exceptions which apply to backward arrows:
   a. No backward arrow may be traveled on if one has already taken a sling on the trip.
   b. No backward arrow may be traveled on if one has already taken a forward arrow.[1]

Path diagrams may involve feedback or reciprocal relations (in which variables are simultaneous causes of each other or exist in a circular feedback loop). In these situations, the covariances between variables involve an infinite number of paths. As discussed in the section on feedback loops (see Chapter 11), we can calculate the predicted average sums of squares and cross-products terms in these situations using matrix algebra.

## Solving for Path Coefficients: Single Predictor Regression

### Standardized and Deviation Single Predictor Regression

To begin, consider a correlation matrix from the first published regression model, Galton's (1886) study of the inheritance of height. A standardized path diagram of the data is shown in the left-hand side of Figure 3.3, and the correlation matrix used to produce the standardized solution is given in the first three rows of Table 3.1.

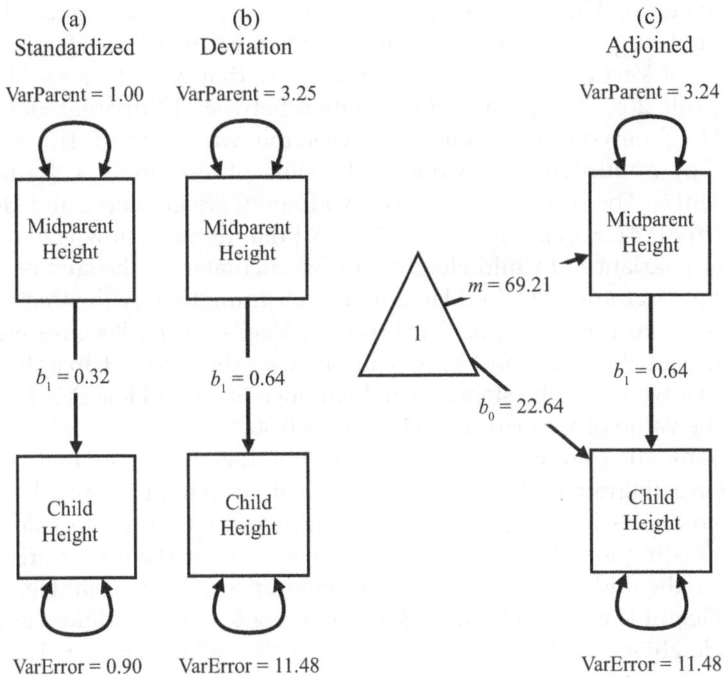


**FIGURE 3.3**
Single predictor models for standardized (a), deviation (b), and adjoined (c) score matrices.

**TABLE 3.1**
Galton Height Data Correlation Matrix

| | | | |
|---|---|---|---|
| | **Correlations** | | |
| | MidParent Height | Child Height | |
| MidParent Height | 1 | 0.32 | |
| Child Height | 0.32 | 1 | |
| | **Variances/Covariances** | | |
| | MidParent Height | Child Height | |
| MidParent Height | 3.2380 | 1.3325 | |
| Child Height | 1.3325 | 2.2407 | |
| | **Average Adjoined Sums of Squares and Cross-Products** | | |
| | Intercept | Midparent Height | Child Height |
| Intercept | 1 | 69.2068 | 66.7459 |
| Midparent Height | 69.2068 | 4792.823 | 4621.339 |
| Child Height | 66.7459 | 4621.339 | 4467.817 |

The value of VarParent is 1, because this variable was standardized. It would not be permissible, for example, to travel to Child Height, "take" the sling of VarError and then return, because that would be a violation of tracing rule 2(b). The predicted correlation between Midparent Height and Child Height involves "traveling" between the two variables. This is accomplished by multiplying the value of the sling of VarParent (1) by the path coefficient $b_1$. The correlation between Midparent Height and Child Height is then $1(b) = 0.32$. Solving for $b_1$ is 0.32/1. All that remains now is to calculate the (unit) variance of Child Height. This is calculated as the sum of permissible trips beginning at Child Height and returning to it by the tracing rules. There are two possible trips: VarError + $b_1$*VarParent*$b_1$. Because we know the values of VarParent and $b_1$, we can calculate the second trip as $0.32^2$ and, given that we know the standardized variance of Child Height is 1, we now know the value of VarError to be $1 - 0.32^2 = 0.90$.

The same diagram can also be used for the deviation score matrix, which results in a diagram that shows the variance of Midparent Height, the unstandardized regression weight, and the conditional variance of Child Height after adjusting for Midparent Height (also known as the error variance), as shown in the middle of Figure 3.3. The covariance of Midparent Height with Child Height is equal to $1.3325 = 3.5^* b_1$, making $b_1 = 0.64$ and the variance of Child Height as $2.2407 = b_1{}^2$Varparent + VarError, which gives us 11.48 as the error variance of Child Height.

## Adjoined Matrix

The situation for a single predictor equation with an intercept is slightly more complicated as shown in Figure 3.3(c), but the same strategy is used of first solving for exogenous variables, then covariances, and then the average sum of squares for the endogenous variables working from the least to most endogenous. The single exogenous variable, the triangle of ones, has the value of 1 from the data and so is not usually explicitly shown. The Midparent Height mean is $1(m) = 69.21$, making $m = 69.21$. ($1^* b_0{}^* b_1$ violates rule 2(b) of going backward on a path after initially going forward.) The variance of MidParent Height is then $4792.82 - 69.21^2(1) = 3.24$. This then leaves us with the problem of how to solve for $b_0$ and $b_1$. We can write out the two average cross-product (CP) trips involving Midparent Height to Child Height and the Intercept to Child Height:

$$CP_{MidparentHeight,ChildHeight} = 69.21b_0 + 4792.82b_1 = 4621.34 \quad (3.1)$$

and

$$\text{CP}_{1,\text{ChildHeight}} = 69.21b_1 + b_0 = 66.75 \tag{3.2}$$

We can solve this by rewriting Eq. (3.2) in terms of $b_0 : b_0 = 66.75 - 69.21b_1$ and inserting that into Eq. (3.1).

$$\begin{aligned}\text{CP}_{\text{MidparentHeight,ChildHeight}} &= 69.2(66.75 - 69.21b_1) + 4792.82b_1 \\ &= 3.488b_1 + 4619.1 = 4621.34\end{aligned} \tag{3.3}$$

and solving for $b_1$.

$$b_1 = \frac{4621.45 - 4619.10}{3.488} = 0.64 \tag{3.4}$$

Now that we know that, we can solve for $b_0$ by going back to Eq. (3.2):

$$\text{CP}_{1,\text{ChildHeight}} = 69.21(0.64) + b_0 = 66.75 \tag{3.5}$$

which gives us $66.75 - 44.45 = b_0 = 22.64$

All that remains now is to solve for the value of $E$. This we do by considering the average sums of squares of Child.

$$\begin{aligned}4467.82 &= 22.64^2(1) + (2)22.64(1)69.21(0.64) \\ &\quad + 0.64^2(3.24) + 0.64^2(69.21) + E \\ &= 4456.33 + E\end{aligned}$$

which gives us $E = 11.49$.

## Building Blocks for Tracing Rules

Although these example SEMs show how regressions are expressed as path diagrams, it helps to build confidence in using the tracing rules to practice with some simple examples. These examples also constitute small modules which can be "snapped into" a larger model to produce desired statistics. SEMs can be expressed in more than one way, especially if the interest is in using a more complicated score matrix, such as the adjoined matrix, to produce a parameter usually associated with a less complicated score matrix such as the standardized score matrix. Three small examples, shown in Figures 3.4(a) and (b), are similar to the unstandardized one predictor regressions shown in Figure 3.3(b) but have the addition of an unmeasured

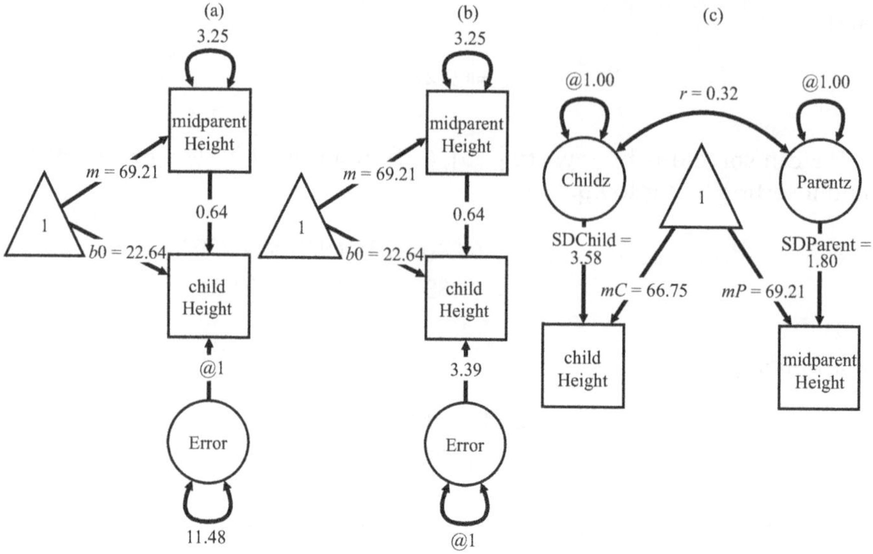


**FIGURE 3.4**
Alternative specification of error variance (a), error standard deviation (b), and correlation coefficient (c) using the adjoined score matrix.

variable, the residual (observed-predicted) calculating the same variance of the residual score. Explicit representation of the error variable is found in some statistical software such as Amos, which does not allow specification of conditional variances on the dependent variable due to the matrix algebra used in the program (the Bentler-Weeks model discussed in Chapter 11). Figure 3.4(b) calculates the standard deviation of the residual. In addition to being perhaps more interpretable than an error variance, this model is useful when the researcher wishes to calculate partial or semi-partial correlations in a path diagram (see Chapter 4) and is a basis for the equal precision measurement model discussed in Chapter 7.

Figure 3.4(c) shows a path diagram based on the adjoined matrix which calculates the variable means, standard deviations, and the correlation between the two variables. Paths from the column of ones to the manifest variable represent variable means. By the tracing rules, the average squared deviation of each variable is equal to sum of the square of the path from the intercept to the manifest variable and the square of the path from the latent variable with unit variance to the manifest variable. The predicted average SSCP matrix from the model is shown in the last three rows of Table 3.1. For example, the average SSCP for Midparent Height is:

$$\text{ASSCP}_{\text{MidparentHeight}} = 4792.82 = 69.21^2 + \text{SDParent}^2 \tag{3.6}$$

which can be solved as: SDParent $= \sqrt{4792.82 - 69.21^2} = 1.80$. This leaves us to calculate the sling between the two (unmeasured) standardized variables. The average cross-product between Child and Parent is then:

$$4621.34 = 69.21(1)66.75 + 1.80(r)3.58 \tag{3.7}$$

which, on solving for $r$, gives $r = 0.32$, which is the correlation coefficient between the variables.

## Note

1 Put slightly differently, if you ignore for a moment the boxes which represent the variables' names, this rule means that arrow heads may not touch during a trip.

# 4

# *Three-Variable Models*

The previous chapter discussed the building blocks of path diagrams and how the tracing rules can be used to calculate parameter estimates from the average sum of squares of a given score matrix. It is a good idea to get some practice using the tracing rules to ensure you understand how the tracing rules work. This chapter introduces small SEMs, which are general linear models you may have encountered in classes in analysis of variance and regression. One way to work with the tracing rules is to use them to solve for the path diagram parameters given the predicted average sums of squares of one of the three score matrices introduced earlier. Other readers may simply use these examples as practice exercises to acquaint themselves with a particular SEM program. These examples also serve as patterns that can be used more generally when the manifest variables in these models are replaced with analogous latent variables. In some models, centering predictor variables to deviation score form is discussed as a way of reducing collinearity among predictors. Examining how testing a parameter's statistical significance in the general linear model differs from statistical significance tests using the Wald statistic in SEM is also explained.

The examples here include models for regression, partial correlation, semipartial correlation, the single factor model, quadratic regression, mediational models, and models from the general ANOVA model such as dummy coding, orthogonal, effect, analysis of covariance, and latent difference scores. All models presented are *just-identified* SEMs, meaning that they contain as many freely estimated parameters as unique elements of the average cross-products matrix under consideration.

## Standardized Two-Predictor Regression Example

Although the path diagram for a two-predictor model has already been introduced, a real-world example provides practice in the tracing rules and introduces the idea that a given parameter estimate in a SEM may be correlated with other parameter estimates, thereby reducing the statistical power of the model.


DOI: 10.1201/9781003595267-4

Experience has taught that it is easiest to begin with the standardized two-predictor regression model given that this involves statements about correlation coefficients. Our example comes from Widaman and Reise (1997) in a study investigating the correlates of smoking (operationalized as past 30-day use). Data were from 661 male Euro-American high school seniors and part of a larger study by Johnston, O'Malley, and Bachman (1995). Three self-report items are considered: (1) Last 30 Days Cigarette Use, (2) Smokers Trying to Be Mature ("In my opinion, when a guy my age is smoking a cigarette, it makes him look like he's trying to appear mature and sophisticated") and (3) Don't Mind if Around Smokers ("I personally don't mind being around people who are smoking").

Correlations between the variables are shown in the top half of Table 4.1, and the corresponding path diagram is shown in Figure 4.1. The predicted correlation matrix in terms of the parameters of the diagram is shown in Table 4.2. Satisfy yourself that the predicted correlation matrix is correct by solving for the parameters: We can see that the estimated variances of the predictor variables are one because the predicted variables are standardized. The covariance between these two variables, $C_{T,D}$, is the correlation (–0.33), again because standardized variables are involved. (Paths going from the predictors to the criterion and back up to the other predictor violate the tracing rules of not being able to go forward and then backward on a single trip.) We can now use these numerical values to solve for the path coefficients *A* and *B*, which represent the standardized amount of gain expected associated with each of the predictor variables if the other predictor variable is held at some constant. As with the case of the adjoined matrix example discussed earlier, we need to

**TABLE 4.1**

Correlations and Covariances from Johnston et al. (1995) Study ($N = 661$)

| Variable | Last 30 Days (C) | Trying (*T*) | Don't Mind (*D*) |
|---|---|---|---|
| | How frequently have you smoked cigarettes during the last 30 days? | In my opinion, when a guy my age is smoking a cigarette, it makes him look like he's trying to appear mature and sophisticated. | I personally don't mind being around people who are smoking. |
| Last 30 Days (C) | 1 | −0.43 | 0.48 |
| Trying (*T*) | −0.43 | 1 | −0.33 |
| Don't Mind (*D*) | 0.48 | −0.33 | 1 |
| | Variance/Covariance Matrix | | |
| Last 30 Days (C) | 1.54 | −0.72 | 0.90 |
| Trying (*T*) | −0.72 | 1.84 | −0.68 |
| Don't Mind (*D*) | 0.90 | −0.68 | 2.28 |

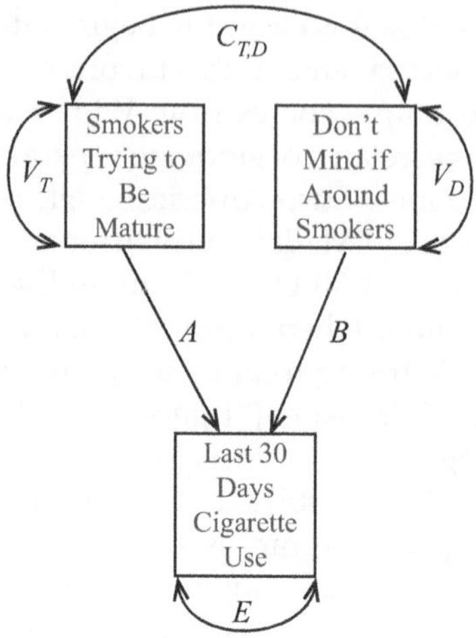


**FIGURE 4.1**
Two-predictor regression model for Widaman and Reise example.

**TABLE 4.2**
Predicted Correlation or Covariance Matrix for Widaman and Reise (1997) $N = 661$

| Variable | Last 30 Days Cigarette Use (C) | Smokers Trying to Be Mature (T) | Don't Mind if Around Smokers (D) |
|---|---|---|---|
| Last 30 Days Cigarette Use (C) | $A^2V_T + B^2V_D + 2ABC_{T,D} + E$ | $AV_T + BC_{T,D}$ | $BV_D + AC_{T,D}$ |
| Smokers Trying to Be Mature (T) | $AV_T + BC_{T,D}$ | $V_T$ | $C_{T,D}$ |
| Don't Mind if Around Smokers (D) | $BV_D + AC_{T,D}$ | $C_{T,D}$ | $V_D$ |

write two equations for this, substituting the known values for the variances and correlation in the diagram:

$$r_{T,C} = (1)A + (-0.33)B = -0.33 \tag{4.1}$$

and

$$r_{D,C} = (1)B + (-0.33)A = 0.43 \tag{4.2}$$

As before, we can solve this problem by substituting Eq. (4.2) into Eq. (4.1), which would give us $A = \frac{0.43 - B}{0.33}$, which, on substituting into Eq. (4.1), gives us:

$$-0.33 = \frac{0.43 - B}{-0.33} - 0.33(B) \tag{4.3}$$

Solving for $B$ gives us 0.37 and, using this value, we can solve for $A$ as −0.30, as shown in Figure 4.2. All that remains is to solve for the value of the variance of the dependent variable, which we know to be 1. By the tracing rules, this is equal to $A^2V_T + B^2V_D + 2AB(C_{T,D}) + E$. The reason that the second to last term is $2AB(C_{T,D})$ is because you travel once around the path beginning with $A$ and in the counterclockwise direction starting with $B$. The sum of the first three terms for the variance of $C$ (i.e., all except for $E$) represents the squared multiple correlation coefficient for the model. $E$ represents the proportion of variability due to error. Solving for $E$ gives you the value of 0.69 shown in Figure 4.2.

## Uses of a Standardized Solution

Remember that a standardized coefficient represents the effect size associated with the unique contribution of a predictor to the criterion. It does not inform you which variable is generally "more important," especially when many predictor variables are present. Some researchers mistakenly examine a regression model with several predictors and then drop all coefficients with nonsignificant regression weights. This results in dropping groups of predictors which may be useful in predicting the criterion but which provide largely redundant information and are highly correlated with each other. A better approach within multiple regression involves considering several

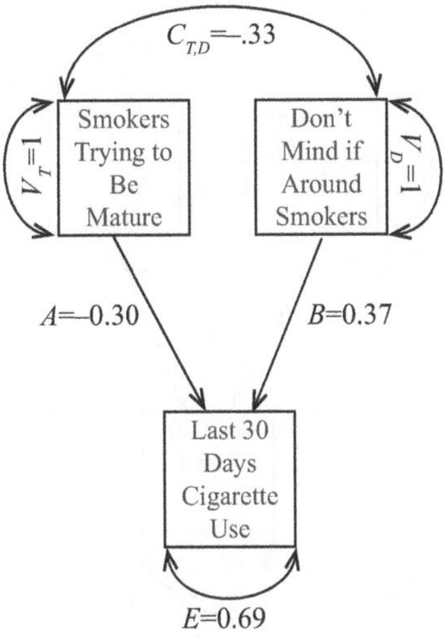


**FIGURE 4.2**
Standardized regression solution.

candidate subsets of regression models. From an SEM perspective, however, there is no reason to consider subsets of regression. If it is reasonable to think that the predictors are correlated with each other because they are indicators of a common unobserved variable, the relationship of this latent variable to the criterion of interest can be modeled instead.

## Unstandardized Two-Predictor Regression

As mentioned in the last chapter, the variance/covariance matrix is the average SSCP matrix of deviation scores and is useful when the researcher is interested in calculating variance components of variables and, within regression, in determining the amount of gain expected in a criterion for each point gain in a predictor, holding all other predictors constant. The variance/covariance matrix for these data are shown on the bottom of Table 4.1. As mentioned before, the elements on the diagonal of this matrix represent variances of the variables.

The corresponding estimated unstandardized parameter values are shown in Figure 4.3. As for the standardized solution, slings associated with exogenous variables are simply equal to their covariance matrix values. The first three terms in the predicted variance associated with the criterion in Table 4.3

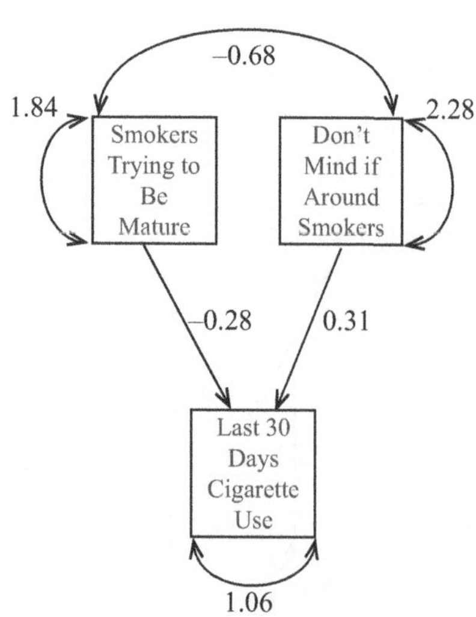


**FIGURE 4.3**
Unstandardized solution.

**TABLE 4.3**

Predicted Average Adjoined SSCP Matrix for Widaman and Reise (1997) Data

| Variable | Last 30 Days Cigarette Use ($C$) | Smokers Trying to Be Mature ($T$) | Don't Mind if Around Smokers ($D$) | One |
|---|---|---|---|---|
| Last 30 Days Cigarette Use ($C$) | $E + A^2V_T + 2ABC_{T,D} + A^2M_T^2 + 2AIM_T$ $+2ABM_TM_D + I^2 + 2IBM_D + B^2V_D + B^2M_D^2$ | $AV_T + BC_{T,D} + AM_T^2$ $+IM_T + M_TM_DB$ | $BV_D + AC_{T,D} + BM_D^2$ $+IM_D + AM_DM_T$ | $AM_T + I + BM_D$ |
| Smokers Trying to Be Mature ($T$) | $AV_T + BC_{T,D} + AM_T^2 + IM_T + M_TM_DB$ | $M_T^2 + V_t$ | $M_TM_D + C_{T,D}$ | $M_T$ |
| Don't Mind if Around Smokers ($D$) | $BV_D + AC_{T,D} + BM_D^2 + IM_D + AM_DM_T$ | $M_TM_D + C_{T,D}$ | $M_D^2 + V_D$ | $M_D$ |
| One | $AM_T + I + BM_D$ | $M_T$ | $M_D$ | 1 |

are equal to the variance of the predicted score on the criterion, and the last term represents the variance associated with the residual score.

## Two-Predictor Regression with Intercept

The path diagram for this same model based on the average adjoined sum of squares and cross products matrix and the estimated parameters for the model are given in Figure 4.4. This diagram is more complicated than the two-predictor unstandardized regression model, but the same general strategy of solving first for most exogenous variables to more endogenous variables works as well, with the predicted average SSCP matrix shown in Table 4.3. The specific steps involved are not outlined here, except to say that solving for exogenous effects first (such as predictor variable means) permits solution of more endogenous effects (such as average sums of squares of predictor variables). For those who like a challenge, the parameters involved in the estimated matrix are given in Table 4.3. The path diagram in Figure 4.4 can be used to predict cigarette use for any two numeric values of the predictor variables.

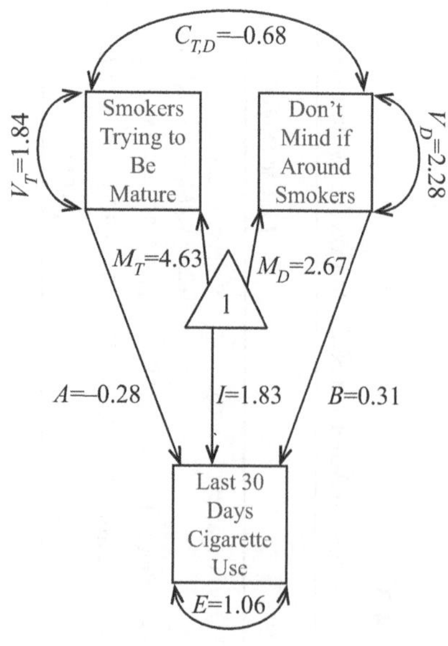


**FIGURE 4.4**
Two-predictor solution with intercept.

## Orthogonality, Redundancy, and Suppression

At this point, some vocabulary is helpful to describe how covariation between predictor variables affects the precision and statistical significance of parameters in a regression model. Predictors are said to be *orthogonal* if the covariance between them is zero. In a two-predictor model, for example, this means that the standardized regression weight associated with a given predictor is equal to its correlation. Estimates of the statistical significance of orthogonal predictors have a smaller sampling variability in their estimated regression weights and therefore, all other things being equal, have more statistical power in tests of the statistical significance of regression weights than similar data sets involving correlated predictors.

A pattern of *redundancy* exists between predictors if the direct path from the predictor to the criterion is the same sign as all compound paths. Conceptually, this means that the information provided by the predictor overlaps with that provided by the other predictors. This pattern has two implications. The standardized path coefficient associated with a predictor is less in absolute value than the observed correlation between the predictor and criterion. Sampling error associated with the regression parameter of one predictor becomes correlated with the sampling error of other predictors, resulting in reduced statistical power to reject the null hypothesis. In the smoking example, such a pattern of overlap exists because the association of a predictor and the criterion consists of two positive quantities (the direct association of DontMind with cigarette use) or the product of two negative quantities (the association of Trying with cigarette use).

In other situations, the sign of one regression weight is opposite in sign of one or more compound paths involving covariances and other predictors. This is referred to as a *suppressor effect*. If regression paths are both the same sign but opposite in sign to the covariance between the predictors, we refer to this as *complementarity*. In both situations, the observed regression weight associated with a predictor variable may therefore be larger in absolute magnitude than one would anticipate based on inspection of the correlation matrix alone. If, for example, the correlation between the two predictors in our example had been positive instead of negative, one can see that the implied correlations in the model based on the path values would have been substantially lower. Suppressor and complementarity effects, like redundancy, have larger standard errors than their orthogonal counterparts because their estimated standard errors are larger.

Because the word "suppression," like "regression," has a rather evil-sounding tone, some individuals believe that finding suppressor effects implies that the regression model has failed in some way, that the data are not suitable for regression, or that the observed data have failed some basic assumption of the regression model. This belief is further reinforced in some situations in which the observed standardized coefficient is greater than 1 or less than −1. If one accepts the analogy that standardized regression coefficients represent correlation-like quantities, they reason, how could a correlation with a criterion be greater than one?

Although it is true that a correlation coefficient cannot exceed the bounds of 1 to −1, it's important to remember that a two-predictor regression model represents a plane of predicted values of $Y$ based on possible values of both predictors (and that three or more predictors represent a multidimensional hyperplane). As such, a regression weight expresses only the tilt of this plane associated with increases in the predictor. Suppressor effects suffer from the same disadvantage as redundancy in that the estimate of sampling variability in the estimated regression weight becomes larger the more correlated the predictors are. For this reason, standardized coefficients in SEMs can also exceed one in absolute magnitude and, unless some other aspect of the model is problematic, such values are completely acceptable.

## Ratio of Explained Variance

Although linear models often summarize statistical significance in terms of average sums of squares, this can also be accomplished using the proportion of a variable's explained variance divided by the total variance of the variable. In linear models, this ratio takes into account the degrees of freedom involved, which gives us an $F$ statistic with $k$ and $N-k-1$ degrees of freedom as:

$$F_{(k,N-k)} = \frac{\dfrac{\sigma^2_{\hat{Y}}}{k}}{\dfrac{\sigma^2_{Y} - \sigma^2_{\hat{Y}}}{N-k-1}} \tag{4.4}$$

## Statistical Significance of Path Coefficients

For single predictor regression the $t$-test of whether the estimated regression weight $b$ is significantly different from zero can be written as:

$$t_{N-2} = \frac{b-0}{s_b} \tag{4.5}$$

where $s_b$ is the standard error of the regression weight, i.e., an estimate of the standard deviation of the regression weight statistic that would result if we were to replicate the study infinitely many times. If we take $\sigma_e^2$ to be the estimated error variance (i.e., $\sigma_Y^2 - \sigma_{\hat{Y}}^2$) $s_b$ can be written as:

$$s_b = \sqrt{\frac{1}{N-2}\frac{\sigma_e^2}{\sigma_X^2}} \tag{4.6}$$

The significance of the $t$-statistic is evaluated relative to the percentile points of the $t$-distribution under the null hypothesis. The corresponding statistical significance test in SEM uses a slightly different statistical test is used called the Wald statistic which divides by $N-1$ instead of $N-k-1$ where $k$ is the number of predictors in the model (one in the case of a single predictor regression models). In data with sufficiently large sample size, this difference should not be appreciable. The Wald test is used more generally in SEMs because of the multiple patterns of relationships in an SEM beyond those considered in the linear model.

## Sampling Distribution of Multiple Predictors

When we consider the statistical significance of path coefficients involving multiple predictors, the sampling distribution of individual regression weights is slightly more complicated. This is necessary because errors of estimation associated with the estimates of the other regression weights in the model introduce another source of imprecision. For the case of multiple predictors, we calculate a variance/covariance matrix of standard errors of regression weights with the formula:

$$C = \frac{\sigma_{\text{Error}}^2\left(\frac{x'x}{N}\right)^{-1}}{N-1} \tag{4.7}$$

where $\left(\frac{x'x}{N}\right)^{-1}$ is the inverse of the variance/covariance matrix of the predictor variables, and $\sigma_e^2$ is again the variance of the errors of prediction. The standard error for any predictor is calculated as the square root of the diagonal entry of C corresponding to that variable. The matrix C is useful in SEMs because it can be inspected to understand dependencies between estimated parameters.

## Correlated Predictor Variables Have Correlated Estimated Path Coefficients

What lessons are we to draw from Eq. (4.7)? As we can see from this value, the estimated standard error associated with a given parameter value is now a function of both the observed variability in the predictor variable and the covariances of that predictor with other predictors in the model. Off-diagonal elements of the inverted matrix will be the negative of the sign of the predictors, so if predictor variables are positively correlated, this means that covariances between the regression weights associated with those predictor variables will be negatively related. This relationship can be thought of in terms of errors of prediction of the regression weights associated with a path model.

As an informal way to understand this, consider a path diagram with positive regression weights and positive covariance between predictors, and that we replicated this study infinitely many times. The path values across replications for the two-predictor model would equal their true population values. Consider the results from any given replication of our study. Observed path coefficients from one predictor in this replication will probably be a little higher or lower than the true population values on the basis of chance. If, in this replication, the estimate of the parameter is an overestimate of the population value, this has implications for the remaining parameters of the model. It will not affect our estimate for the covariance between predictors because that value corresponds simply to the observed covariance between the predictors in the data. The estimate of the regression weight associated with the other predictor, however, is affected and will result in an underestimate of the other predictor variable relative to its population value to "make up the difference" in the unsuccessful estimation of the covariance of the predictor with the criterion. (Similarly, an underestimate of one regression weight will result in an overestimate of the regression weight associated with another predictor).

Similarly, in cases of suppression (assuming positive regression weights), a positive covariance obtains between standard errors with the regression weights due to the down-adjustment of the covariance of the regression weight with the criterion. In cases where the predictors are orthogonal to each other, the corresponding standard errors associated with the regression weights will also be orthogonal, i.e., there will be no dependencies between parameter estimates from such a model.

Correlations between predictor variables, regardless of whether they are part of a pattern of redundancy or suppression, have the net effect of increasing the values of $C$ for the correlated predictors. One can take this statement as given or, for those with matrix algebra expertise, recall that the inverse is defined as the adjoint divided by the determinant. High correlations between predictors result in lower values for the determinant, resulting in higher values for the inverse. Multiplying the inverse by the error variance associated with the criterion results in higher values of $C$.

**TABLE 4.4**

C and $R_C$ for Widaman and Reise Data

| Predictors | *A* (Trying) | *B* (Don't Mind) |
|---|---|---|
| *A* (Trying) | 0.0012598 | −0.0007341 |
| *B* (Don't Mind) | −0.0007341 | 0.0018646 |
| | $R_C$ | |
| *A* (Trying) | 1 | −0.48 |
| *B* (Don't Mind) | −0.48 | 1 |

When considering larger SEMs, the matrix of correlations of parameter estimates can be used to localize where patterns of dependencies exist between parameter estimates. For the data in the Smoking example, *C* and the correlation matrix of the estimates, $R_C$, are shown in Table 4.4. As can be seen from the values of the correlations between parameter estimates, although the values are somewhat correlated, they are not highly correlated with each other. Recall that the sign of the regression weight is negative, so errors of estimation on the DONTMIND variable are correlated with higher negative values on the TRYING variable.

## Third Variable Explanations

### Partial Correlations

In many situations, the bivariate correlation between two variables can be misleading because both variables are associated with other "third" variables. For example, consider a data set which sought to determine if collegiate alcohol use was related to college grades (Wood, Sher, Erickson, & DeBord, 1997). In this study, an attempt was made to determine the relationship between first-semester cumulative grade point average (GPA) and a self-report measure of alcohol quantity and frequency of consumption. The bivariate correlation between these two variables was −0.27, indicating that more alcohol-involved students tended to have lower cumulative GPAs.

It could be, however, that this association is misleading given that other variables are associated with both alcohol consumption and freshman cumulative GPA. For example, less academically able students (as operationalized by standardized admissions tests such as the ACT or prior academic achievement such as high school GPAs) are less likely to succeed in college and may be also more likely to consume alcohol while in college (due to high school alcohol consumption or frustration with the college experience,

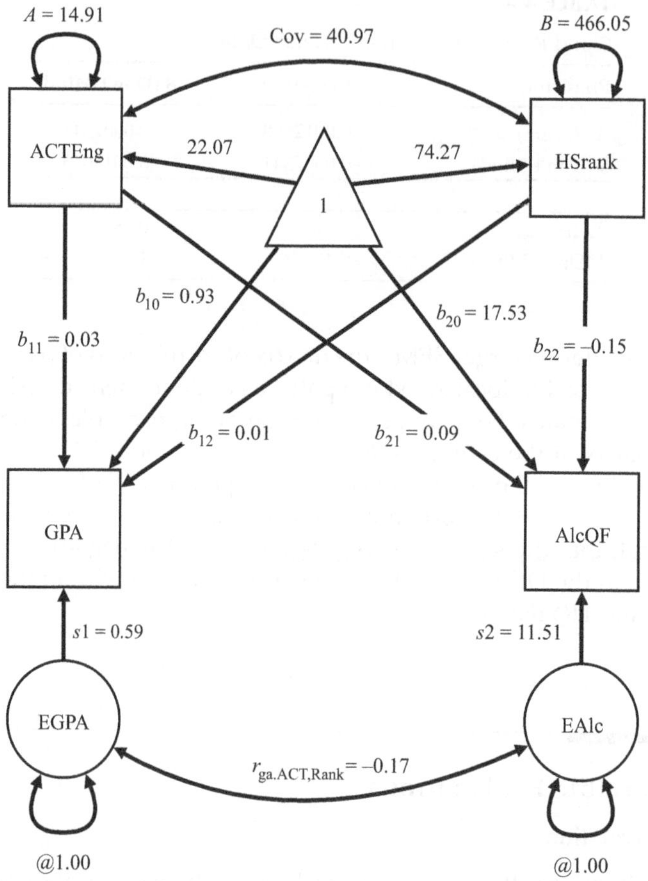


**FIGURE 4.5**
Partial correlation between GPA and alcohol use adjusted for effects of ACT English and high school rank.

for example). In such situations, the researcher may be interested in assessing the degree of association between two variables after adjusting scores for such confounders or "third variable" explanations. Figure 4.5 is a path diagram showing the partial correlation between freshman cumulative GPA and alcohol quantity/frequency consumption. In this diagram, unstandardized regression coefficients are reported, but since the slings associated with error variances are one, the sling joining the errors corresponds to a correlation. The partial correlation between Freshman GPA and Alcohol Quantity/Frequency is shown at the bottom as $r_{GA.ER}$, which is interpreted as the correlation between grades and alcohol after partialing out the effects of English and high school class rank.

We can see from the diagram that two two-predictor regression models are shown in the figure. The regression weights $b_{11}$ and $b_{12}$ correspond to the unstandardized regressions weights predicting GPA from English and high school class rank, respectively, and the regression weights $b_{21}$ and $b_{22}$ represent these same predictors with Alcohol Quantity/Frequency as a criterion. From this diagram we can see that high school class rank (HSrank) appears related to both undergraduate GPA and Alcohol Quantity/Frequency, with individuals who had better high school class ranks less likely to consume alcohol and more likely to perform better during the freshman year of college. ACT English, on the other hand, appears related most strongly to undergraduate GPA. As a result, HSrank appears to be the stronger third-variable explanation for the association between collegiate alcohol consumption and undergraduate GPA, but this association does not completely explain the association.

The value $s_1$ represents the standard deviation of errors of prediction associated with predicting GPA from English and HSrank, while $s_2$ represents the standard deviation of errors of prediction predicting alcohol quantity/frequency from English and HSrank. If standardized versions of the variables were used, these would have represented the square root of 1 minus the squared multiple correlation of these regression models. Because the deviation score matrix was used, these quantities represent the standard deviations of the error variances associated with these two dependent variables.

Provided that the regression assumptions of the model are met, this correlation represents an estimate of the correlation one would expect if a sample of individuals were observed who all had the same values on the predictor variables of ACT English and high school GPA.

## Semi-Partial Correlations (Part Correlation)

In other settings, a researcher may wish to adjust only one of the variables in the model and leave the other unadjusted. Such situations occur when the researcher is interested in examining if an association remains between an adjusted variable with some other unadjusted variable. Because a regression weight performs this adjustment as well, the statistical significance of a semi-partial correlation is the same as that for its corresponding regression weight. First, the absolute value of the semi-partial correlation of a predictor with a criterion is always less than or equal to that of the partial correlation in which both variables are adjusted. No causal assumption about the variable is made; however, the semi-partial correlation is simply a standardized magnitude of association. The path diagram for a semi-partial correlation treating freshman GPA as a criterion is shown in Figure 4.6.

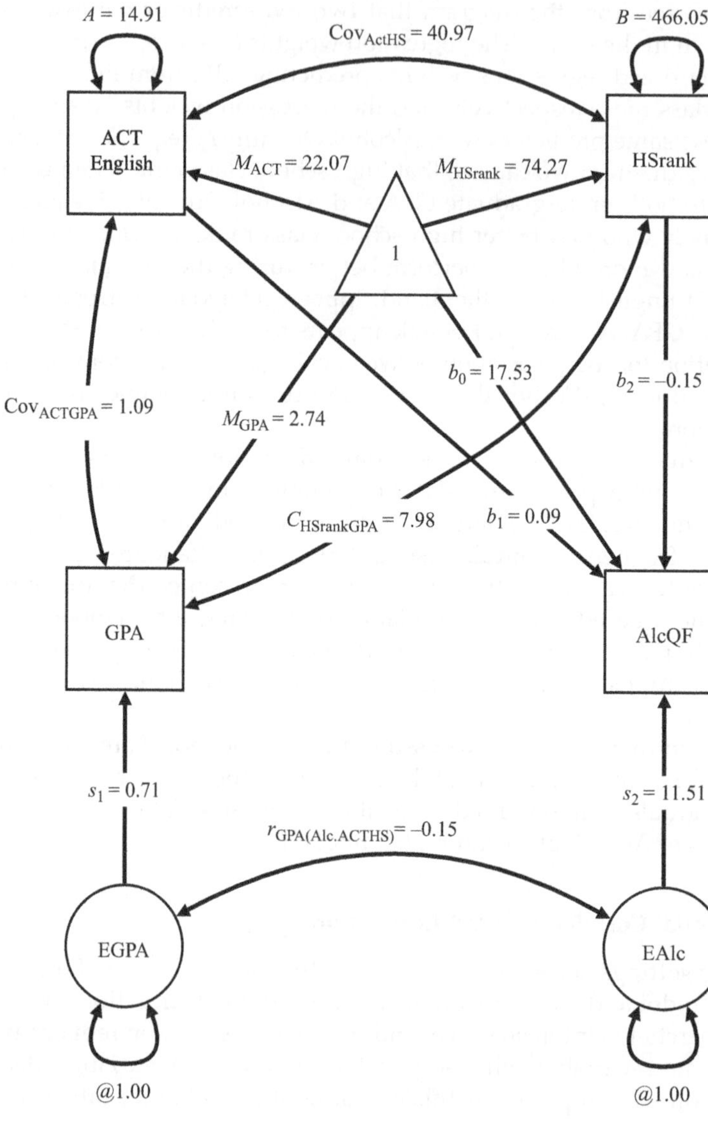


**FIGURE 4.6**
Semi-partial correlation between alcohol use (adjusted for ACT English and high school Rank) with high school GPA.

Notice that the directional arrows of Figure 4.5 involving freshman GPA have been replaced by slings, making the path $s_1$ the standard deviation of the criterion variable. In this model, $s_1$ is an estimate of the standard deviation of freshman GPA.

## Three-Indicator, One-Factor Model

Although much more will be said about latent variables later, it is helpful at this point to consider the path diagram for a single-factor, three-indicator model as shown in Figure 4.7. This diagram differs from those usually shown for confirmatory models in that it is based on the average adjoined SSCP matrix. However, because manifest variable intercepts are estimated, the resulting parameters describe variances and covariances between the variables. One could just as easily have employed centered versions of the data and modeled the data without intercepts. The estimated paths would be the same. Similarly, if standardized versions of the variables were employed, the resulting parameter estimates would look as shown in Figure 4.8.

In both models, an unmeasured latent variable $F$ is thought to underlie all patterns of covariation in the data and a substantial portion of the variability in each manifest variable as well. So long as the matrix is of full rank (meaning, informally, that no combination of two variables completely predict the third), it is possible to decompose a three-variable study in terms of a single latent variable with estimated intercepts.

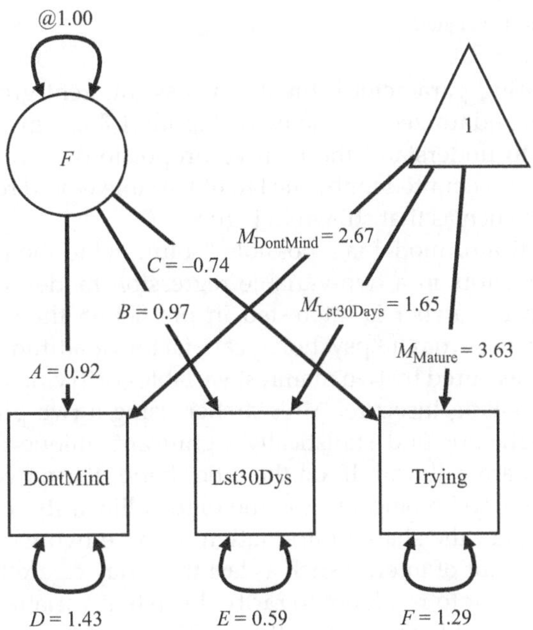


**FIGURE 4.7**
Unstandardized one-factor model.

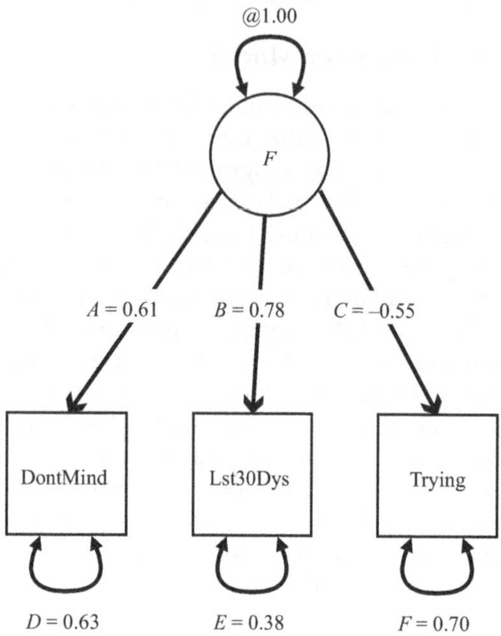


**FIGURE 4.8**
Standardized one-factor model.

When comparing parameter estimates across different groups, examination of the unstandardized estimates of Figure 4.7 are most appropriate. Those seeking to understand the relative proportions of variability in the manifest variable accounted for by the latent variable would refer to the standardized values such as that shown in Figure 4.8.

This three-indicator model is a possible "solution" to the problem of collinearity of predictors in a two-variable regression model. It could be, for example, that a researcher is interested in predicting the behavior of last 30-day cigarette use from a "psychological" factor of attitudes about smoking, which are measured by two manifest variables of Trying to Look Mature and Smokers Looking Insecure. A researcher using a two-predictor regression model might not find statistically significant unique effects because these predictors are collinear. If, on the other hand, these two variables are indicators of the psychological factor, the high collinearity results in higher factor loadings, and the shared information of these two variables, in turn, predicts the behavior of interest such as last third-day cigarette use.

It is also important to recall not to rarify this latent variable. A latent variable is simply a variable that "is not there." The "latent variable" may actually be the net effect of several variables or a single manifest variable. For example, some research suggests that tobacco use is tied to socioeconomic status (SES). It could be, for example, that the associations between manifest variables due to the latent variable would be far less dramatic if a manifest variable corresponding to SES were included as a predictor of the manifest variables (i.e., is a covariate), or if all individuals under study came from the same SES level. Even if the research question at hand does not deal with SES, it may be advisable to include such variables in the model in order to better assess the relationships between the variables of interest. Failing that, a discussion of such unincluded variables are important points to mention when discussing the limitations of a model.

To summarize, a latent variable model may be a "solution" to the problem of collinearity of predictor variables, a way to estimate whether one or more missing variables can predict a number of manifest variables, or a way to assess whether three variables would be indicators of a "common" latent variable such as tobacco-related attitudes/behaviors. Third-variable associations due to unincluded variables, constructs, subgroup differences, or methods of assessment constitute conceptual limitations of the model.

## Quadratic Regression

In some situations, a quadratic regression line is thought to underlie the data. In this case, the second predictor in the regression consists of squaring the predictor variable and entering it as another predictor variable. The example data set considered here is a rather simple one, but I include it here because it introduces the idea of centering data to reduce collinearity between predictors. Consider the data presented in Figure 4.9, which shows the relationship between fuel efficiency and mileage in data taken from 15 cars that were assessed at speeds ranging from 20 to 55 miles per hour (data taken from Davis & Boundy, 2018).

These data show a strong curvilinear trend. It is inadvisable, however, to simply square the speed variable and enter this into a regression equation, as the squared predictor will be highly correlated with the original data. In these data, for example, the correlation between speed and its square is 0.99. As a result, the correlations between the regression estimates for such a large regression and the likelihood of identifying either predictor as statistically significant is low. If the predictor variable is centered, however, the correlation between the linear and quadratic predictors is greatly reduced and will usually be modest except in very extreme situations. The path diagram for a polynomial regression is presented in Figure 4.10. The fact that the estimated variance of the speed variable is equal to the variance of the square of speed is not accidental, but due to the fact that the speed variable is uniformly distributed. This is not usually the case unless the study has uniform levels of

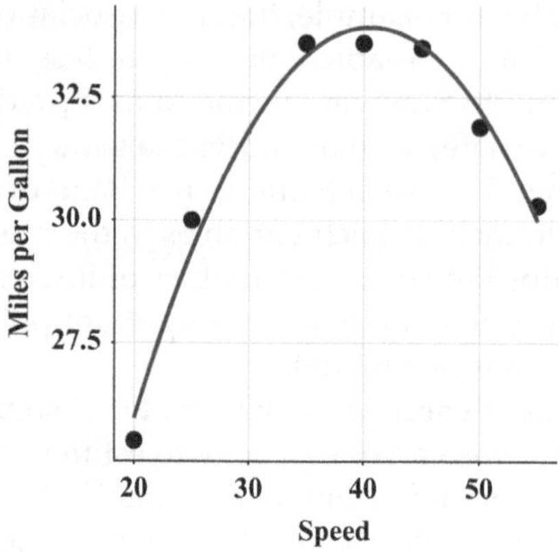


**FIGURE 4.9**
Quadratic regression example.

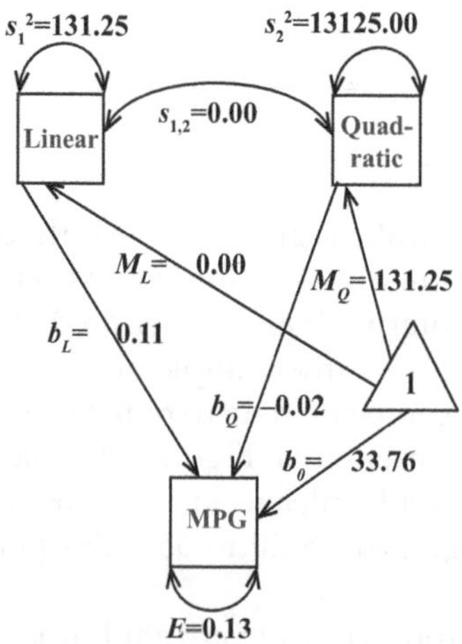


**FIGURE 4.10**
Path diagram for quadratic regression.

the criterion as a design feature. The average MPG is calculated, as for any two-predictor model, as the intercept plus the sum of the compound paths involving the means of the predictors multiplied by the regression weights – in this case, 33.76 + 131.25(−0.02).

If we replace MPG with $Y$ and the Speed variable by $X$, the regression model fit by this diagram corresponds to the equation:

$$Y = 0.11X - 0.02X^2 + 33.76 \quad (4.8)$$

Calculating simple derivatives of a quadratic function such as this allows the researcher to determine the location of the maximum (or minimum) of the predicted regression line. This is useful for determining, in this example, the speed at which mileage is highest. Recall that the first derivative of a function is calculated by multiplying each term in Eq. (4.8) by its exponent and reducing the exponent of that term by one. In this example, this would yield:

$$\frac{dY}{dX} = 0.11476 - 0.03781X = 6.0705 \qquad (4.9)$$

which means that the maximum/minimum occurs 3.03 units above the mean speed in the study (which was 37.5), meaning that the optimal speed occurs at 40.53 mph. In order to determine whether this critical point on the curve is a maximum (highest possible value) or minimum, we calculate the derivative of 4.9, which is then −0.0189. The fact that this quantity is less than zero implies that the function is a maximum at this value.

## Mediational Models

In some situations, theory assumes a causal ordering among predictors of a criterion. The questions of interest then focus on the proportion of criterion variation accounted for by the direct and compound paths relating a predictor to the criterion. Some individuals refer to these types of models as "mediational" effects (Baron & Kenny, 1986) because the research questions focus on the degree to which the relationship between a predictor and criterion is "mediated" by intervening variables. To consider such models, let's return to the Widaman and Reise data set again, but this time focusing on the predictor variables asking the boys whether they thought that people who smoked looked "mature and sophisticated" (Mature) or "cool, calm, and in control" (Cool). The dependent variable remains cigarette consumption in the past 30 days. Suppose that a researcher wanted to test the hypothesis that the reason that the Cool variable predicted cigarette use was because being cool gave the appearance of looking mature and sophisticated. As a result, the researcher is interested in knowing the strength of the unique effect of Cool predicting cigarette use and the effect of Cool on cigarette use via the compound path through mature. The path diagram corresponding to this model is given in Figure 4.11. This figure represents two regression models. A single predictor regression model is shown at the top predicting the Mature variable from the Cool variable. The second model is a two-predictor model predicting cigarette use from both the Cool and Mature variables.

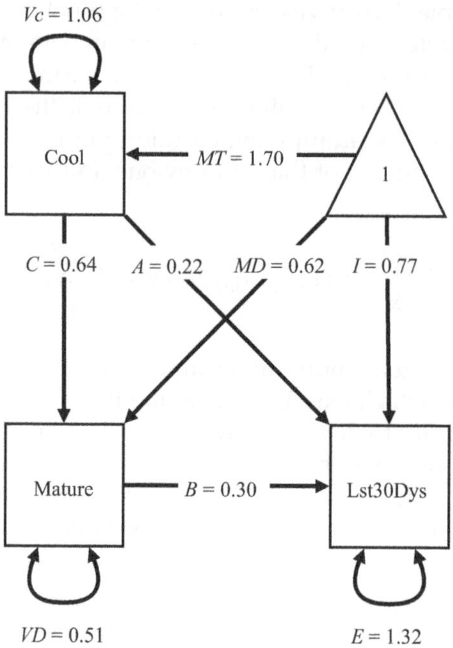


**FIGURE 4.11**
Mediational model of last 30-day consumption of mature by cool.

Before considering the direct and compound effects in the model, it is first helpful to discuss what does and does not change when comparing a standard two-predictor model to this model, which assumes a causal link between the predictors. By the tracing rules, there is no covariance trip going from Cool to Mature that involves cigarette use, as this would involve going forward and then back, which is not allowed. As a result, the covariance between Cool and Mature is $V_C A$, meaning that $A = \text{CoV}(\text{Cool}, \text{Mature}) / V_C$, i.e., $A$ represents the usual single variable regression weight and differs from the usual covariance only by division by the variance of Cool.

In standard parlance of mediational models, coefficient $A$ is referred to as the unique effect of Cool on cigarette use, and it becomes a natural question as to whether the causal structure assumed between the aforementioned predictor variables results in different estimates of the association between Cool and cigarette use, by using the tracing rules for the model just described to solve for the predicted covariances and comparing them with the predicted covariances from a two-predictor regression model. However, it should be clear that the regression weights from the two models will be identical – i.e., the causal ordering of the predictor variables has no effect on the parameter estimates associated with the criterion so long as the causal ordering among the more exogenous variables completely explains all covariation between the predictors. The "unique" effect of Cool on the criterion is therefore equal to the unstandardized regression weight in a two-predictor model. There is no need, therefore, to consider separate regression models to estimate unique effects under various scenarios of causal ordering if the researcher is interested only in unique effects. These are readily available from the regression analysis alone.

A second question, however, concerns whether the compound path predicting cigarette use from Cool via Mature is statistically significant. In this situation, this compound path is equal to $V_C AB$, which is equal to 0.19. The total effect, then, is the sum of all indirect (compound paths involving more than one arrow) and direct effects (single path effects) in the model. In this case, 0.19 + 0.22 = 0.41.

The statistical significance of indirect effects cannot be tested using the Wald statistic because the sampling distribution of products of path coefficients cannot be thought of as normally distributed. For this reason, bootstrap sampling is often used to determine the standard error associated with the statistical significance of mediational effects. A bootstrap estimate of the confidence interval approximates the sampling distribution of an estimator by repeatedly resampling from the observed data with replacement. Similarly, if the same model is estimated based on standardized estimates, the direct, indirect, and total effects associated with Cool are 0.18, 0.16, and 0.34, respectively. We would conclude from this that there are small but significant indirect and direct effects on the prediction of cigarette use in the data. See Cole and Maxwell (2003) for a description of the conceptual issues and misconceptions associated with mediational models such as those shown in Table 4.5.

**TABLE 4.5**

Total, Direct, and Indirect Effects

| | | | 95% Confidence Interval[a] | | | | |
|---|---|---|---|---|---|---|---|
| **Effect** | **Estimate** | **SE** | **Lower** | **Upper** | ***Z*** | ***p*** | **% Mediation** |
| Indirect | 0.193 | 0.0386 | 0.118 | 0.268 | 4.99 | <.001 | 47 |
| Direct | 0.217 | 0.0583 | 0.102 | 0.334 | 3.72 | <.001 | 53 |
| Total | 0.409 | 0.0443 | 0.323 | 0.496 | 9.23 | <.001 | 100 |

[a] Based on 1,000 bootstrap samples.

## Categorical Independent Variables

In Chapter 3, a single predictor regression model functioned as a *t*-test between two groups using a predictor variable consisting of 1's and 0's to code for group membership. It is also possible to consider other linear models as special cases of two predictor models. These models include dummy, effect, and orthogonal coding strategies for three-group one-way designs, analysis of covariance, and a main-effects-only model for a 2 × 2 experimental design. Extending the model to three predictor variables enables us to consider analysis of a 2 × 2 experimental design with interaction in addition to the four-group one-way experimental design. Although the discussion here involves simple manifest variables, it is possible to extend the linear model to analogous latent variable models. To begin, consider the following example data set taken from Pedhazur (1997), shown in Table 4.6.

### ANOVA Model

One way to analyze such data is with the ANOVA model (sometimes called the Cell Means model) in which the means of each group are estimated using dummy variables *D*1, *D*2, and *D*3, which are scored as one if an observation is taken from the respective group and zero otherwise (Kutner, Nachtsheim, Neter, & Li, 2005). The resulting regression weights will then consist of

**TABLE 4.6**

Coding Strategies for One-Way Analysis of Variance

| | | Dummy Coding | | | Orthogonal Coding | | Effect Coding | |
|---|---|---|---|---|---|---|---|---|
| *Y* | Group | *D*1 | *D*2 | *D*3 | *O*1 | *O*2 | *E*1 | *E*2 |
| 1 | 1 | 1 | 0 | 0 | −1 | 0 | 1 | 0 |
| 2 | 1 | 1 | 0 | 0 | −1 | 0 | 1 | 0 |
| 3 | 1 | 1 | 0 | 0 | −1 | 0 | 1 | 0 |
| 4 | 1 | 1 | 0 | 0 | −1 | 0 | 1 | 0 |
| 5 | 1 | 1 | 0 | 0 | −1 | 0 | 1 | 0 |
| 4 | 2 | 0 | 1 | 0 | 0.5 | 1 | 0 | 1 |
| 5 | 2 | 0 | 1 | 0 | 0.5 | 1 | 0 | 1 |
| 6 | 2 | 0 | 1 | 0 | 0.5 | 1 | 0 | 1 |
| 7 | 2 | 0 | 1 | 0 | 0.5 | 1 | 0 | 1 |
| 8 | 2 | 0 | 1 | 0 | 0.5 | 1 | 0 | 1 |
| 7 | 3 | 0 | 0 | 1 | 0.5 | −1 | −1 | −1 |
| 8 | 3 | 0 | 0 | 1 | 0.5 | −1 | −1 | −1 |
| 9 | 3 | 0 | 0 | 1 | 0.5 | −1 | −1 | −1 |
| 10 | 3 | 0 | 0 | 1 | 0.5 | −1 | −1 | −1 |

estimates of the mean for the groups ($M_{G1}$, $M_{G2}$, and $M_{G3}$), as shown in Figure 4.12(a). Means of the three dummy variables ($M_1$, $M_2$, and $M_3$) are equal to 1/3, given that one-third of the observations in each dummy variable are coded as ones. The error variance of $Y$, $V_E$, represents an estimate of the common standard deviation across all three groups.

Wald tests of these path coefficients test whether the mean of each group is significantly different from zero and will closely resemble those of the $t$-statistics for a regression with no intercept; but, because they divide by the square root of $N-1$ instead of the square root of $N-k-1$, they will be slight overestimates of the proper $t$-statistic. We can also use the tracing rules to calculate the overall $R^2$ for the model (0.75), which could then be used to reproduce the $F$-statistic for the model from a standard analysis of variance.

Notice that the negative covariances between dummy variables mean that the regression model shown here demonstrates a pattern of suppressor effect if all three means are greater than zero. This will always be the case for dummy coded effects, with more categories being associated with weaker correlations. We can calculate the correlation between dummy vectors in this example as $\frac{-0.11}{0.22} = -0.50$.

One can also think of these models in terms of the matrix algebra involving the criterion vector $y$ and the $X$ matrix of predictors involved from Chapter 2. The criterion vector of $Y$ is:

$$y = \begin{bmatrix} 1 \\ 2 \\ 3 \\ 4 \\ 5 \\ 4 \\ 5 \\ 6 \\ 7 \\ 8 \\ 7 \\ 8 \\ 9 \\ 10 \\ 11 \end{bmatrix} \text{ and the X matrix of predictors is: } \begin{bmatrix} 1 & 0 & 0 \\ 1 & 0 & 0 \\ 1 & 0 & 0 \\ 1 & 0 & 0 \\ 1 & 0 & 0 \\ 0 & 1 & 0 \\ 0 & 1 & 0 \\ 0 & 1 & 0 \\ 0 & 1 & 0 \\ 0 & 1 & 0 \\ 0 & 0 & 1 \\ 0 & 0 & 1 \\ 0 & 0 & 1 \\ 0 & 0 & 1 \\ 0 & 0 & 1 \end{bmatrix}.$$

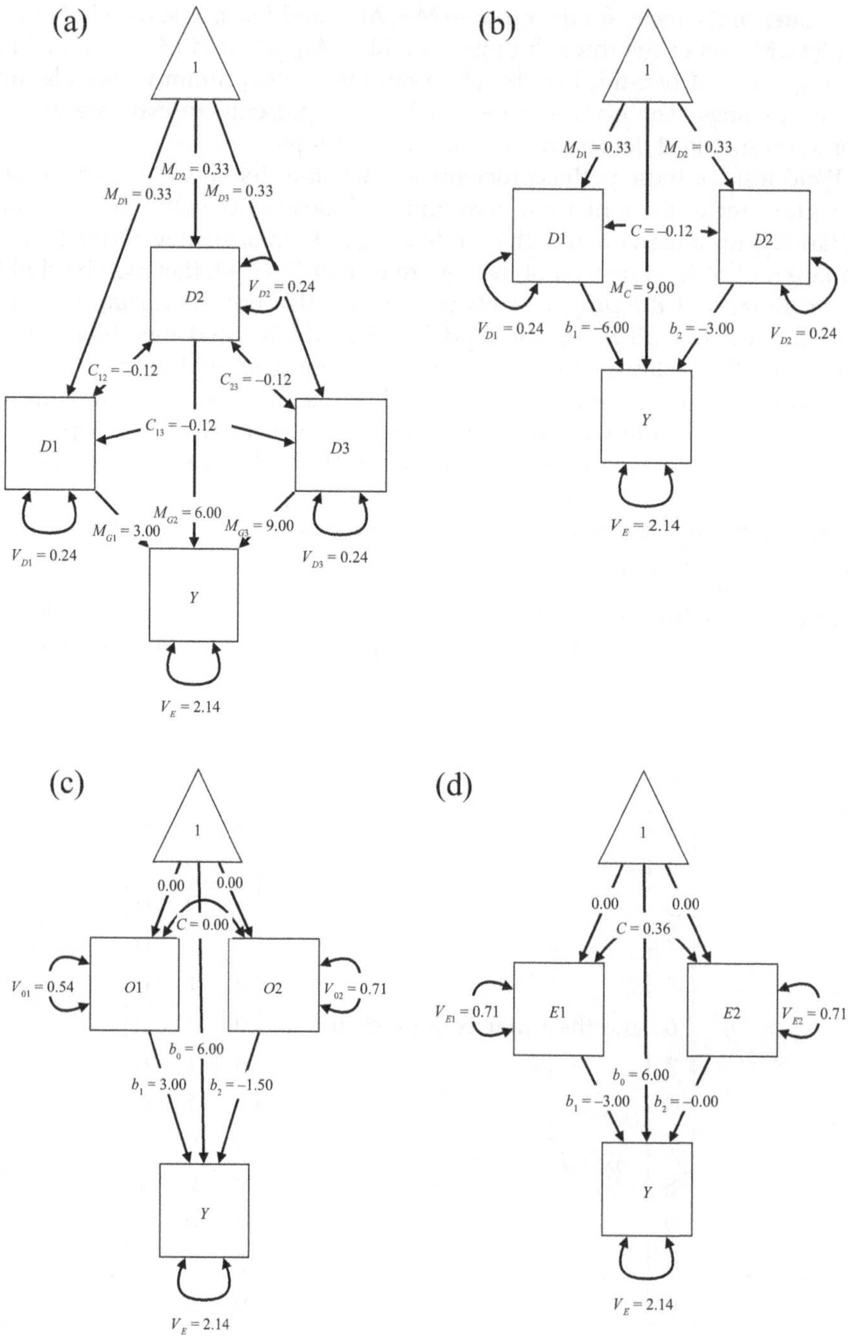


**FIGURE 4.12**
Coding strategies for ANOVA (a), dummy (b), orthogonal (c), and effect coding (d).

The column of ones across all rows used in Chapter 2, though, is not present. The three dummy coded vectors are pointed at by the column of ones in the diagram. This means that the paths from the column of ones to each of the dummy variables represents the mean of each dummy variable, which is 1/3. Slings associated with the dummy coded variables represent variances and covariances because they are pointed at by the column of ones.

## Dummy Coding Relative to a Control Group

Alternatively, we may be interested in comparing the means between groups. For example, if the first group were to be treated as a control group and the other two groups were treatments, we might be interested in knowing whether the estimated treatment for a given group was systematically higher or lower than the control group. In this situation, only two vectors are necessary, and the freely estimated intercept in the diagram corresponds to the mean of the control group. Figure 4.12(b) shows the path diagram for such a model. The regression intercept in this model corresponds to the mean of the control group.

If we examine the path coefficients, we see that the pattern presented is one of suppressor effect if the two other groups are both above or below the control group. The statistical significance of the path coefficients now answers the question of whether the mean for a given group (Group 2 for $b_1$, or Group 3 for $b_2$) is different from the mean for Group 1. Had Group 2 been considered as the control group, a model comparing performance of Groups 1 and 3 to 2 would demonstrate a pattern of redundancy because all direct and compound paths would be the same sign. As before, $R_{Y.12}{}^2$ can be calculated by dividing the variance explained by the predictors by the total variance of the dependent variable. Statistical significance tests of the individual regression weights from the Wald test are similar to those from a standard regression except for the slightly less conservative standard errors. The matrix algebra representation of the regression is the same as before, except that we include the column of ones and only the two dummy coded vectors.

## Orthogonal Coding Contrasts for Three Groups

Alternatively, one could ask whether the average of Groups 2 and 3 is different from the control group and, in addition, whether evidence exists that Group 2 is different from Group 3. These two questions are orthogonal contrasts, meaning that the covariance is zero between the vectors used to answer these questions. Specifically, the first hypothesis is tested by constructing a predictor variable, which is −1 for Group 1 and 0.5 for observations in Groups 2 or 3 as shown in Table 4.6. Note that this coding strategy is merely calculating a mathematical average across the rows of the observations; i.e., it is calculating the grand mean centered sum, $O_1 = \frac{\bar{X}_1 + \bar{X}_2}{2} - \bar{X}_3$, and using the tracing rules to enable us to calculate the variance and mean of this variable within the path diagram. The second contrast is scored zero if the observation is from the first group, 1 if it is from the second group, and −1 if it is from the third group, i.e., the mean difference of the centered means of Groups 2 and 3. The resulting path diagram is shown in Figure 4.12(c). The intercept in this model now represents the grand mean, or the average across all observations in the data set. The means of the three groups are calculated relative to this by multiplying the coding for that group in the two effect coded predictors. For example, Group 1 was coded −1 on *O*1 and zero on *O*2. As a result, the predicted mean for that group is 6 + 3(−1). Group 2 received a 0.5 on *O*1 and 1 on *O*2, so the predicted mean for that group is: 6 + 3(0.5) + −1.5(−1) = 9. For Group 3, the predicted mean is: 6 + 3(0.5) + (−1.5)(1) = 6.

The same statistics can be calculated as before (model $R^2$, statistical significance of paths, etc.) The advantages of this approach lie in the fact that each variance component can be directly calculated as the square of the respective regression weight. If one were to use a standardized score matrix instead, the intercept is, of course, zero, and the square of the standardized coefficients (0.75 and −0.43 for *O*1 and *O*2, respectively) are equal to the proportion of variance in *Y* accounted for by the hypothesis.

## Effect Coding for Three Groups

Effect coding is another strategy of coding data with a system of 1's, 0's, and −1's, but this time with the objective of comparing the means of individual groups to the overall mean, sometimes referred to as the grand mean. As with orthogonal coding, two coded vectors are possible, with the estimated intercept representing the grand mean. In Figure 4.12(d), the means of Groups 1 and 2 are compared to the grand mean. As shown in Table 4.6, the predictor variable $E_1$ designed to compare Group 1 to the grand mean is coded as 1 for individuals in Group 1, 0 for those in Group 2, and −1 for those in Group 3. The second variable $E_2$ is coded as 0 for Group 1, 1 for Group 2, and −1 for Group 3. No statistical test for an added variable comparing Group 3 to the grand mean can be estimated, because the mean of Group 3 is known once you know how far Groups 1 and 2 differ from the grand mean.

As can be seen in the figure, the intercept, $b_0$, corresponds to the grand mean of the data, effect coded vectors $E_1$ and $E_2$ have a positive covariance. The correlation between these two predictor variables is 0.5 in this example. Thus, if the two coded vectors describe values above the grand mean, a pattern of redundancy is present in the estimates of the regression weights, and if one is below the grand mean but not the other, a pattern of suppressor effects is present. In this figure, we can see that Group 1 is estimated to have a mean three points below the grand mean of 6, and Group 2 is estimated to be equal to the grand mean. As for the dummy, orthogonal, and ANOVA models, the estimated error variance of $Y$, $V_E$, is 2.

## Dependent *t*-Test

It is possible to specify an SEM to conduct a dependent *t*-test using the small structural components mentioned in Chapter 3. As an example of this, consider a data set considered by Charles Darwin, who was interested in knowing if cross-fertilization resulted in improved growth in plants (as discussed by Fisher, 1966). He selected 15 individual plants that were either self-fertilized or cross-fertilized and recorded their growth. The data are shown in Table 4.7.

In Figure 4.13, the height of the self-fertilized plants have an estimated mean of 140. The increase in height associated with cross-fertilization is shown as 21.53 because it expresses the gain over 140 communicated by the height factor, which has paths to both self-fertilized and cross-fertilized plants. The standard deviations of the self- versus cross-fertilized plants is

**TABLE 4.7**

Darwin's 1876 Data on the Effects of Cross-Fertilization

| | Fertilization | |
|---|---|---|
| **Plant** | **Cross** | **Self** |
| 1 | 188 | 130 |
| 2 | 96 | 163 |
| 3 | 168 | 160 |
| 4 | 176 | 160 |
| 5 | 153 | 147 |
| 6 | 172 | 149 |
| 7 | 177 | 149 |
| 8 | 163 | 122 |
| 9 | 146 | 132 |
| 10 | 173 | 144 |
| 11 | 186 | 130 |
| 12 | 168 | 144 |
| 13 | 177 | 102 |
| 14 | 184 | 124 |
| 15 | 96 | 144 |

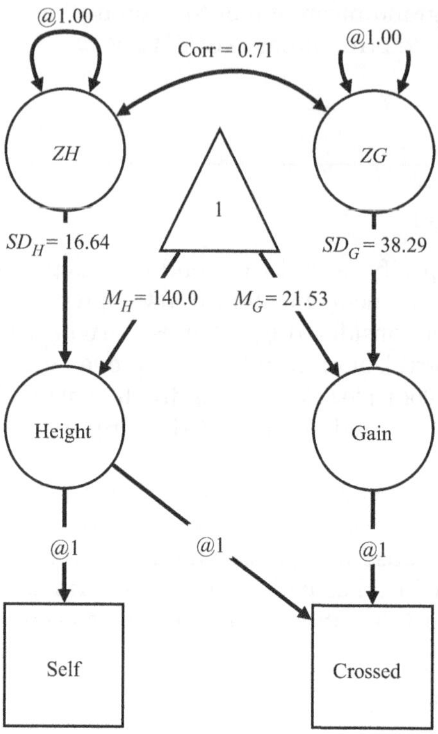


**FIGURE 4.13**
Dependent *t*-test for Darwin cross-fertilization data.

shown by the paths from the *ZH* and *ZG* latent variables and the correlation between the self- and cross-fertilized plants is estimated as 0.71.

## Analysis of Covariance for Two Groups

One may think of this as a type of blocking of the experimental design in which each block contains only one individual. In some situations it is not possible to ascertain individuals across conditions, but a good indicator related to inter-individual differences is available. If the indicator is continuous in nature, if the relationship between the indicator and the criterion is linear, and if the functional relationship across all groups is well characterized by a common slope, analysis of covariance (ANCOVA) provides a more powerful test of inter-individual differences than available by a simple *t*-test of group differences alone.

As a small example of analysis of covariance involving two groups, consider the following data set. Weight loss medications such as Ozempic (semaglutide) work by reducing levels of glucagon in the bloodstream. In this fictional example, the initial levels of glucagon and the kilos of lost weight using the weight loss drug are examined. There is some question as to whether men experience more weight loss using the drug than women. A scatterplot of the data is shown in Figure 4.14. As can be seen, the data from men largely occur above the regression line and that of women fall below it. If a simple *t*-test comparing the means of men and women is conducted, no significant difference between men and women is found $\left(t(8) = -1.14, \underline{p} = 0.29\right)$ and a correlation between glucagon and weight loss is found $\left(\underline{r} = 0.68, \underline{p} = 0.03\right)$.

Neither of these statistics, however, seems to capture the differences between men and women in the scatterplot or the strength of the association between the predictor and criterion in men and women separately. If we dummy code sex of participant as 1 = Male and 0 = Female, the data from the study look like what appears in Table 4.8.

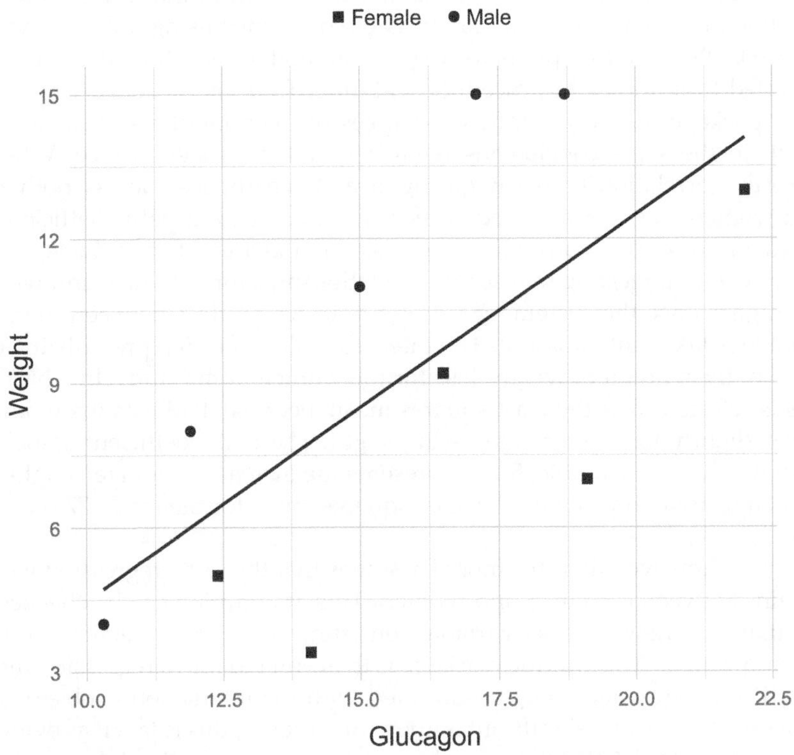


**FIGURE 4.14**
Scatterplot of Ozempic data.

**TABLE 4.8**
Example Weight Loss Data

| Glucagon | Glucagon Centered | Weight | Sex |
|---|---|---|---|
| 10.3 | −5.41 | 4 | 1 |
| 11.9 | −3.81 | 8 | 1 |
| 15 | −0.71 | 11 | 1 |
| 17.1 | 1.39 | 15 | 1 |
| 18.7 | 2.99 | 15 | 1 |
| 12.4 | −3.31 | 5 | −1 |
| 14.1 | −1.61 | 3.4 | −1 |
| 16.5 | 0.79 | 9.2 | −1 |
| 19.1 | 3.39 | 7 | −1 |
| 22 | 6.29 | 13 | −1 |

The path diagram corresponding to an analysis of covariance for these data is given in Figure 4.15. This analysis is useful in that it replicates the solution one obtains from an analysis of covariance using a GLM package for ANCOVA or a two-predictor regression model. As before, the *t*-statistics reported from the GLM procedure will differ from the Wald statistics of an SEM package due to the inflated degrees of freedom of the Wald test but result in the same conclusions regarding statistical significance. Whereas the raw correlation between glucagon and weight loss across both men and women was 0.68 as noted earlier, the standardized path coefficient for glucagon from the two-predictor regression model was 0.89, a substantially larger value suggesting that separate relationships for both men and women are higher once the systematic difference in weight loss between men and women is taken into account. The intercept value of 6.35 represents the estimate of the amount of weight loss for the women in the study. In ANCOVA this is referred to as the least squares mean, because it adjusts the women's mean slightly for the differences in weight. The path coefficient associated with the dummy variable, 5.41, represents the increased average weight loss of men relative to women. The least squares mean for men is 11.77 (i.e., 5.41 + 6.35).

Notice, however, that this model assumes that the same regression weight obtains between both men and women. This assumption can be checked by calculating a new product variable consisting of centered versions of both sex and weight. This product variable is then entered into a regression model and, if the regression weight associated with the interaction is observed to explain a statistically significant amount of variance, this is taken as evidence that the standard ANCOVA model is not appropriate for the data.

If this product variable is significant, some authors (such as Kutner et al., 2005) recommend transforming the data so that it is no longer significant.

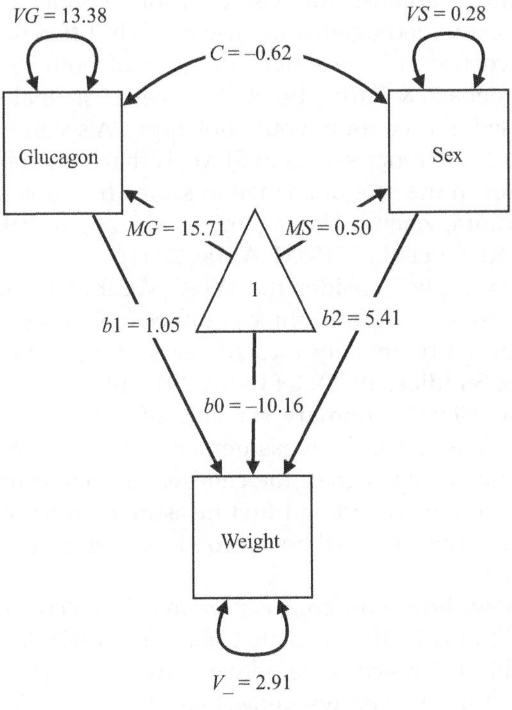


**FIGURE 4.15**
Path diagram for simple analysis of covariance.

In other cases, however, the differential effect may be a substantive question in its own right, sometimes referred to as a "moderator effect" (e.g., Baron & Kenny, 1986). Preferred ways of dealing with such interaction effects are to use the Latent Moderated Structural (LMS) Equations approach (available in the Mplus software package; Klein & Moosbrugger, 2000), Bayesian inference with product variables (available in the Blavaan program; Merkle & Rosseel, 2018), or use of a multigroup model in which one of the variables is categorized.

## Difference Scores

In longitudinal studies, researchers may be interested in the degree of change in scores across measurement occasions. Although this could take the form of simply calculating a difference score (say, Time 2 – Time 1), a path diagram can also be used to model the difference score as a latent variable. Again, in the case of manifest variables, there is no reason to prefer an SEM to do this, and some have cautioned against the use of such difference scores because growth scores compound measurement error of both the test and retest assessments (Cronbach & Furby, 1970). If, however, an analogous latent variable model is used, this caution would not apply. A second reason to understand the logic of difference scores in SEMs is that it then becomes possible to model changes in the rate of change in scores (i.e., acceleration or deceleration; e.g., Grimm, Zhang, Hamagami, & Mazzocco, 2013; Hamagami & McArdle, 2007; Kievit et al., 2018; McArdle, 2009).

As a simple example, consider the WISC Vocabulary data of 273 Black and White children drawn from three counties in Georgia who were tested from the ages of approximately 6 to 10 years of age (Osborne & Lindsey, 1967; Osborne & Suddick, 1972). Of these, 204 children had data on all four measurement occasions. Summary statistics of the raw score of children's vocabulary from the first three measurement occasions are shown in Table 4.9 along with the average age of the children at each testing occasion. The difference between the second and first measurement occasions (Diff1) and the difference between the third and second measurement occasions (Diff2) are shown as well.

Figure 4.16 shows how a difference score may be calculated using an SEM without having to explicitly calculate a manifest variable difference score. Specifically, if the difference score between two measurement occasions is Diff = Time 2 − Time 1, then we know that Time 2 = Diff + Time 1. In the diagram, the unobserved manifest variable is then modeled as a latent variable with no indicators. As can be seen, the mean of the difference score, 7.77, matches the value associated with the variable if it had been calculated as an added variable in the data. In this diagram, the variance of the WISC variable at the first measurement occasion is 40.46 and the mean is 18.78. The average growth at Time 2 is 7.77, and the variance of the difference scores is 17.96. This means that, on average, children increased their scores by 7.77 from 18.78 over the period of 0.88 years.

**TABLE 4.9**

Three Measurement Occasions and Difference Scores

| | Wisc Vocabulary Scores | | | Difference Scores | |
|---|---|---|---|---|---|
| | **Wisc_1** | **Wisc_2** | **Wisc_3** | **Diff1** | **Diff2** |
| Mean | 18.78 | 26.55 | 35.98 | 7.772 | 9.43 |
| Standard Deviation | 6.376 | 7.256 | 7.763 | 4.248 | 4.123 |
| Minimum | 3.54 | 12.19 | 15.72 | −1.61 | −0.52 |
| Maximum | 34.96 | 47.21 | 56.15 | 20.14 | 18.56 |
| Age | 6.066 | 6.946 | 8.80 | 0.88 | 1.85 |

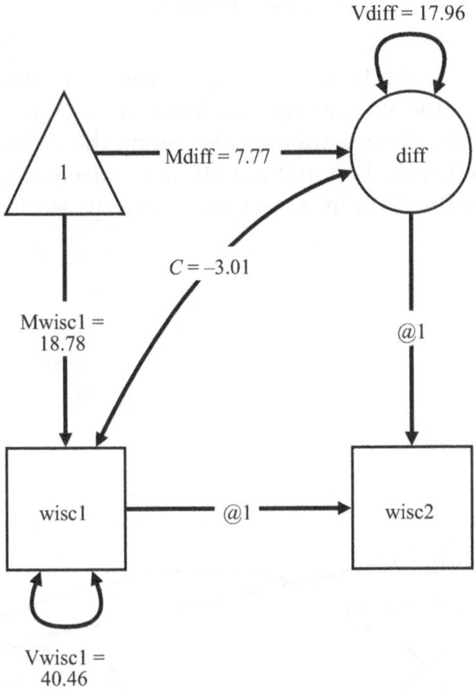


**FIGURE 4.16**
Simple latent difference score.

## Putting Time in the Model

If we were to think about difference scores as representing how fast the children's growth in vocabulary is increasing, we could say that scores were increasing 7.77 per 0.88 years. It might be more straightforward to talk about this change in terms of the anticipated growth over a single year based on time between assessments. It is easy to use path diagrams to calculate the estimated per-year gain in vocabulary scores. To do this, the estimated annual gain is given by

$$D_1 = \frac{T_2 - T_1}{\text{Age}_2 - \text{Age}_1} \tag{4.10}$$

Where $T_1$ and $T_2$ represent the Time 1 and 2 scores, respectively, and $\text{Age}_1$ and $\text{Age}_2$ represent the respective average chronological age at the two times. Rearranging this so that the difference score $T_2$ appears on the left-hand side gives us:

$$T_2 = (\text{Age}_2 - \text{Age}_1)D_1 + T_1 \tag{4.11}$$

We can then model this by fixing the path from the latent difference score to be equal to the chronological age difference in the two measurement occasions. In the case of three measurement occasions, then, it becomes possible to test whether the estimated annual growth in children's vocabulary changes from the second to the third measurement occasion, as shown in Figure 4.17.

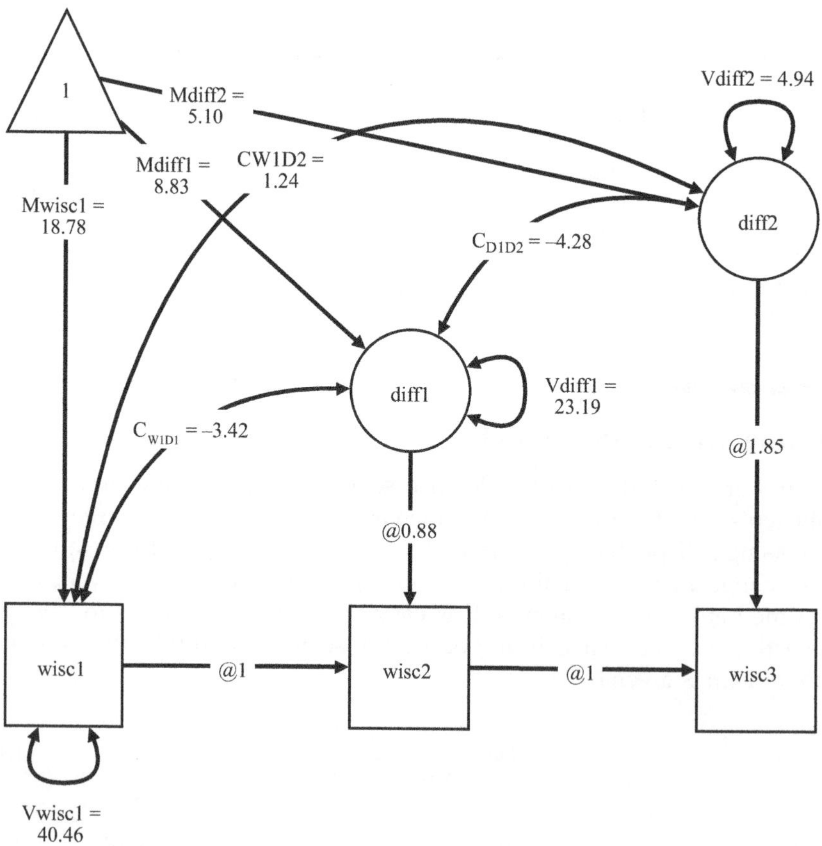


**FIGURE 4.17**
Latent differences in terms of annual growth scores.

## Acceleration of Growth

Just as we might think of estimated annual growth as "how fast" vocabulary is changing, we might also ask whether the rate of growth – i.e., the acceleration or difference in differences – is increasing or decreasing over time. Just as a simple difference score may be thought of as a velocity (i.e., a change from one place to another such as miles per hour), second differences may be thought of as acceleration (i.e., a change in the rate of acceleration; Grimm et al., 2013). Figure 4.18 calculates this difference in difference scores using

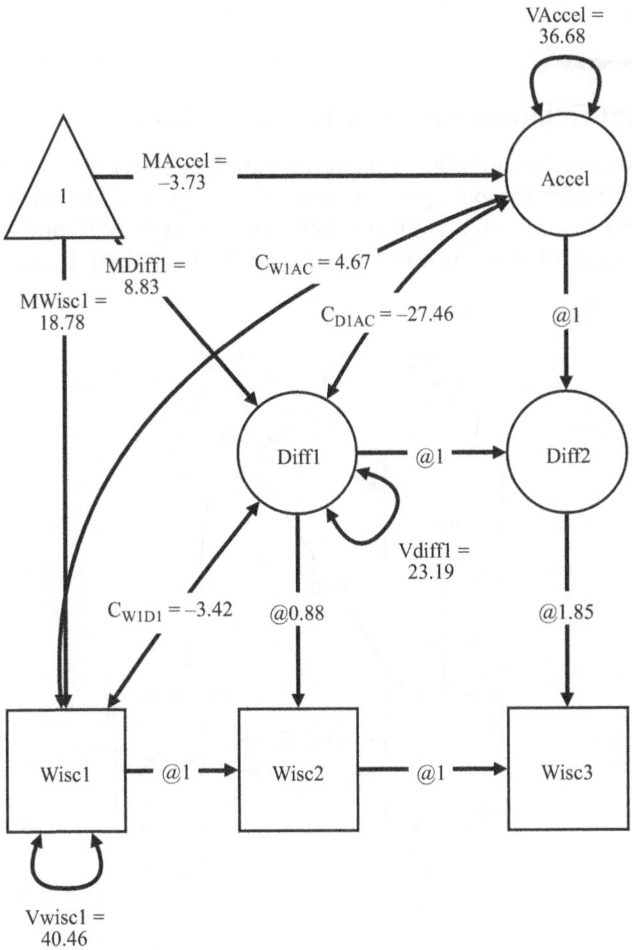


**FIGURE 4.18**
Latent difference scores and acceleration.

the same logic as before, but this time applied to difference scores. If, for example, it is thought that growth in children's vocabulary was accelerating from the second to third testing from that observed between the first two measurement occasions, the difference in differences would be expected to be positive. Conversely, if the rate of growth were slowing down, we would expect the acceleration to be negative.

In Figure 4.18, the second difference, Accel, is −3.73, reflecting the fact that the average growth between the second and third measurement occasions, calculated as 5.10 in Figure 4.18, is less than that observed for the difference between the first and second measurement occasions, which was 8.83.

## Reciprocal Relationships and Feedback Loops

It is not the case that only one directional path can exist between pairs of variables. Figure 4.19, for example, estimates the reciprocal feedback loops associated with educational aspirations between boys (REASP) and their friends (FEASP) in a study by Duncan, Haller, and Portes (1968). In this model, the

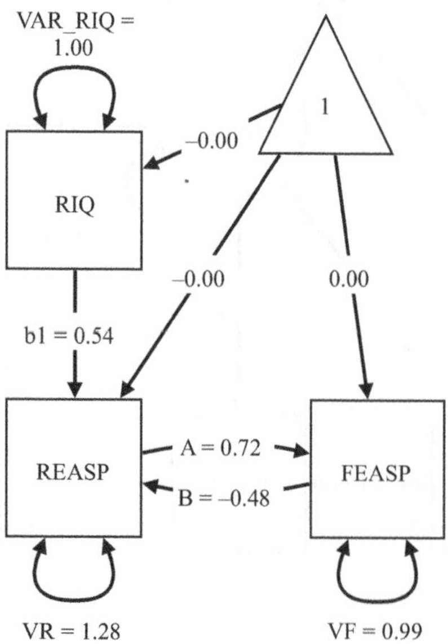


**FIGURE 4.19**
Reciprocal relationships of educational aspirations between students and friends.

boys' IQ test scores (RIQ) are thought to affect only the boys' educational aspirations and not that of their friends. This condition, in which a variable has an association with only one of the downstream variables and not another, is termed an "instrumental variable" (Wright, 1928). While we may think that an instrumental variable is reasonable in controlled experiments, where a variable determining random assignment to an independent variable exists, in correlational studies the reasonableness of the instrumental variable is often asserted rather than proven. Even when a genetic marker constitutes the instrumental variable linked to a behavior, for example, it may be that factors such as assortative mating may invalidate a feedback loop model of individuals and their spouses.

In this model, boys' educational aspirations appear to positively affect those of their friends ($b = 0.72$), but that friends' educational aspirations negatively affect the boys' educational aspirations ($a = -0.48$). In addition, boys' IQ tends to positively affect their educational aspirations ($b_1 = -0.54$). The data that are reported are a correlation matrix, as shown in Table 4.10.

Arriving at these values using the tracing rules is a bit complicated by the fact that the tracing rules describe an infinite series of products according to the tracing rule paths. To see this, consider for the moment just the two variables of REASP and FEASP. To make the presentation a bit easier, consider the path from respondent to friend, $b$, and from friend to respondent, $a$. In Chapter 11, we will talk about the fact that we can calculate effects by writing the arrows in our model in a square matrix called $A$ and, using matrix algebra, use the inverse of this matrix to calculate the predicted average cross-products matrix of a model. For this model, we could write this "arrow" matrix $A$ as:

$$A = \begin{bmatrix} 0 & a \\ b & 0 \end{bmatrix} \tag{4.12}$$

**TABLE 4.10**

Correlations between Respondent IQ (RIQ), Respondent Educational Aspirations (REASP), and Friend Educational Aspirations (FEASP); $N = 329$

| Variable | RIQ | REASP | FEASP |
|---|---|---|---|
| RIQ | 1.00 | 0.40 | 0.29 |
| REASP | 0.40 | 1.00 | 0.37 |
| FEASP | 0.29 | 0.37 | 1.00 |

Now, calculating the inverse of this matrix $A$ involves calculating the sum of $I + A + AA + AAA$ ... until the product of $A$'s becomes zero. Unfortunately, because a feedback loop is involved, the product of $A$'s never becomes zero. Carrying out this multiplication for the first four terms of $A$'s gives:

$$\begin{bmatrix} 1+ab+a^2b^2+a^3b^3+a^4b^4+\cdots & a+a^2b+a^3b^2+a^4b^3+\cdots \\ b+ab^2+a^2b^3+a^3b^4+\cdots & 1+ab+a^2b^2+a^3b^3+a^4b^4+\cdots \end{bmatrix} \tag{4.13}$$

Which we can rewrite as:

$$\begin{bmatrix} \sum_{i=0}^{\infty} a^i b^i & \sum_{i=1}^{\infty} a^i b^{i-1} \\ \sum_{i=1}^{\infty} a^{i-1} b^i & \sum_{i=0}^{\infty} a^i b^i \end{bmatrix} \tag{4.14}$$

It might seem like this problem is unsolvable, but it is a well-known fact that $\sum_{i=0}^{\infty} a^i b^i = \frac{1}{1-ab}$ provided that $ab < |1|$ (Hall & Knight, 1899). This is the term associated with the effects for the diagonal. Off-diagonals of (4.14) are then $a\frac{1}{1-ab}$ and $b\frac{1}{1-ab}$. We can use matrix algebra in this example to generate the predicted correlation matrix between the two variables of respondent and friend educational aspirations. From the diagram, we can see that the variance for Friend educational aspirations uncorrected for the feedback loop with respondent aspirations is 0.99. The variance of Respondent educational aspirations is $0.54^2 + 1.28 = 1.57$. If we call a matrix of arrow coefficients $A$, and the variances of the two variables $S$, we can calculate the predicted correlation using the formula ASA' (see Chapter 11):

$$\begin{bmatrix} \frac{1}{1-ab} & a\frac{1}{1-ab} \\ b\frac{1}{1-ab} & \frac{1}{1-ab} \end{bmatrix} \begin{bmatrix} \sigma_R^2 & 0 \\ 0 & \sigma_F^2 \end{bmatrix} \begin{bmatrix} \frac{1}{1-ab} & b\frac{1}{1-ab} \\ a\frac{1}{1-ab} & \frac{1}{1-ab} \end{bmatrix} = \begin{bmatrix} 1 & 0.37 \\ 0.37 & 1 \end{bmatrix} \tag{4.15}$$

This replicates the correlation between the respondent and friend educational aspirations shown in Table 4.10.

## The Broader Picture

Although many examples have been considered here, a few basic "takeaway" messages of the past two chapters are helpful.

### It's All a Structural Model

First, the statistical models of regression and analysis of variance are types of structural models. These models, however, are all "just-identified," meaning that there are exactly as many parameters in the model as there are unique elements of the corresponding average SSCP matrix or covariance matrix. Although this observation makes the point that there is nothing "distinct" about an SEM versus more traditional models, it also has the practical benefit that you may wish to estimate a more traditional model as an SEM if you desire to produce estimates in the presence of missing data.

### Categorical Coding Schemes

Second, tricks for coding variables have also been introduced, and the choice of which one to use depends on the questions you wish to ask of the data. Dummy coding (use of a system of ones and zeros to code categorical variables result in negative covariances between predictors. Effect coding, consisting of positive and negative 1 and zero, introduces a pattern of positive covariances between the predictor variables. The magnitude of the covariances decreases with an increased number of categories being coded.

### Correlations Between Parameter Estimates

Third, estimates of structural model parameters are often correlated. Orthogonal coding schemes result in uncorrelated parameter estimates and have greater statistical power for that reason. Patterns of redundancy and suppressor effects, however, have correlations between parameter estimates, i.e., that an estimate which is erroneously high on one parameter due to sampling error may introduce bias in other parameters of the model.

### Path Diagram Tricks

Statistics that are more easily calculated in the covariance or standardized score matrix can be calculated in a path diagram that uses a more cumbersome score matrix such as the adjoined or deviation score matrix. Specifically, a variance can be calculated by specifying the unit vector as a cause of the variable, a standard deviation can be calculated by specifying that a latent variable with unit variance causes the manifest variable with a freely estimated path, and a correlation can be calculated as the sling joining two variables with unit variance.

### Alternate Explanations

Many of the models considered here can be proposed as alternate explanations of other models in the chapter. One can, for example, think of a three-variable factor model as an alternate explanation for a regression model. Specifically, there may be a latent variable (or simply a missing third variable), which underlies all associations between the variables. Such third-variable explanations can be a simple variable such as sex or gender, SES, or educational level. The unincluded variable may be a method factor, such as the assessment strategy used, or a more psychological construct. In reality, no single missing variable is responsible, but a latent variable can represent the net effects of several unincluded variables. Alternately, it is likely that any relationships between variables, such as attitudes toward problem behaviors like alcohol or risky sex, operate in a feedback loop with the behaviors themselves. Such considerations do not invalidate the use of any given model but do constitute a limitation of the favored model when interpreting the results.

# 5

# *Assumption Checking*

As discussed in the Introduction, it is your job to disbelieve your model and be skeptical about claims made from your data. Although statistics books often stress that the analyst should examine the reasonableness of a proposed model based on the data, this advice is only rarely accompanied by explicit guidelines of what such careful examination might entail. This chapter discusses assumption checks for standard linear models such as regression and analysis of variance. The goal is to evaluate whether a small number of observations has an outsized effect on the magnitude or statistical significance of parameter estimates and whether the functional form assumed by the model between variables is well characterized by the chosen statistical model. Visual representation of data such as scatterplots, LOESS regressions, and univariate distributions are sufficient to investigate such counterclaims. When dealing with larger data sets or highly multivariate data, however, adjustments need to be made to the visual presentation of the data, and diagnostic fit measures may be useful as well.

You may have already learned about assumption checking in previous study and, if so, this chapter can be omitted. The assumption checks described here are appropriate for the models involving only manifest variables, as in the previous chapter, but can also be used in latent variable models using factor scores. Additional assumption checks for latent variable models will be discussed in a subsequent chapter after latent variable structural models have been introduced.

## Assumption Checking via Scatterplots

Anscombe (1973) provided perhaps one of the simplest cautionary tales of the perils of not looking at plots of the data when evaluating a model. In light of present-day concerns about replication, one can say that even well-replicated findings can be misleading or, conversely, that replications may exist but are not evident unless further investigations have been considered. This demonstration illustrates that one can never move directly from a statistical model to probability statements about parameters estimated from the data. Visual inspection of the data is informative. The example data are

DOI: 10.1201/9781003595267-5

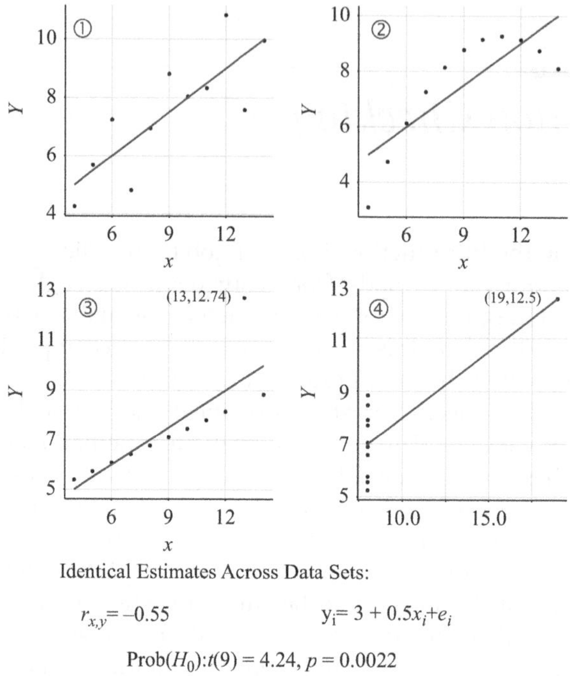


**FIGURE 5.1**
Anscombe data quartet example.

small and widely available on the internet. Anscombe presented four bivariate example data sets with 11 observations each (shown in Figure 5.1). The respective summary statistics for the predictor and criterion variables were identical across the data sets ($\bar{X} = 9$; $\sigma_X = 3$; $\bar{Y} = 7.5$; $\sigma_Y = 1.9$) Further, the correlation coefficients between the predictor and criterion were the same (0.82), and, as a result, the regression lines predicting $Y$ from $X$ were also identical across the examples ($Y = 3 + 0.5X$) as well as statistically significant ($t(9) = 4.24$, $p = 0.0022$). If one did not check assumptions, one might think that the data replicate the same phenomenon. A graph of the data, however, reveals that each data set produces quite different conclusions about whether and how the two variables are related.

The first example describes what we usually think of in linear regression with individuals at lower levels of the predictor tending to have lower scores on the criterion. The relationship isn't perfect, with errors of prediction above and below the regression line. The second example, however, shows a strong relationship between the predictor and criterion but that a straight line doesn't characterize the nature of the relationship. Errors of prediction are systematic; low and high values of the predictor are associated with negative errors on the criterion, and values in the midrange are associated with positive errors of prediction. The third example also shows a perfect relationship between the predictor and criterion; the correlation between the two variables would be perfect were it not for the one extreme data point, (13, 12.74). Notice that the unstandardized regression weight resulting from omitting that observation would be different ($b = 0.35$).

The final example shows that the entire correlation is due to the presence of a single data point, ($X = 19$, $Y = 12.5$), which is extreme on both the predictor and the criterion. The remaining observations show variation on only the criterion. If this extreme individual were not in the data, there would be no regression weight, as there would be no variation in the predictor.

## Terminology

At this point, it is helpful to introduce some terminology. First, we will refer to the shape of the mathematical relationship between predictors and the criterion as the "functional relationship." For example, in the second Anscombe example, we may say that the functional relationship between the predictor and criterion is not well captured by a linear regression and that a quadratic model may be more appropriate. Polynomial curves are not the only type of functional relationship, however, and other possibilities such as exponential, logarithmic, or logistic curves are possible.

Second, as demonstrated, parameter estimates and, possibly, statistical significance levels may appear substantially different with or without a single or small number of individuals. These individuals are described as "influential." Anscombe's fourth example is an example of a single highly influential observation. Finally, observations may be said to be "outliers" if the errors of prediction associated with the observations are dramatically larger than for the remaining observations. Anscombe's third example shows an individual observation that is an outlier, although it is also somewhat influential as well, given that the regression slope is somewhat steeper than it would be if based on the remaining observations. Outlying observations, while possibly also being influential, reduce the statistical significance of model parameters and the overall significance of the model. As discussed next, the remedies for an influential observation may be different from those for an outlying observation.

## LOESS Regression

LOESS regression is a method of generating a smooth curve through data points based on subsets of regression as proposed by Cleveland (1979). Under this model, predictor values are arranged in increasing order. A subset size for the data is chosen (window-size), and a local linear regression is calculated based on the first through window-size observations. The process is then repeated but based on the second through window-size +1 observations. The process is then repeated until all observations have been included and the resulting regression line is produced by smoothing the results from the several regressions. More complicated versions of LOESS regression take into localized patterns using local polynomial regressions rather than linear regressions (Cleveland & Devlin, 1988) or robust LOESS regressions that attempt to adjust for influential observations (Loader, 1999). Although more complicated algorithms of LOESS regression involve a greater computational burden, this is generally not a problem given advances in computing resources, although some caution is appropriate when attempting to fit multivariate LOESS regression models (Figure 5.2).

Although a useful tool, LOESS regression requires interpretation and, especially in small samples or in regions where data are sparse, evident departures from a functional relationship may be artifactual. Consider LOESS regressions for the Anscombe examples. The first example contains a less than perfectly straight line, which reflects sampling and the small sample size used. The second example correctly identifies a regression line suggestive of a quadratic equation, while the third example suggests a possible piecewise regression for the data in which one regression has no errors of prediction while the others contain errors of prediction. The final example, in which variability in the predictor variable was due to a single observation,

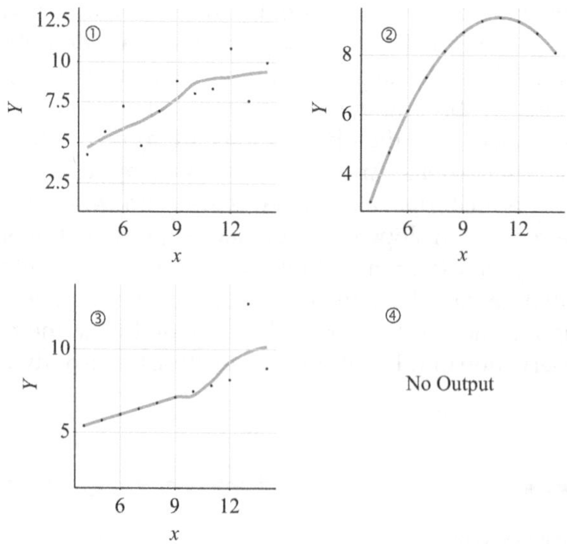


**FIGURE 5.2**
LOESS regressions for Anscombe data.

produced no output given that no variance in the predictor variable is observed without the influential observation.

In some situations, LOESS regressions may suggest a useful transformation of the predictor or criterion variables, which may result in a better estimate of the functional form between a predictor and criterion. In other cases the functional form suggested by a LOESS regression suggests a relationship that does not correspond to a parametric curve. For example, one illustrative analysis was based on a sample of 1,904 freshmen who reported drinking at least one drink in the past 30 days. The data were taken from the study described by Littlefield, Sher, and Wood (2010). The predictor variable in this small example is the number of drinks that the person would usually drink on any one occasion. The criterion of interest is a scale that measures how strongly the student believed that alcohol would enhance an experience. The LOESS regression plot for these data is shown in Figure 5.3.

The scatterplot has been jittered to facilitate a better understanding of the data. As can be seen, the LOESS regression notes a marked bend in the data, with individuals who usually consume five or more drinks having roughly the same levels of enhancement beliefs, while below the five-drink level, a relatively linear relationship appears to exist. This type of pattern may lead the researcher to explore a variety of different models, depending on the context of the study. Given that consuming five drinks in a sitting is often defined as a "binge" drinking episode, one strategy may be to rescale the data so that those who answered five or more drinks receive a score of five. Alternatively,

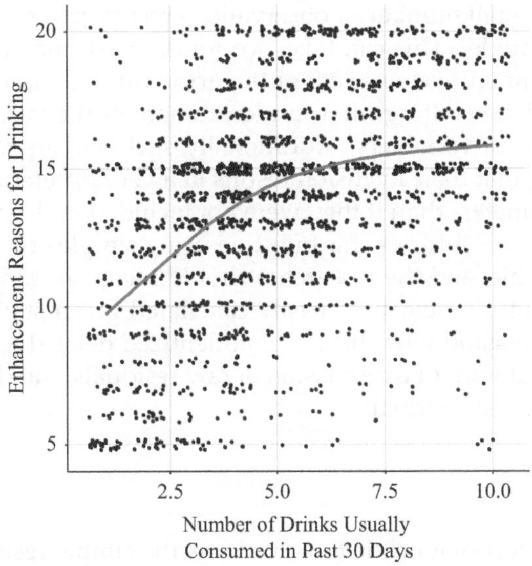


**FIGURE 5.3**
Example LOESS regression relating number of drinks to enhancement reasons for drinking.

the researcher may decide to explore a piecewise regression model by analyzing individuals above and below the five-drink level separately. In large data sets, it is possible to consider scoring the drinks variable as an ordered categorical variable, reflecting the belief that the categories of drinking are representative of some underlying continuous dimension. If a parametric form is desired for the data, it is possible that the researcher could estimate a constrained exponential curve, provided that the researcher believes the earlier sections of the curve to be well typified by a slowly increasing exponential process.

In any event, the LOESS regression suggests that assuming a linear functional relationship is not warranted and polynomial approximations such as the quadratic are also inappropriate. This discovery regarding the nature of the functional relationship between these two variables is important regardless of whether this association is the prime hypothesis under consideration, or whether it is part of a general statistical adjustment (such as what might occur if a partial correlation adjusting for consumption was desired), or whether it is part of a larger latent variable model.

## Residual Analysis

We now turn to regression diagnostics designed to assess the case where an observation or small number of observations results in a misspecification of the regression model. This can take two forms. First, the outlying observation can result in an overestimate of the error variance associated with the regression model, resulting in an underestimate of the variance accounted for by the predictor variable(s). Additionally, such outliers can also be influential, meaning that their inclusion results in a very different estimate of the regression parameters than if they were not included in the analysis.

Often, researchers address this question using simple errors of prediction. This is inadvisable, and the researcher should focus on studentized deleted residuals instead. These can be easily calculated given statistical software. In order to understand why this is so, studentized deleted residuals are presented as the endpoint of a discussion of raw residuals, standardized residuals, and studentized residuals.

### Raw Residuals

Perhaps the easiest diagnostic to consider is the simple residual, calculated as the observed minus the predicted value ($e = Y - Xb$). Unfortunately, raw residuals are somewhat difficult to interpret across studies, as the size of the residual is dependent on the scale of the criterion and the degree to which predictors explain the criterion. For this reason, many researchers look to standardized residuals, which have a mean of zero and standard deviation of one.

### Standardized Residuals

Standardized residuals can be readily calculated from the raw residual by dividing by the standard deviation of the residuals, $s_{y.x}$. In formula form, $z_e = \frac{e}{s_{y.x}}$. Some individuals use a general cutoff of 1.96 or 2.58 to count an observation as significantly outlying. Relying on the absolute value of the residual to define an extreme residual value is ill-advised, however, as such values are more likely to occur the farther away the observation is from the centroid (i.e., the mean of all predictor variables and criterion). This is the reason, for example, that confidence intervals widen the farther points are considered from the average of the predictor, as shown in Figure 5.4 for Anscombe's first data example.

### Studentized Residuals

The studentized residual follows from the definition of the standard associated with that score. One point of connection with the leverage (or "lever"), to be introduced momentarily, is that $s_e = s_{y.x}\sqrt{1-\text{lever}}$. (Note that we are studentizing the raw residual value, not the standardized residual.) The studentized residual is then calculated as: $\frac{e_i}{s_{y.x}\sqrt{1-\text{lever}}}$, where $e_i$ is the raw residual associated with observation $i$.

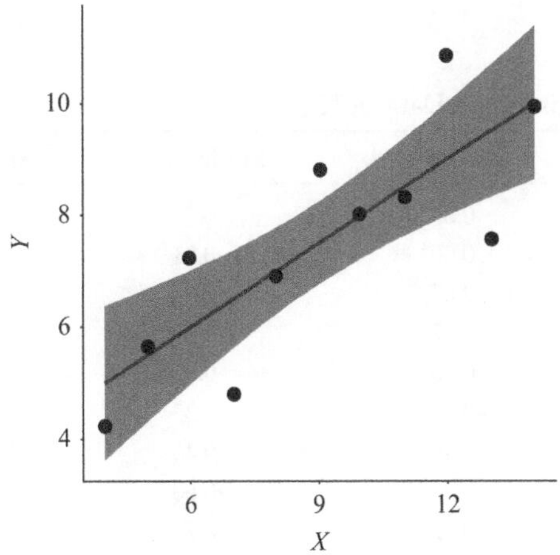


**FIGURE 5.4**
Confidence intervals for Anscombe's first example.

The resulting pattern of residuals adjusts for the distance of the predictor variable from the mean. Yet another difficulty, however, in using the studentized residual is that model misfit associated with a single extreme observation may result in significant residuals for all remaining observations in the data set. As a result, there is an alternative (but somewhat computer-intensive) approach to calculating studentized residuals based on all other observations in the data set except the individual observation under consideration. The result is the studentized deleted residual.

## Studentized Deleted Residuals

If it is the case that a single observation is unusually aberrant in terms of their $Y$ values, it makes the most conceptual sense to calculate the residual for observations based on all data except the observation under investigation. Doing so prevents a "slightly off" regression equation from being used on the conforming observations and also prevents those individuals who are members of the group for which the single regression line actually holds from being falsely identified as individuals with significantly deviant criterion values. The studentized deleted residual, usually denoted as $t_i$, is given as $t_i = \frac{e_i}{\sqrt{\text{MSE}_i(1-h_{ii})}}$, where $e_i$ is the error of prediction, $\text{MSE}_i$ is the mean square error of the regression based on all data but the $i$th observation, and $h_{ii}$ is the lever associated with the $i$th observation (explained in the next section).

As an example, Table 5.1 shows the standardized residuals, studentized residuals, and studentized deleted residuals for the third Anscombe example.

**TABLE 5.1**

Residuals for Anscombe Dataset #3

| Observation | Standardized | Studentized | Studentized Deleted |
|---|---|---|---|
| 1 | 0.33 | 0.38 | 0.36 |
| 2 | 0.00 | 0.21 | 0.00 |
| 3 | 0.07 | 0.07 | 0.07 |
| 4 | −0.07 | −0.07 | −0.07 |
| 5 | 0.00 | 0.00 | −0.19 |
| 6 | −0.33 | −0.33 | −0.31 |
| 7 | −0.46 | −0.46 | −0.44 |
| 8 | −0.59 | −0.60 | −0.57 |
| 9 | −0.72 | −0.76 | −0.74 |
| 10 | 2.76 | 3.00 | 13.54 |
| 11 | −0.99 | −1.13 | −1.16 |

In this example, all three approaches to residuals reveal that observation 10 has a much different residual score. The value of the studentized deleted residual for observation 10, though, is much higher than the studentized residual, given that the regression line without observation 10 describes a perfect line. In practice, more than one studentized deleted residual is often large enough to be concerning, and side analyses are necessary to determine the behavior of the model under different deletion or adjustment scenarios.

## Influence Analysis

Although examination of aberrant residuals in a model may help identify individuals or subgroups who are exceptions to the general model, influence analysis investigates whether an individual or subgroup result has a significant effect on the model. This may take the form of a global effect across all model parameters, the fitted values of the model, or it may be limited to individual parameters of the structural model. Other influence diagnostics such as Cook's $D$ attempt to quickly identify observations that may be influential or outlying.

### Leverage (Lever)

The lever, which is often symbolized as $h_i$, is one such general measure of the influence of an observation on the parameters of a structural model. For example, the fourth example in Anscombe's examples appears to be due entirely to the observation of an aberrant predictor variable. (The third example, by contrast, contains a value that has an unusual criterion variable score.)

Conceptually, leverage views the observed regression weight as an aggregate of several individual "votes" as to what the regression weights of the structural model are. Each "vote," however, is not counted equally. Some votes may be very influential in determining the observed regression weight, while other votes may not "count" at all. Figure 5.5 shows these "votes" graphically relative to the centroid (the point corresponding to the means of $X$ and $Y$). In the first example, we can see that the slope associated with the most extreme data points reflect "votes" for the value of the slope of the resulting regression line. Data points closer to the centroid represent votes for a greater variety of values. Some are steeper than the final slope, and some are even votes for no association. This first case constitutes a "normal" regression situation, as the individual "votes" about the regression weight value seem a uniformly "reasonable" aggregate of the individual votes with no particular pattern of slope "votes."

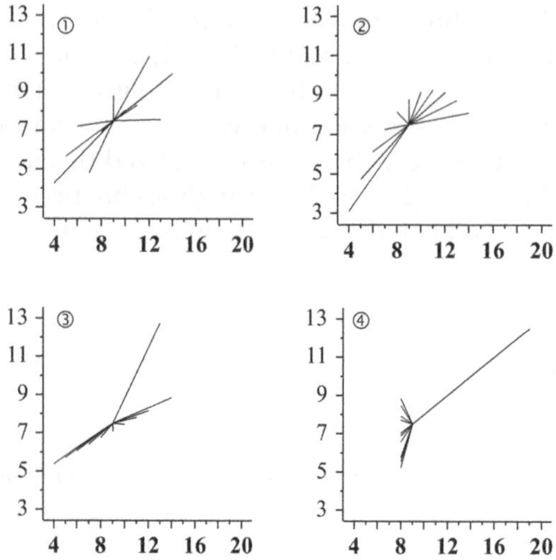


**FIGURE 5.5**
Anscombe data regressions as aggregation of slopes associated with each observation.

In the second case, though, the pattern of "votes" looks systematic and reflects the misspecification of the functional form between the predictor and criterion in this example. No particular observation, however, appears to have an outsized effect on the regression slope. Example 3, however, does show an outsized effect for one observation, with the remaining observations constituting, in aggregate, votes for a much lower slope. Similarly, Example 4 contains an individual that determines the positive value of the regression slope with the remaining observations "voting" for an equal number of positive and negative slopes – i.e., no association between the predictor and criterion.

The lever is a diagnostic associated with each individual observation considered jointly across all the predictor variables of a structural model. A slight rearrangement of the regression model can be used to demonstrate how the lever is calculated. For the simple case of bivariate regression, consider the formula for an unstandardized regression weight:

$$b = \frac{\sigma_{xy}}{\sigma_x^2} = \frac{\sum_{i=1}^{N}(X_i - \bar{X})(Y_i - \bar{Y})}{\sum_{i=1}^{N}(X_i - \bar{X})^2} \tag{5.1}$$

If we multiply both numerator and denominator of this equation by $\sum_{i=1}^{N}(X_i - \bar{X})$ and rearrange the terms of summation, this gives us:

$$b = \sum_{i=1}^{N} \left( \frac{(X_i - \bar{X})}{\sum_{i=1}^{N} (X_i - \bar{X})^2} \right) \left( \frac{Y_i - \bar{Y}}{X_i - X} \right) \tag{5.2}$$

Notice that this equation describes the overall regression weight in terms of contributions of each individual observation. The first term in parentheses is a weight term that determines how much of a "voice" the individual has in determining the overall regression weight based solely on the individual's variation from the average $X$ score. To make this a bit more explicit, some people write these two components to the lever in two parts as follows:

$$b = \sum_{i=1}^{N} h_i \left( \frac{Y_i - \bar{Y}}{X_i - X} \right) \tag{5.3}$$

Notice that the fraction in parentheses in Eq. (5.3) is, in some sense, an estimate of an unstandardized regression weight in that it represents the amount of change in the criterion given change in the predictor. The weight that this vote receives, $h_i$, can be expressed as a squared $z$-score representing the distance of each observation from the average predictor value.

#### Multivariate Leverage

When more than one predictor variable is involved, obviously, the formulae just discussed are not much help because the graph can take into account only one or two predictor variables at a time. Fortunately, however, the multivariate extension of the level using matrix algebra is straightforward and involves calculation of a matrix often termed $h_i$ or the "hat matrix." The term "hat matrix" comes from the relationship of this statistic to the formula for predicted criterion scores or $\hat{Y}$. Recall from Chapter 2 that the predicted scores for a multiple regression in matrix form $\hat{Y} = Xb$ can be written out as $\hat{Y} = X(X'X)^{-1} X'Y$. Observe that the $X(X'X)^{-1} X'$ part of this equation describes a diagonal matrix with as many rows and columns as there are individuals in the data set. The diagonal of this matrix, usually termed $h_{ii}$, contains the lever values, or weights for the votes about the regression slopes in the model for each observation in the data set. (Two subscripts are needed to designate the leverage for each individual based on the hat matrix, owing to the fact that both the column and row of the square hat matrix must be indexed.) More formally:

$$h_i = \text{diag}\left( X(X'X)^{-1} X' \right)^{-1} X' \tag{5.4}$$

where diag() is a matrix algebra operator that selects the vector of the diagonal elements of the matrix. Leverage values are primarily designed to detect the presence of individuals who have unusual values across the predictor variables. Researchers using leverage commonly compute leverage based on either the adjoined or the deviation score matrix, and you should consult your statistical software manual to determine which statistic is being calculated when you request these statistics. If X is taken to be a deviation score matrix, the diagonal elements of the hat matrix represent the squared deviations of that individual from the mean divided by the variance of the predictors. In the multivariate case, this variance is represented by the sum of squares and cross products matrix instead of a simple sum of squares.

An additional complication in interpreting statistical software results is that some packages report a centered leverage index, which adjusts the average leverage value by a factor of $1/N$; i.e., $h_{ii}^{*} = h_{ii} - \frac{1}{N}$. Because lever values vary as a function of sample size and number of predictors, no numeric cutoff can be recommended. A general guideline for significant leverage, however, is twice the average leverage across observations.

## Mahalanobis Distance

Many statistical software packages, especially those dealing with structural equation modeling, often report a statistic closely related to the leverage termed the Mahalanobis distance, except that this statistic is calculated based on all variables in the model, rather than just those specified as exogenous in the model. Weisberg (1985) points out that if the Mahalanobis distance is computed based on only exogenous variables, the distance can be written as: $D_M = (N-1)h_{ii}^{*}$. In addition to being closely related to the idea of leverage (though in a slightly different metric), the Mahalanobis distance is a general measure of distance between points in space. It obeys a few commonsense properties we would expect a distance measure to have, such as the distance between any point and itself being 0, distance being > 0 only if the two observations are different, and the distance between observations A and B being less than or equal to the distance from A to C, and C to B, a relationship known as the triangle inequality. The Mahalanobis distance is equal to the regular "Euclidean" distance if and only if all the variables are standardized and have zero correlations. As with the lever, no clear bright line cutoff exists for determining whether an observation is influential. If multivariate normality is assumed, the significance of a given observation can be used by comparing the percentile points of a $\chi^2$ distribution with degrees of freedom equal to the number of variables used to calculate the distance.

### DFFITS

DFFITS is a measure of how much how much one particular observation has on the fitted value $\hat{Y}_i$. One easy way to remember this is to recall that the letters "DF" stand for the difference between the fitted value of $\hat{Y}_i$ for the *i*th case when all cases are used in fitting the regression function, and the predicted value $Y_{i(i)}$ is the predicted value obtained when the *i*th case is not used in fitting the regression function. The formula is:

$$(\text{DFFITS})_i = \frac{Y_i - Y_{i(i)}}{\sqrt{\text{MSE}_i h_{ii}}} \tag{5.5}$$

Notice that the denominator is an estimate of the standard deviation of $Y_i$, but that it uses the mean square from the regression in which the *i*th case was omitted for estimating the variance. The numerator is therefore standardized by this denominator. Conceptually, this amounts to fitting a regression line based on the data and then removing an observation and refitting the model. If the predicted value for that observation changes significantly, this is taken as evidence of influence on the estimated regression. Generally speaking, DFFITS that exceed 1 for small to medium data sets are thought to be influential, and values of $2\frac{\sqrt{p}}{n}$ are thought to be influential for larger data sets where *p* is the number of predictors in the model and *n* is the sample size.

### DFBETA and DFBETAS

DFBETA is a measure of how influential the *i*th case is on each of the regression coefficients of a model. It is similar to DFFITS, except that DFFITS assesses the influence of an observation on the remaining predicted values, while DFBETA focuses on the influence associated with the regression weights. The formula for $\text{DFBETA}_{(i)}$ is the difference between a given regression weight and the regression weight that would be observed if the *i*th observation is excluded (i.e., $(\text{DFBETA})_{ij} = b_j - b_{j(i)}$ for each of the *j* regression weights and *i* observations. This DFBETA, however, is not scaled, making it difficult to compare across regression weights and across samples. The scaling of DFBETA is accomplished by dividing DFBETA by the estimated standard error of this difference. If we take $c_{kk}$ to be the *k*th diagonal element of $(X'X)^{-1}$ and *p* the number of predictors in the model, the formula for DFBETAS that results is:

$$(\text{DFBETAS})_{ij} = \frac{b_j - b_{j(i)}}{\text{MSE}_{(i)} c_{kk}} \text{ for } k = 0, 1, \ldots p \tag{5.6}$$

Generally speaking, for small samples, a DFBETAS value that exceeds $\frac{2}{\sqrt{N}}$ in absolute value is indicative of a significant effects while, for larger samples, a value that exceeds 1 in absolute value is taken a significantly large.

### Cook's *D*

Cook's *D* is designed to assess how influential a given observation is on all of the remaining predicted values of the criterion. (This is in contrast to DFFITS, described earlier, which looks at the influence of a particular observation's predicted value, $\widehat{Y}_i$, or other measures such as the Mahalanobis distance and the lever, which look at how influential predictor variables are relative to the rest of the population.) The formula for Cook's *D* is:

$$D_i = \frac{\sum_{j=1}^{N}\left(\hat{Y}_j - \hat{Y}_{j(i)}\right)}{p\text{MSE}} \tag{5.7}$$

The numerator is similar to for DFFITS, except that all predicted values in the data, rather than only that associated with just one observation, are considered as part of the influence of the observation. The denominator, as before, serves to standardize the measure. If multiple criterion variables are considered (as happens, for example, in cases of factor analysis, where the manifest variables are thought to be predicted by the latent variable), Cook's *D* can be expressed by the usual matrix algebra:

$$D_i = \frac{\left(\hat{Y}_j - \hat{Y}_{j(i)}\right)'\left(\hat{Y}_j - \hat{Y}_{j(i)}\right)}{p\text{MSE}}, \tag{5.8}$$

with $\hat{Y}$ and $\hat{Y}_i$ representing vectors of predicted scores based on all the data and based on all but the *i*th observation, respectively.

It is also possible to think of Cook's *D* as a measure of how influential an observation is in terms of the regression coefficients of the model under consideration. It is possible to express Cook's *D* as:

$$D_i = \frac{(b - b_i)'X'X(b - b_i)}{p\text{MSE}}, \tag{5.9}$$

with $b$ and $b_{(i)}$ representing vectors of the regression weights obtained under the complete data set, with the *i*th observation deleted, and with $X'X$ being the usual sum of squares and cross-products appropriate to the model (i.e., adjoined if the regression intercept $b_0$ is estimated, or the deviation version of $X$ if the model does not include estimation of the intercept). As a result, some analysts prefer to use Cook's $D$ first to determine if an observation is generally influential within a regression model, which can then be followed up with an examination of DFBETAS to see whether the influence associated with the observation is particularly pronounced for a subset of the regression weights.

Finally, Cook's $D$ may be thought of as a combination of the lever and the studentized deleted residual. Specifically:

$$D_i = \frac{t_i^2}{p}\left(\frac{h_{ii}}{1-h_{ii}}\right), \tag{5.10}$$

where $t_i$ represents the studentized deleted residual and $h_{ii}$ represents the lever, as defined earlier. Generally speaking, significantly influential observations for Cook's $D$ are values that are greater than $\frac{4}{N}$, which is indicative of a possibly influential/outlying observation, and values greater than 1, which represents a strong candidate for further scrutiny.

**Covratio**

The covariance ratio (Covratio) is designed to assess the degree to which a given observation affects not only the variances (and thereby the standard errors) associated with estimated regression weights), but also the magnitude of the covariances of estimates of regression weights (which can have implications for the patterns of observed statistical significance of the observed regression weights, for example). The formula for the covariance ratio is:

$$\text{COVRATIO}_i = \frac{\text{MSE}_{(i)}\left(X'_{(i)}X_{(i)}\right)^{-1}}{\text{MSE}\left(X'X\right)^{-1}} \tag{5.11}$$

Although the covariance ratio is not generally used as often as the lever, DFFITS, DFBETA, or Cook's $D$, values outside the interval $1 \pm \frac{3p}{N}$ are thought to be influential (where $p$ is the number of predictors and $N$ is the sample size). Significant covariance ratio values have the potential to identify observations that may interfere with the statistical significance of path coefficients.

## Summary

Quite a bit of ground has been covered regarding assumption checking, and this highlights the need for a conscientious analyst to explore not only the single statistical model anticipated for the data, but a variety of side analyses. Exploring such models is no vice if the goal is to refute a favored model. However, the remedies for such violations can be different, depending on the sample size and reasons for the assumption violation. Although we have discussed remedies for cases when the functional form of an association is not linear, commentary on remedies for the various diagnostics that flag individual or a small subset of observations have not yet been addressed.

First of all, such observations may indicate a problem with the validity of the data point(s), such as failure to understand instructions, or possibly cultural or racial differences in the meaning of words. In these situations, the observations are probably best removed from the data and the presence of such observations in the data noted as a limitation of the study. This may be particularly useful in small sample datasets where the estimated error variance in the model is substantially larger including such observations than when they are excluded. In other situations, aberrant reports, such as an extremely high number of sexual partners or substance use reports, may be legitimate data values. The researcher may wish to winsorize scores on the observation (i.e., replacing the data value with a value slightly larger than the next highest values in the data) to explore whether doing so results in significant changes in model conclusions. This has the advantage of not removing data that have been gathered; but it also serves to produce parameter estimates that are more likely to be replicated, given that the researcher is unlikely to observe such aberrant values upon replication. In large data sets, it may be possible to conduct side analyses composed of just these observations if it is found that they constitute an identifiable subgroup in the study, such as race, age, or characteristics such as being a first-generation college student.

In the end, however, the question of which analysis to include as the main result of the study and which analyses are suitable for reviewers' supplements or archival analyses involves tradeoffs. If the analysis based on the original raw data is similar to that of the side analyses, it would seem reasonable to report that analysis. To the extent that other analyses provide different insights, however, winsorizing, piecewise regressions, and analysis of the data as ordered categories are replicable by other researchers and provide a more nuanced but defensible summary of the data.

# 6

# *Vector Algebra*

> For everything works through innate forces shown by lines, angles, and figures.
>
> **– Roger Bacon (1267)**

## Introduction

Vector algebra is a geometric approach to representing relationships in data that is often presented in discussions of exploratory factor analysis. Unfortunately, given the need to discuss the large topic of structural equation modeling, vector algebra is often overlooked as a way to present structural models in general. Vector algebra representations of multiple regression and factor models promote an understanding of how such models are related. The easiest classes of vector algebra models to understand are based on the deviation score matrix or variance/covariance matrix because the lengths of vectors involved have a ready interpretation as standard deviations. As we will see, though, extensions of vector algebra to standardized and adjoined score matrices are straightforward.

To give a summary of this material, a vector is a line segment with an arrowhead. The length of this line can represent a standard deviation, represent an average sum of squares, or be of length one, depending on whether the score matrix under investigation is a deviation, adjoined, or standardized score matrix. Simple algebra on a variable can be expressed as multiplying a vector's length by a constant. If two deviation or standardized vectors are joined at their endpoints without changing the orientation, the cosine between the two variables represents the correlation between the two variables. Vectors can be arranged in the form of a right triangle to explain a regression model. Such representations restate an endogenous variable in terms of an approximation of one or more exogenous variables and an unexplained component (the error variable). Vector representations are also used to understand relationships between variables in both exploratory and confirmatory factor models.

To begin, the basics of vector notation are introduced, and how new vectors representing the addition or subtraction of other variables is described.

DOI: 10.1201/9781003595267-6

A vector representation of regression demonstrates geometrically why the most optimal rescaling of an exogenous variable to predict a criterion involves an error vector, which is uncorrelated with the predictor variable.

## Vector Algebra Preliminaries

Vector representations express variation and relationships in data using single-headed arrows. The length of this arrow represents an average sum of squares of one of the three types of score matrices. Relationships between two variables are expressed by the angle between the variables' vectors. Vector algebra can also represent basic algebra involving variables or simple structural models.

### What Is a Vector?

Some individual numbers in statistics represent single quantities such as weight, length, volume, pressure, temperature, or time and are called "scalars." Sometimes, however, it's necessary to represent force, velocity, or acceleration. In statistics, we think of our variables as representing an amount of some force, for example, and we may be interested in describing the joint behavior of multiple variables or forces in the data.

A vector is simply a line with an arrow at one end, as shown in Figure 6.1. The arrowhead of a vector is sometimes called its terminus, and the beginning of the vector is called its "initial point" or "tail." The variability of a variable is represented by the length of this line, often called the "norm" of the vector, When we use vectors to describe the deviation score matrix, the length of the vector associated with a variable is simply the standard deviation. When vectors are used to describe standardized variables, the length of standardized variables is always 1. For completeness, it's also possible to use vectors for the adjoined matrix. In this case, however, the norm of the vector represents the average sum of squares ($\frac{\sum X^2}{N}$) associated with the variable.

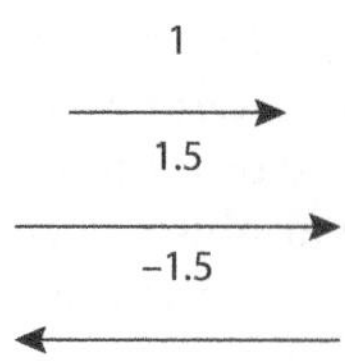


**FIGURE 6.1**
Representation of a vector and multiplication by a constant.

As with multiple regression, vector representations based on the adjoined matrix include representations of both variability and overall level or intercept associated with the model.

Thinking in vector terms takes a little getting used to, given that most often we are used to thinking of variables as being anchored in some grid, such as the $x-y$ coordinate system found in plots of regression lines. The utility of vector algebra, however, comes from the fact that one can slide different vectors around to perform mathematical operations and to determine other statistics of interest, much in the same way as one might place one overhead transparency on top of another without changing the orientation of either. In other words, this "sliding around" can be done in any fashion, subject to the constraint that one may not change the direction of the vector while sliding it around.

## Multiplication by a Constant

"What happens to the standard deviation of a variable if each score in a data set is multiplied by a constant?" is a question often raised in introductory statistics courses. In vector algebra, this is calculated by multiplying a given vector by the desired constant (let's call it $c$). The result is a vector (some of which may overlap or be overlapped by the original vector) with length equal to the constant, $c$, multiplied by the original length of the vector as shown in Figure 6.1.

Multiplication by a constant is used in regression and structural equation models in two ways. First, variables can be standardized or set to some other metric by multiplying some well-chosen constant. If, for instance, one wanted to standardize a particular variable whose vector length was $k$, simply multiplying that variable by $1/k$ produces a new vector with length 1. A second use of multiplying a vector by a constant occurs when producing useful statistics such as predicted scores, which will be described later in this chapter.

## Correlation Between Variables

The correlation between two variables is expressed by the cosine of the angle between the vectors. For example, if the vectors representing two variables point in the same direction, this denotes a perfect positive correlation between them because the cosine of $0° = 1$. Similarly, a perfect negative correlation means that the vectors representing the variables point in opposite directions because the cosine of 180° is $-1$. A zero correlation is represented when the vectors are at right angles to each other (cosine of $90° = 0$). Such variables are said to be "orthogonal." Less than perfect positive correlations are expressed by an angle less than 90 degrees. Negative correlations are expressed by an interior angle greater than 90 degrees and less than 180 degrees.

To determine a correlation between two vectors, we slide one of them until their initial points touch and examine the interior angle between the two vectors. For example, in the vectors in Figure 6.1, we can see that multiplying a vector by −1.5 results in a new variable that is perfectly negatively correlated with the original. This negative correlation would be shown as two arrows pointing in opposite directions joined at their initial points.

More precisely, the correlation between two variables is represented by the cosine, a trigonometric function. For example, a squared correlation of 0.5 represents an association between two variables that is midway between perfect correlation and zero correlation. As such, the angle between the two vectors representing the two variables is therefore 45°.

## Visualizing Effect Sizes as Vectors

Figure 6.2 shows the graphical relationships between variables under a variety of scenarios of correlations and squared correlations. To find any of these angles ($\theta$) associated with a desired correlation or squared correlation, we use the arccos function ($\cos^{-1}$). For example, if we wanted to calculate the angle (in degrees) between two vectors that represents a 0.95 correlation, we would calculate: $\theta = 360\frac{\cos^{-1}0.95}{2\pi}$. Similarly, if we were interested in the angle associated with a squared correlation of 0.95, we would calculate: $\theta = 360\frac{\cos^{-1}\sqrt{0.95}}{2\pi}$.

## Addition of Vectors

Just as in "regular" math, vectors can be manipulated to accomplish operations such as addition and subtraction. When vectors are added or subtracted, a new vector is created to represent the newly created variable. Graphically, two vectors are added by sliding the initial point (the dot) of one variable to the terminus of the other. The result is the vector formed by the initial point

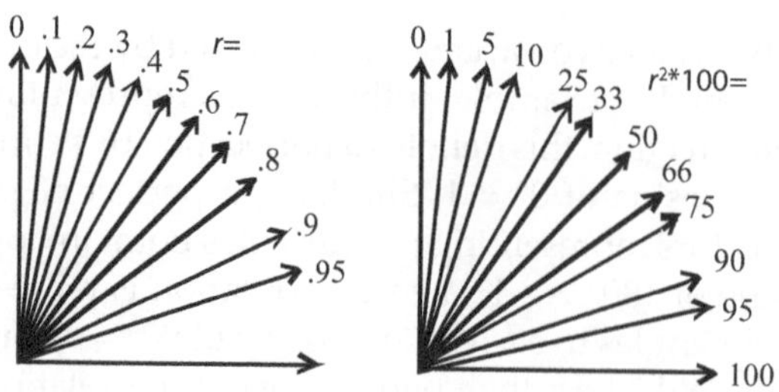


**FIGURE 6.2**
Correlations and effect sizes (squared correlations) represented as vectors.

of the first variable and the terminus of the other. The format of a vector representation allows us to observe several properties about the newly created variable. First, the length of the variable represents the standard deviation of the variable, which is the sum of the two variables. As shown in Figure 6.3, a little experimentation with two vectors demonstrates that variables that are positively correlated with each other will have a sum with a large standard deviation, those that are negatively correlated will have a small standard deviation, and those that are uncorrelated will have a standard deviation equal to the square root of the sums of the variances of the other two variables.

Second, the angle of the new variable with the other vector describes the correlation between the component vector and the sum. From this we can see, for example, that if two variables have the same variability (same standard deviations), the angle between either variable and the sum is half that of the angle between the two vectors. If you call the original angle $\theta_1$, the cosine of the angle of vector $B$ (or $A$) with the sum C is $cos\frac{\theta_1}{2} = \pm\sqrt{\frac{1+cos\,\theta_1}{2}}$, where the $\pm$ is used to keep the same sign as the original cosine of $\theta_1$. So, for example, if you know that two variables with equal standard deviations are correlated 0.7, each variable will be correlated $\sqrt{\frac{1+0.7}{2}} = \sqrt{0.85} = 0.92195$ with the sum. By contrast, if the two original vectors are at right angles $(cos(\theta_1)=0)$, each vector will be correlated $\sqrt{\frac{1}{2}} = 0.7071$, which, as you can see from Figure 6.3, corresponds to a 45° angle.

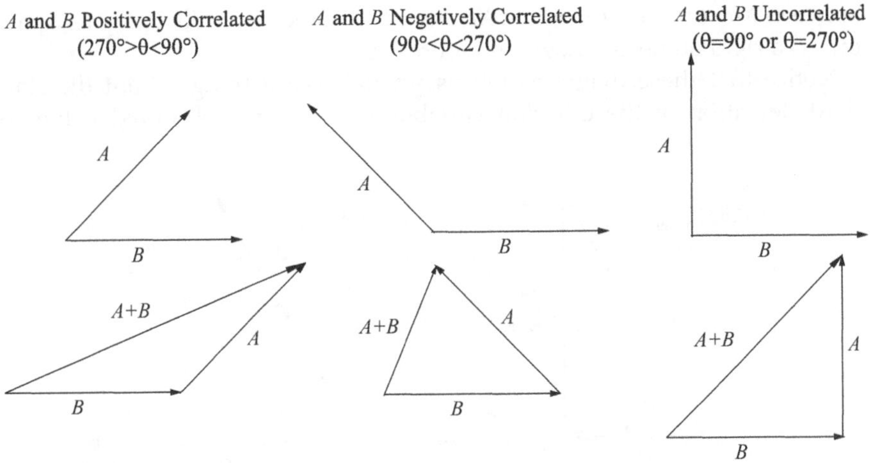


**FIGURE 6.3**
Addition of vectors positively and negatively correlated as well as uncorrelated.

## Vector Representation of Regression Models

It is quite easy to represent regression models with vector algebra, and understanding the models from this geometric perspective may help you understand why statisticians seem to "talk backwards" on occasion when they say phrases such as "the regression of $Y$ on $A$ and $B$" when the "=" way of expressing an equation, sometimes causing the novice to think that a "regression of $A$ and $B$ on $Y$" would make more sense. As we will see, vector representations also help us to understand how a regression weight is a simple rescaling of a predictor to the "best possible" approximation of a desired criterion and how the assumption of error variance being uncorrelated with the predictor plays a critical role. Although we have already discussed how the regression of $Y$ on $X$ is different than the regression of $X$ on $Y$, vector algebra can also explain how regression models are different from factor models, even though both models stem from an examination of "relatedness" between two variables.

### One Variable Regression Using Vectors

Recall from the earlier discussion in this chapter that multiplying any vector, such as a predictor variable, merely lengthens or shortens the length of the vector. $Xb$, or the predicted value of $Y$ based on $X$, is simply a multiplier chosen to be the best possible approximation of $Y$ in terms of $X$. This best possible approximation is defined as that multiplier of $X$ that results in the shortest length of the line from $Xb$ to the terminus (or arrowhead) of $Y$. Looking at this relationship graphically, we can see that the shortest distance between $Xb$ and $Y$ occurs when $Xb$ and the distance from $Xb$ to the terminus of $Y$ is a right angle, as shown in Figure 6.4.

Notice that these diagrams tell us several useful things about the standard deviation of the criterion variable and how the observed values of

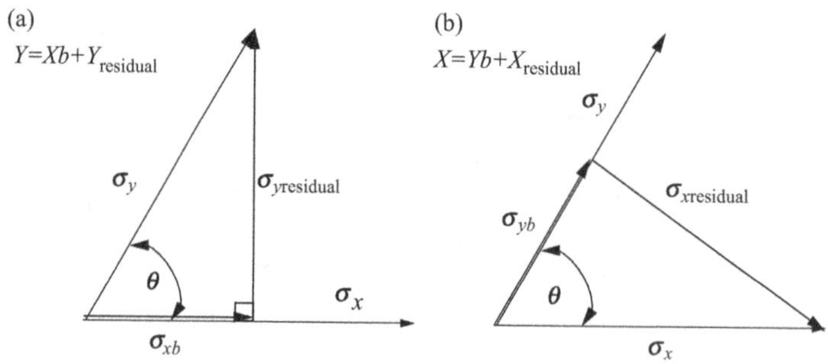


**FIGURE 6.4**
One variable prediction: $Y$ on $X$ (a) and $X$ on $Y$ (b).

the predictor are optimized to produce an estimate with the least amount of squared error of prediction. Consider the diagram at the left first. The observed standard deviation of the predictor variable is given by the length of the vector labeled $\sigma_x$. Multiplying the vector $\boldsymbol{x}$ by the regression coefficient $b$ results in a new version of the predictor which has a smaller standard deviation. (Depending on the initial length of the vector $\boldsymbol{x}$, this could just as easily result in a larger standard deviation in other situations). The length of $\boldsymbol{xb}$ represents the standard deviation of the predicted $y$ values. From the diagram we can see that any other error vector which is not orthogonal to the predicted vector $Xb$ will result in a larger error standard deviation. By the Pythagorean theorem, we can now observe that a vector representation of a single regression represents the partition of the variance of $Y$ as:

$$\sigma_y^2 = \sigma_{\hat{y}}^2 + \sigma_{\text{error}}^2 = b^2\sigma_x^2 + \sigma_{\text{error}}^2 \tag{6.1}$$

The extension to two (and more) predictors is straightforward using the same geometry. Consider the three-dimensional graph shown in the two-predictor regression model, in which regression weights $b_1$ and $b_2$ are so chosen that the sum falls orthogonally to the terminus of $Y$ (Figure 6.5).

In regression the two predictor variables are first weighted by their respective regression weights (as described in multiplying vectors by a constant) and then summed to create a new variable, which is a predicted variable. Graphically, two vectors are added by sliding one of the vectors so that the end point of one vector touches the arrowhead of another vector. The added vector then corresponds to the end point of the first vector and the arrowhead

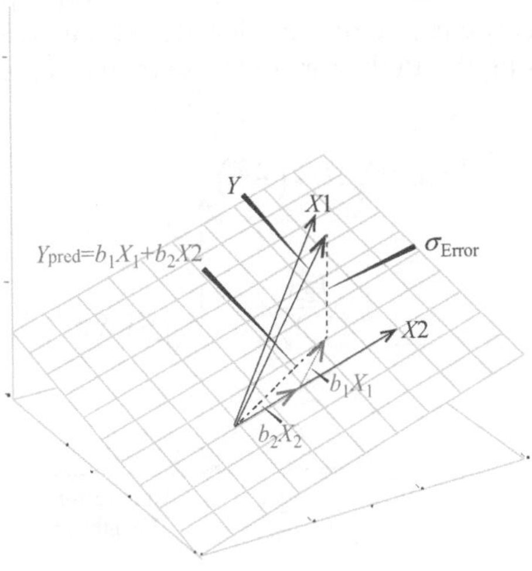


**FIGURE 6.5**
Two-predictor regression model.

of the second vector. Care must be taken so that the orientation or direction of the vector isn't changed while sliding it around. This means that $b_1X1$ must be parallel to the original $X1$ variable, and that $b_2X2$ must be parallel to $X2$ as well.

The resulting vector, which was the sum of two weighted versions of the original variables, has length corresponding to the standard deviation of the predicted criterion score, $Y_{pred}$, as shown in the dotted line at the bottom. This, then, shows that the same right-triangle decomposition – with the criterion representing the hypotenuse, and the predicted $Y$ score, the standard deviation of functioning, as the two legs of the triangle, the right angle between the predicted criterion score, and the error – makes the error standard deviation as small as possible for any possible combinations of the predictor variables.

## Calculating the Standard Deviation from the Adjoined Matrix

As you may have guessed, the same right-angle triangle mathematics can be used to calculate the standard deviation and mean from vectors as well. We define the unit vector as a vector of length 1 and then regress the average sum of squares onto the unit vector. The multiplier of the unit vector corresponds to the mean, and the distance between the end point of the average sum of squares vector and the mean corresponds to the standard deviation of the variable.

Based on Figure 6.6, some commonsense relationships can be seen. For example, the larger the mean, the smaller the estimated standard deviation, and vice versa. Using the Pythagorean theorem, we can see that the average

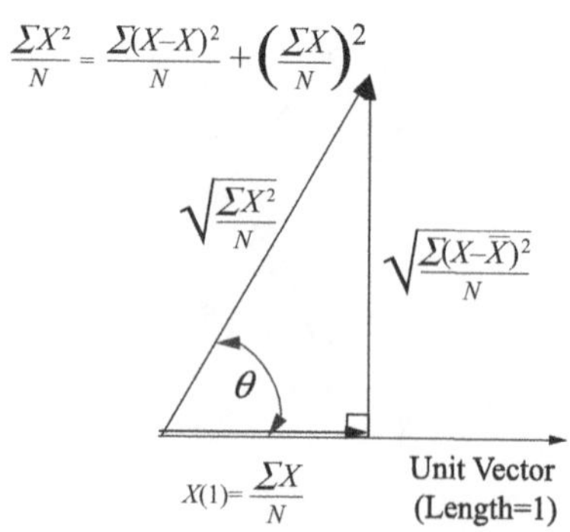


**FIGURE 6.6**
Calculation of mean and standard deviation from adjoined data.

sums of squares is equal to the squared mean plus the variance – or, to put it another way, that the estimated variance is equal to the average sums of squares minus the square of the mean. This is sometimes referred to as the "computational" formula for the variance.

Another observation based on this diagram is that analysis of variances and covariances represents a decrease in the dimensionality of the vectors involved by 1, and we accommodate this reduction in dimensionality by dividing our estimates of standard deviations by $N-1$ instead of $N$. As mentioned earlier, in practice, the effect of this adjustment is minimal given larger sample sizes.

## Relationships to Factor Analysis

Vector algebra figures heavily in the specification of factor models, and especially in the re-expression of initial models to more conceptually interpretable solutions. Although a detailed explanation of rotation in factor analysis is taken up later (Chapter 16, "Factor Rotations"), it helps to motivate the use of vectors in factor analysis by considering two brief examples.

### Vector Representations in Exploratory Factor Analysis

Exploratory factor analysis differs from the confirmatory models considered in later chapters in that it assumes that the average cross-products matrix (almost always a correlation matrix) has been adjusted for measurement error prior to analysis. Again, for reasons described later, initial exploratory factor decompositions consist of patterns of positive and negative values for all but the first factor. (If the manifest variables are positively correlated, which is often the case, this first factor consists of all positive values.) The notion of "simple structure" in factor analysis, then, corresponds to rotating variable vectors to a new pair of axes such that the manifest variables have vector values that are large in absolute magnitude on only one of the dimensions.

As a vector example of this, one of the classic data sets for factor analysis consists of eight physical measurements on 305 girls (Harmon, 1976). The top half of Figure 6.7 shows the unrotated factor loadings for a two-factor solution for these variables. Note that the unrotated vectors contain substantial values on both factors – i.e., that each vector is represented by two values that are not close to zero. Note also that the way that we read a coordinate system is in terms of the vertical and horizontal distance of the terminus of the vector; i.e., we are thinking of each vector relative to two unmeasured predictor variables. This observation makes the case that latent variables represent one or more unmeasured predictor variables.

Rotation to simple structure is an attempt to define a new set of reference axes for the vectors such that each vector has a substantial value for only one of the coordinates. Varimax rotation, shown at the bottom of the figure, involves rotating these vectors (i.e., spinning the reference axes) to a best-fitting solution that still preserves the orthogonal nature of the two factors. Notice in the plot on the bottom of Figure 6.7 that each vector now has one substantial nonzero loading on one factor and a much smaller loading on the other factor. This is a graphical representation of the idea of simple structure, which is introduced in Chapter 15.

Oblique rotation, by contrast, relaxes the assumption that the latent factors must be uncorrelated. Promax rotation, for example, is one such oblique approach, as shown in Figure 6.8. As can be seen, the reference axes were estimated to be substantially correlated ($r = 0.48$, corresponding to a 61° angle between the two factors). Each of the vectors in the solution, however, now has a substantial value on one factor and a near-zero value on the other factor. As a result, each manifest variable is more easily interpreted as being a component of one and only one factor. In addition, as is often the case,

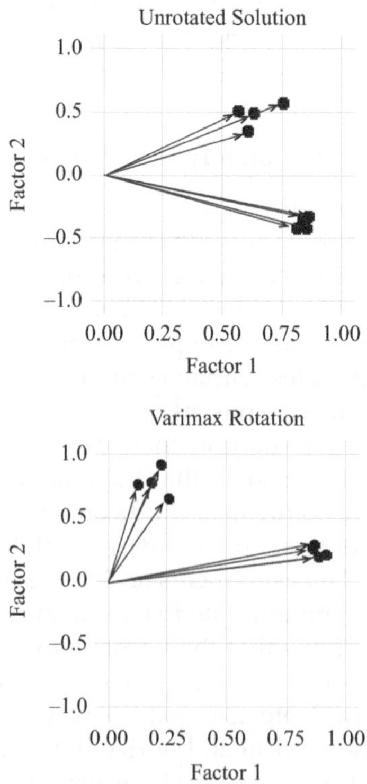


**FIGURE 6.7**
Unrotated and orthogonally rotated factor solutions.

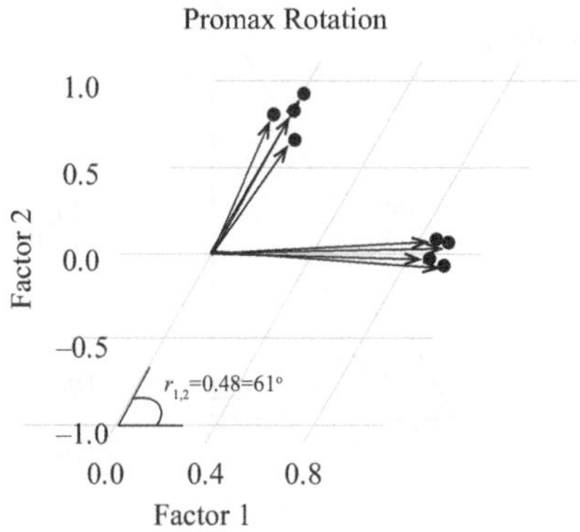


**FIGURE 6.8**
Promax rotation example.

it is reasonable to think that the latent variables are related to each other. In the case of physical measurements of individuals, such as this, it seems reasonable to think that there may be two dimensions (one reflecting length of arms and legs in this example, and the other reflecting girth), but that these two dimensions are somewhat correlated (i.e., some individuals are generally larger than others).

## Vector Representations in Confirmatory Factor Models

The same general strategy for representing latent variables as axes with manifest variables represented as vectors can also be applied to confirmatory models. The largest difference between the two approaches is that exploratory factor analysis assumes that the variables are represented only in terms of their variation after removing the effects of measurement error. In confirmatory factor models, however, the error variance is included in the model. As the latent variable is exogenous, the error variance associated with a manifest variable is represented as the distance between the manifest variable vector orthogonal to the latent variable $\eta$, as shown in Figure 6.9. The diagram representing a single variable is the same as for the single regression predictor shown in Figure 6.4, but the names of the variables have been changed. Instead of $X$, the manifest variable predictor, the latent predictor variable is designated as $\eta$, which is sometimes assumed to be the unit vector. Instead of the letter $b$ to denote a regression weight, the Greek letter $\lambda$ is used The length of the vector representing the factor is the square root of the variance in a path diagram, the multiplier of the factor that yields the best (and orthogonal) estimate of the manifest variable is the factor loading, and the vertical distance between this value and the tip of the manifest variable vector is the square root of the error variance.

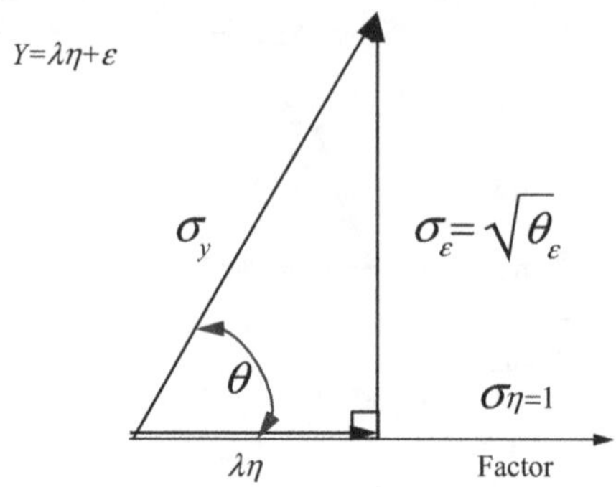


**FIGURE 6.9**
Vector representation of single variable in CFA.

The next chapter returns to a presentation of the problem of measurement error in predictor variables using path diagrams, but it also uses a vector diagram approach to explain how the problem of regression attenuation due to measurement error can be understood for single predictor regression models.

# 7

# *Reliability Models*

Up to this point, we have spoken of errors of prediction as a single property of one or more dependent variables. We have attached no substantive interpretation to errors of prediction, other than to note that one may reduce the magnitude of errors of prediction by including other variables, which account for variability in the criterion. In doing so, greater statistical power results in tests associated with path coefficients. We have assumed in the models so far that observed variation represents a measure of "true" variability uncontaminated by errors or imprecisions of measurement.

However, in many situations, it seems reasonable to distinguish between errors of prediction, due to the measurement properties of the variable, and unexplained variability, which is unexplained by other variables in the model. Quantifying these two properties of a variable would inform a researcher of the relative merits of searching for more accurate ways of assessing the construct of interest versus searching for covariates which might further improve the total explained variation in a model.

Recall in Chapter 3 that we spoke of a single predictor model as a partitioning of the variability of a dependent variable into two parts: one part being the variability due to a predictor (or predictors), the variance of the predicted dependent variable scores; and the other being the residual or unexplained variation. Such a partition was shown in path diagram notation on the left-hand side of Figure 3.3 on page 31, for example.

## Reliability and Attenuation

A full discussion of classical test theory is rightly the object of a course in its own right. Lord, Novick, and Birnbaum (1968) is a classic text in the area, for example. In this chapter, however, enough basics are supplied to point the reader to the primary literature, show how corrections associations that are attenuated due to unreliability can be specified, and preview the advantages of models with multiple measures of constructs as a superior analytic strategy.

DOI: 10.1201/9781003595267-7

## The Reliability Coefficient

The analogy from the model of single variable regression carries over nicely to ideas of reliability. Recall our having earlier noted that the ratio of predicted criterion scores to the observed criterion scores is one way to calculate a squared multiple correlation coefficient. This followed from the equation (and path diagram, if you wish) for the regression:

$$Y_i = \beta X_i + \varepsilon_i = \hat{Y}_i + \varepsilon_i, \tag{7.1}$$

which expresses the regression model for the *i*th individual. Greek letters are used for the coefficients in the model given that they reflect parameter values and not empirical estimates. Given the assumption that the errors of prediction are orthogonal to predicted scores, it then follows that:

$$\sigma_Y^2 = \sigma_{\hat{Y}}^2 + \sigma_\varepsilon^2 \tag{7.2}$$

In similar fashion, we can specify that the observed variability of a given variable, $Y$, can be partitioned into variability due to measurement error and the "true score" uncontaminated by measurement error, i.e.:

$$\sigma_Y^2 = \sigma_T^2 + \sigma_E^2, \tag{7.3}$$

where $T$ indicates the true score value and $E$ denotes the error variable. In just the same way as the squared multiple correlation coefficient, the ratio of predicted scores to observed scores, represents the proportion of variability due to the criterion, we can construct an analogous quantity, $\alpha$, which represents the proportion of observed variability in a variable due to the true score, i.e.:

$$\alpha = \frac{\sigma_T^2}{\sigma_Y^2} = \frac{\sigma_T^2}{\sigma_T^2 + \sigma_E^2} \tag{7.4}$$

This quantity is also known to some as the intra-class correlation coefficient, which will be explained when repeated measurement models are discussed. In just the same way that we can multiply a squared multiple regression weight by the observed variance of a criterion, we can also use coefficient alpha (if we are supplied one) to specify a model in which not one, but two latent variables are specified, namely the true score variable with variance equal to $\sigma_Y{}^2(\alpha)$ and an error variable with variance equal to $\sigma_Y{}^2(1-\alpha)$. This permits us to specify a path diagram such as that shown on the left-hand side of Figure 7.1. Only one free parameter can be specified in

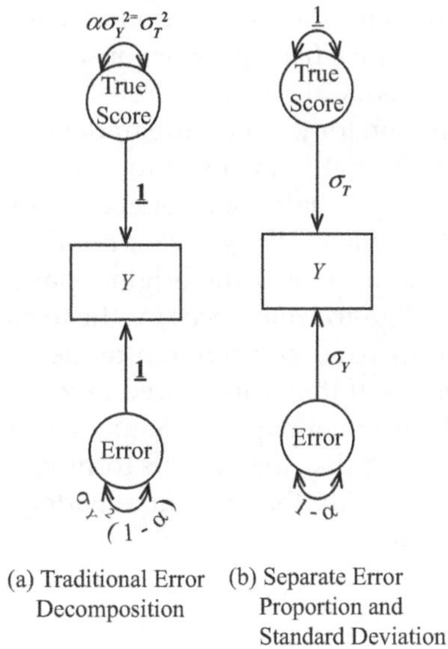


**FIGURE 7.1**
Correction for attenuation of a single predictor. (a) Traditional error decomposition. (b) Separate error proportion and standard deviation.

this model, as it is a statement of the behavior of a single variable. For reasons that will be explained later when we outline problems with use of standard corrections for attenuation, the error variance of this model is usually constrained to a particular value (i.e., the observed variance of the manifest variable multiplied by the reliability coefficient), and the remaining variance component is freely estimated.

Alternatively, it is a little less work to employ the model on the right, in which the error variance is constrained to $1-\alpha$ and the path from the error term to the manifest variable is fixed to the standard deviation of the variable. This makes it a little easier for the researcher to change the value of the reliability if desired for side analyses.

Additionally, it is often useful in such models to specify the variance of the true score variable as 1, with a freely estimated path reflecting the standard deviation of the true score values. These models are equivalent under the tracing rules.

## Estimation of Reliability

As mentioned, the estimate of $\alpha$ cannot come from the variance/covariance of the data, as such a model would be underdetermined. Various estimates of reliability are appropriate, given the properties of the construction of the instrument, the scoring, and the assumption of whether the items used to construct the instrument are a fixed pool or constitute only a random sample of items that might be used. Calculation of reliability requires access to some type of scoring information for the measure under consideration, such as the dichotomous or Likert-format items used to calculate a score, ratings from multiple judges, or even split-half-scale versions of the scale.

The papers by McGraw and Wong (1996; minor corrections in McGraw and Wong, 1996a) are an update of the original paper by Shrout and Fleiss (1979), and all three are "go-to" references for the formulae for the intra-class correlation coefficient for fixed and random models in which items (raters) are nested versus crossed. If items are scored as zero/one, estimates of reliability are found in the original paper by Kuder and Richardson (1937), and it is noted only in passing that the K-R 18 formula presented yields more accurate estimates of reliability than the more widely used (and more easily computed) KR-21 formula.

## Correction for Attenuation for a Single Variable

Even though these few observations are straightforward, it helps to consider an example. Consider the data in Table 7.1, which are the means as well as the correlation and covariance matrix for a study of three variables.

Suppose, for example, that a researcher includes the ACT English subtest variable but wishes to correct the model for the measurement imprecision in the variable. In this example, the variance of this variable was 14.9. Let's

**TABLE 7.1**

Summary Statistics for Study of GPA, ACT English Subtest, and Alcohol Quantity/Frequency

| | **Alcohol Quantity/ Freq. (AlcQF)** | **ACT English (ACTE)** | **High School GPA (GPA)** |
|---|---|---|---|
| | Correlation Matrix | | |
| AlcQF | 1.000 | −0.106 | −0.273 |
| ACTE | −0.106 | 1.000 | 0.400 |
| GPA | −0.273 | 0.400 | 1.000 |
| | Covariance Matrix | | |
| AlcQF | 142.398 | −4.874 | −2.307 |
| ACTE | −4.874 | 14.907 | 1.092 |
| GPA | −2.307 | 1.092 | 0.500 |

further assume that the coefficient α value associated with the ACT English subtest was found to be 0.85.

The usual decomposition, corresponding to the traditional and simplified versions, is given in Figure 7.2. In this model, the variance associated with the true score is freely estimated, while the error variance to the model is fixed to $2.235\left(=0.15^{*}14.9\right)$. The value has been underlined in the diagram to emphasize that this value is fixed, not freely estimated. Alternatively, if the simplified version with unit variance is desired, the values on the right of Figure 7.2 are specified. The path from error to $Y$ is fixed at the standard deviation of the observed variable ($3.86=\sqrt{14.9}$, and the variance of the error variable is fixed to $0.15=1-\alpha$. Again, fixed values in the diagram have been underlined. Although equivalent to the usual way of specifying measurement error, this method is preferable, given that it shows, with a bit more transparency, the measurement assumptions of the model by explicitly showing the error proportion and observed standard deviation.

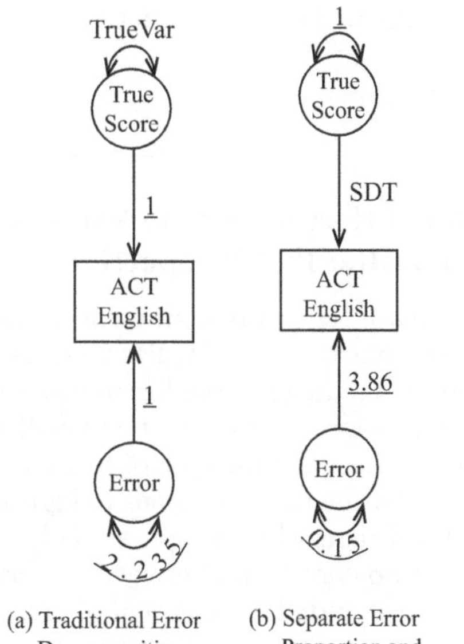


**FIGURE 7.2**
Single variable correction for attenuation example. (a) Traditional error decomposition. (b) Separate error proportion and standard deviation.

## Models that Correct for Attenuation Due to Unreliability

The presence of measurement error in variables has important implications for the power, unstandardized, and standardized coefficients of a regression model. Specifically, the effect of measurement error on estimated correlations is to reduce the absolute value of the correlation between variables. In the case of regression models, the effect of measurement error in dependent variables is to increase the error variability in the model. This increase in error variability will not bias the estimate of the unstandardized regression weight but will affect the estimates of standardized regression weights and will decrease the multiple $R^2$ for the model, leading to lower power.

The situation is more serious, however, when predictor variables contain measurement error. In the case of single predictor models, measurement error in the predictor variable will result in unstandardized and standardized regression weight estimates that are systematically lower in absolute magnitude relative to a model in which the predictor variable had no measurement error. For the case of multiple predictors, however, unstandardized regression weight estimates may be smaller in absolute value than their uncontaminated counterparts, or they may be larger in absolute magnitude due to changes in the estimated covariance between predictors in the model.

## Measurement Errors Effect for Single Variable Regression from a Scatter Plot Perspective

The argument that increased measurement error in the dependent variable has no effect on the estimated unstandardized regression weight, but that measurement error in the predictor variable can lead to bias in estimation of the unstandardized regression weight, can be visualized via an illustrative scatterplot. Consider the effect of adding error to the criterion variable in the scatterplot as illustrated by the two solid lines in Figure 7.3. We will observe the same unstandardized regression weight as with the original data, provided we take care that no correlation between the error and the predictor variable occurs. Even if we added a great deal of error to the criterion variable, say on the order of 10 points, we will always observe the same variance in the predictor variable as with the original data, and the average gain in expected criterion will be the same as in the original. The correlation between the two variables would, of course, be lowered, but the same unstandardized regression weight would be observed.

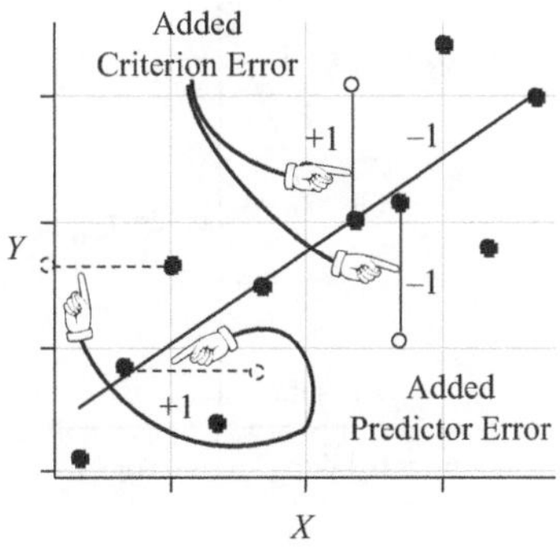


**FIGURE 7.3**
Scatterplot illustration of increased measurement error in criterion or predictor.

If, however, we add error to the predictor variable *X*, this will lead to a slightly different observed variance in *X* and will tend to flatten out the regression line if large amounts of error are added. Consider, for example, how flat the regression line would look if the amount of randomly added horizontal error would be 10. The resulting regression line would look much flatter than the original regression.

## Measurement Error's Effect for Single Variable Regression

It helps to first consider the effect of measurement error in the case of a single predictor regression model. In a nutshell, we can observe that increased measurement error in either the predictor or criterion of a variable will reduce the observed correlation and therefore, the standardized regression estimate. To see this, let us first represent this phenomenon using vector notation and then again using path analysis. Figure 7.4 gives the vector representation of a single variable regression using a standardized score matrix (i.e., the manifest variables *X* and *Y* have unit length). Let's assume a 0.40 correlation (i.e., a 66° angle). In standardized metric, the length of the *X* and *Y* vectors is one, so the standardized regression weight which appropriately shortens the *X* variable to get the best possible version of *Y* on the *X* axis is 0.40, the cosine of the 66° angle as shown in Figure 7.4.

To represent the fact that the predictor variable *X* is measured with measurement error, it becomes necessary to construct a second triangle, as shown

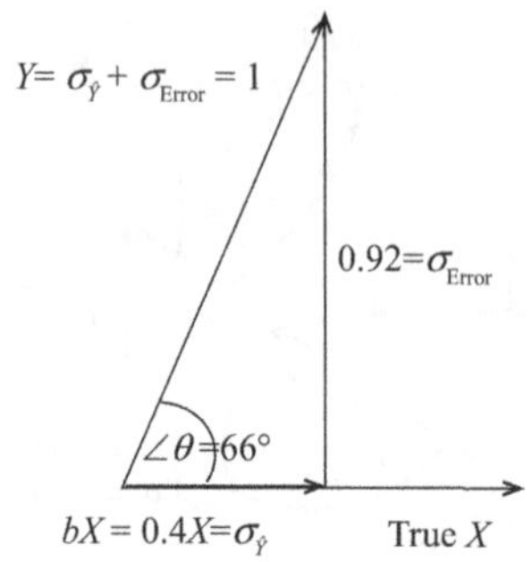


**FIGURE 7.4**
Single variable regression in vector form.

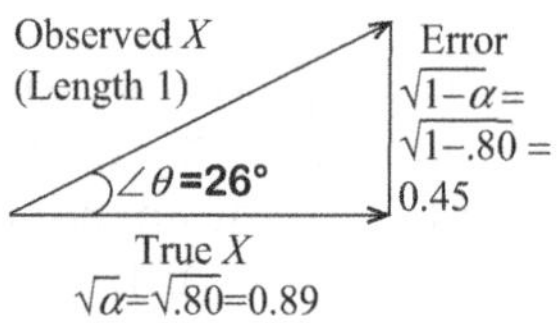


**FIGURE 7.5**
Measurement error associated with X.

in Figure 7.5. In this diagram, the hypotenuse represents the observed variable $X$ and has a length of 1. The proportion of variability in $X$ that is due to true score is 0.80, representing an angle of 26°. The two sides of this triangle represent the standard deviation of the reliability, $\sqrt{\alpha}$, and error, $\sqrt{1-\alpha}$. Because error is assumed to be uncorrelated with true score, the triangle associated with these two vectors is a right triangle.

To understand how correction for attenuation works, we now combine Figures 7.3 and 7.4. Because measurement error is also assumed to be uncorrelated with the $Y$ variable, the triangle in Figure 7.4 is placed perpendicular to both $Y$ and the true score of $X$, i.e., Figure 7.4 stands up off the page of Figure 7.3. The effect of correction for attenuation is then seen by comparing the regression weights needed to express the regression of $Y$ on the observed $X$ and the regression of $Y$ on the true score of $X$. We can see from this, for example, that if the $X$ variable were completely reliable, the angle between Observed $X$ and True $X$ would be zero, and the regression weights of the two models would be identical. However, the longer the length of the error vector (i.e., the lower the reliability), the shorter the vector of True $X$ and the larger the angle between the $Y$ vector and the observed $X$ variable. Seen from above, the regression weight associated with observed vector $X$ will still cover the same proportional length as the regression weight associated with True $X$. The regression weight multiplied by the length of the True $X$

vector is $b\sqrt{\alpha}$. The multiplier needed to reach the same proportion on the unit length of the observed X is $r_{X,Y}$. As a result, $r_{X,Y} = b\sqrt{\alpha}$, meaning that the uncorrected regression weight will always be smaller than the true regression weight by a factor of $\sqrt{\alpha}$.

## Correction for Attenuation via Path Diagram

Finally, it is also possible to consider the effect of correction for attenuation due to unreliability via a path diagram. Figure 7.6 shows how the standardized regression coefficient predicting GPA from Alcohol Q/F would be corrected for attenuation, assuming that the internal consistency of the alcohol variable was 0.78. In this diagram, the predictor variable is, as in Figure 7.2, split into two latent variables composed of true score and measurement error. The error variance is fixed to $0.22(1-\alpha)$, and the path from the true score to the alcohol variable is free to vary to keep the number of free parameters equal to the number of unique elements of the correlation matrix. From this we can see that the correlation between Alcohol Q/F and GPA is 0.88(b), and, since the observed correlation between these variables is –0.273, we can calculate that the standardized regression weight associated with AlcQF is –0.31.

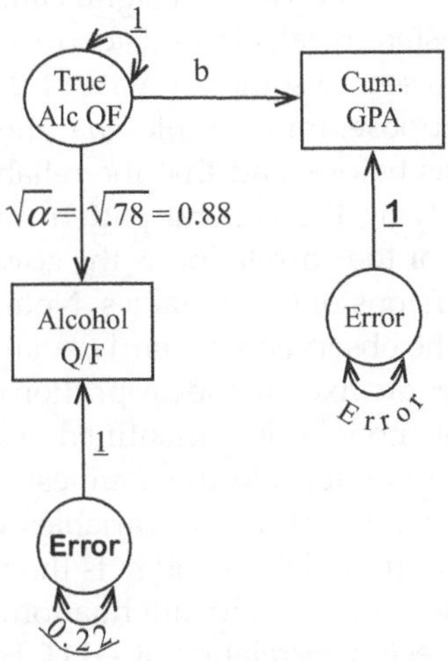


**FIGURE 7.6**
Path diagram for single predictor correcting for attenuation of predictor.

## Regression

We can always conclude that the regression weight corrected for attenuation will always be larger in absolute magnitude than its uncorrected counterpart for single predictor models. However, this is not the case for multiple predictor models, as a trade-off exists between a resulting increase in the corrected regression weight due to attenuation and an increase in the corrected correlation between predictors, which would result in higher redundancy in the estimated regression model and a lower estimate of the unique contribution of a variable to the criterion.

## Estimation of Correlation Corrected for Attenuation Due to Unreliability of Measurement

On occasion, researchers may be interested in the construct discrimination between two variables. One might, for example, calculate a correlation between two variables and observe that it is less than perfect. However, a critical skeptic might reply that a correlation less than perfection was to be expected, given that both measures contained measurement error. An observed low correlation could simply be an artifact of poor measurement of one or both constructs. Duplicating this pattern of correction for attenuation due to unreliability, we can then estimate the correlation between two variables had they been measured with no measurement error.

Suppose, for example, that we are interested in the relationship between ACT English subtest score and a quantity/frequency measure of alcohol consumption in a study of college freshmen. The observed correlation between these two measures is −0.11, indicating a significant, but low negative correlation. If we were interested in calculating a correlation corrected for attenuation, we would need to secure some estimate of the internal consistency of these two measures. Suppose, for example, that the estimated reliability for ACT English was 0.85 as before, and that the reliability of the Alcohol Q/F measure was 0.78. Applying the general pattern on the right-hand side of Figure 7.2, the estimate of the correlation is the covariance between the unit variance "error free" versions of the variables. Note, in the diagram, that the error paths are fixed to the observed standard deviations of the manifest variables, and error variances are fixed to the proportion of variability due to error (i.e., $1-\alpha$). Note that this model is just-identified, with free parameters being the paths from the latent variables to the manifest variables (which, it turns out, correspond to $\sigma_X\sqrt{\alpha}$ for each of the variables where, for each variable, $\sigma_x$ is the observed standard deviation and $\alpha$ is the reliability coefficient). As can be seen, the correlation corrected for attenuation, −0.13, is not appreciably different from the uncorrected correlation of −0.11. However, had the internal consistency of both measures been, for example, 0.25, the corrected correlation would have been −0.42, a dramatically different value (Figure 7.7).

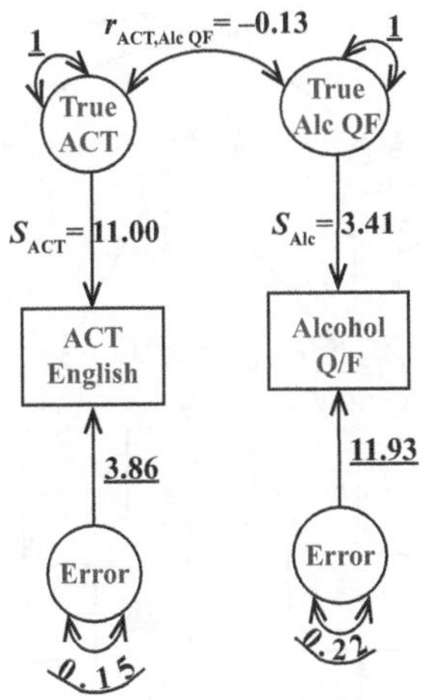


**FIGURE 7.7**
Correlation corrected for attenuation due to unreliability.

## Partial Correlations Corrected for Attenuation

The same strategy of substituting corrected for attenuation versions of an observed manifest variable can be used to produce analogous estimates for many of the models presented in Chapter 4. For example, consider the possibility that a researcher might be interested in knowing the partial correlation between GPA and Alcohol Q/F when both variables are corrected for ACT English subtest score. The observed correlation between GPA and ACT English is, for these data, −0.27.

It is not possible in this situation to use the right-hand path diagram for all three elements of this model, since the variances of the true-score versions of Alcohol and GPA variables are endogenous variables. As a result, the path coefficients from the true score versions to the manifest variable versions of these variables are fixed to one, and the variances are freely estimated. This enables the specification of the variances of the variables to be partitioned into proportions due to the third variable (ACT English) and unaccounted-for variation. As for other models, this unaccounted-for variance is, in turn modeled by a latent variable with unit variance and freely estimated path. In this model, the true score standard deviations (10.37 and 0.48 for Alcohol Q/F and GPA, respectively) represent the standard deviations of the true scores associated with these variables after these variables have been adjusted for the effects of the true score of ACT. The estimated partial correlation of −0.31 is only modestly larger than the unadjusted bivariate correlation reported in Table 7.1 of −0.273 (Figure 7.8).

It is also possible to insert the corrected for attenuation version of a manifest variable into a traditional regression model. Frequently, when considering

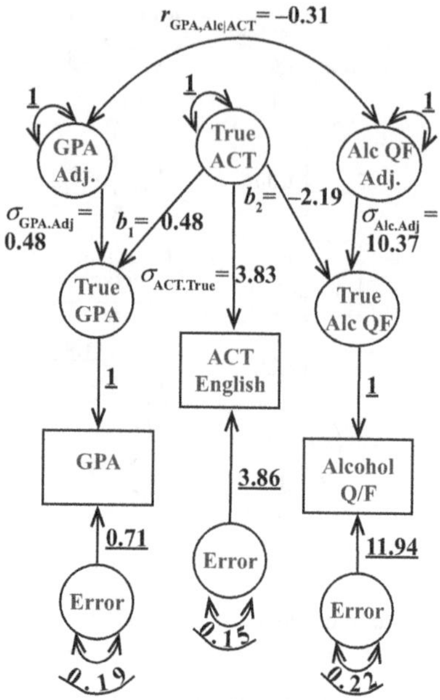


**FIGURE 7.8**
Partial correlation between GPA and AlcQF adjusted for ACT correcting all variables for attenuation.

regression models corrected for attenuation, interest focuses on comparison of standardized paths across variables or on comparison of corrected and uncorrected standardized estimates. For this reason, the two-predictor model considered here will be based on the standardized score matrix. Doing so allows us to write the model in a slightly more compact fashion, as it is not necessary to include an error path fixed to the observed standard deviation of the variables involved (Figure 7.9).

In this model, as in Figure 7.4, endogenous latent variables must have freely estimated variances and, as a result, the paths from the latent variable True AlcQF to the manifest variable of AlcQF has a path fixed to one. The covariance between True ACT E and True GPA, 0.48, is a correlation coefficient, given that the two variables have unit variance. This value is somewhat higher than the observed correlation of 0.40 based on uncorrected versions of the variables. Although the path coefficients in this diagram have exogenous latent variables with unit variance, the endogenous latent variable of True AlcQF does not (having a variance of 0.78). As a result, it becomes necessary to use the formula in Eq. (7.3) to arrive at standardized values. Doing so yields standardized path estimates of 0.04 and −0.33 for True GPA and True ACT E, respectively. These are largely similar to the standardized regression weights of 0.00 and −0.28 for the uncorrected data.

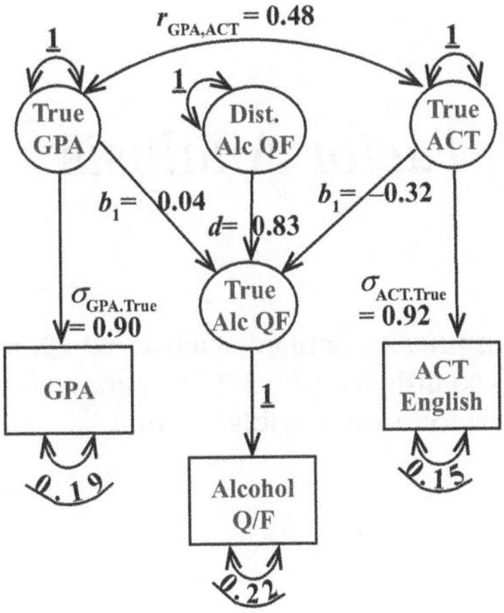


**FIGURE 7.9**
Corrected standardized two-predictor regression.

## Summary

In this chapter, a particular pattern, the correction for attenuation, was inserted into a part of a model to yield the desired model. Although in our examples, the effects of these corrections were modest, they can result in dramatic differences from their uncorrected counterparts if the variables involved are less internally consistent, or if a highly multivariate model is considered such as a regression or partial correlation involving several variables. Given the potential for such differences, one may wonder why corrections for attenuation are not more routinely used in structural equation models. Unfortunately, internal consistency estimates such as coefficient alpha are not unbiased estimates of reliability. They are instead lower bound estimates of the true reliability, meaning that a researcher who observes a coefficient alpha estimate of 0.75 knows that the true reliability is likely some higher value. Although such a conservative estimate may be useful if a researcher wants only to report a measure's psychometrics, within the context of structural equation models, this bias is problematic in that it will lead the researcher to overcorrect for measurement error, resulting in estimates of standardized and unstandardized relationships that are too large in absolute magnitude for single predictor models and that could be over- or underestimates in regression models, depending on whether the association between predictor variables is overcorrected more than the unique association with the dependent variable. Fortunately, such estimation bias is not present in the case of confirmatory factor models in which multiple indicators of the construct are available. Such models constitute a simultaneous estimation of path coefficients and error variability, and they are the topic of the next chapter.

# 8

# *Confirmatory Factor Analysis*

> Why does this magnificent, applied science, which saves work and makes life easier, bring us so little happiness? The simple answer runs: Because we have not yet learned to make sensible use of it.
>
> **—Einstein (1931, p. 12)**

## Introduction

This chapter introduces the idea of confirmatory factor models (CFAs), alternative types of confirmatory factors, and the specifics of how SEM programs arrive at parameter estimates. Latent variables involving more than three variables contain fewer model parameters than elements of the corresponding variance/covariance matrix, meaning that such model parameters cannot be algebraically derived as was done in previous chapters. The conditions under which SEM programs can arrive at a single set of parameter estimates are discussed, as well as procedures to assure that the solution associated with a model is a single set of numbers. The criteria for determining whether parameter estimates constitute a good approximation of the data involve a discussion of SEM loss functions such as least squares loss and maximum likelihood. Critical examination of a confirmatory factor model involves assessment of possible pitfalls in estimation and assumption checking of the model. As this is a relatively long list of discussion points, it helps to begin by first examining the results of a single CFA model and then to discuss how the points just described apply to this example.

## Data Example

The data for our consideration come from a larger prospective study of entering first-year students at the University of Missouri. The study is detailed in Sher and Rutledge (2007). In our example, we consider five variables related to alcohol consumption, shown in Table 8.1.


DOI: 10.1201/9781003595267-8

**TABLE 8.1**

Measures of Drinking Behavior

| Variable Name | Text |
|---|---|
| Plus5 | How many times in the past 30 days have you consumed five or more drinks in a single sitting? |
| AlcF3M | In the past three months how many drinks do you usually consume on any one occasion? |
| TypTh | How many drinks did you consume last Thursday? |
| TypFr | How many drinks did you consume last Friday? |
| TypSA | How many drinks did you consume last Saturday? |

Although missing data were rather minimal, for simplicity, we consider the data based on individuals who completed all five questions, and so the data from 2,557 individuals are considered. For these data, the variance/covariance matrix and the correlation matrix are shown in Table 8.2. From this table, we can see that some variables, such as the number of drinks consumed on Saturday, have larger variances than others, such as the number of times five or more drinks were consumed in the past three months. From the correlation matrix, we can see that the average correlation between variables is about 0.55, although there is some variation in correlations across variables.

**TABLE 8.2**

Observed Variance/Covariance Matrix and Correlation Matrix for Drinking Data ($N = 2,557$)

| | Alc. Use 3 Mo. | # Times 5(+) Drinks | Typical # Drinks Thurs. | Typical # Drinks Fri. | Typical # Drinks Sat. |
|---|---|---|---|---|---|
| Variance/Covariance Matrix | | | | | |
| Alc.Use 3 Mo. | 8.956 | 2.138 | 3.44 | 6.677 | 7.74 |
| # Times 5(+) Drinks | 2.138 | 0.931 | 1.593 | 2.507 | 2.699 |
| Typical # Drinks Thursday | 3.44 | 1.593 | 7.587 | 4.499 | 4.807 |
| Typical # Drinks Friday | 6.677 | 2.507 | 4.499 | 14.629 | 9.862 |
| Typical # Drinks Saturday | 7.74 | 2.699 | 4.807 | 9.862 | 17.538 |
| Correlation Matrix | | | | | |
| Alc.Use 3 Mo. | 1 | 0.645 | 0.417 | 0.583 | 0.618 |
| # Times 5(+) Drinks | 0.645 | 1 | 0.523 | 0.592 | 0.582 |
| Typical # Drinks Thursday | 0.417 | 0.523 | 1 | 0.427 | 0.417 |
| Typical # Drinks Friday | 0.583 | 0.592 | 0.427 | 1 | 0.616 |
| Typical # Drinks Saturday | 0.618 | 0.582 | 0.417 | 0.616 | 1 |

## Unstandardized and Standardized Single Factor Models

Consider a path diagram representing a single factor confirmatory factor model for these data. Figures 8.1 and 8.2 show the estimated values for a single factor model for these data based on the deviation and standardized score matrices, respectively. In both models, a latent variable, named "Alcohol Use," is thought to explain variation and covariation of the manifest variables. Note that in both models, the variance of this latent variable was constrained to the value one for reasons which will be explained shortly. Let us begin the conversation about these two models by considering what these models could mean in light of the topics of the preceding chapters.

As mentioned earlier in the three-variable factor model in Chapter 4, we can think of this model as a series of five single predictor regression models in which the predictor variable was not assessed. Path coefficients in Figure 8.1 associated with the latent variable can be thought of as unstandardized regression coefficients. In Figure 8.1 for every point increase in the latent variable of Alcohol Use, a gain of 3.19 reported drinks for last Saturday is expected, as well as a gain of 2.4 drinks typically consumed in one sitting during the past three months. The unaccounted-for variation of last Saturday's drinks, however, is much larger than that of the typical consumption in one sitting

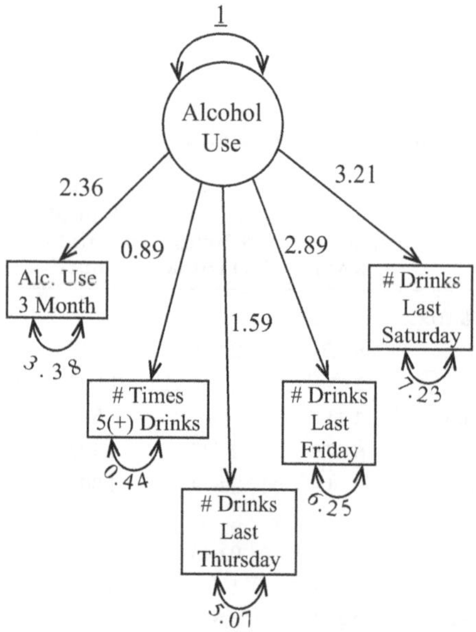


**FIGURE 8.1**
Unstandardized single factor model with unit variance.

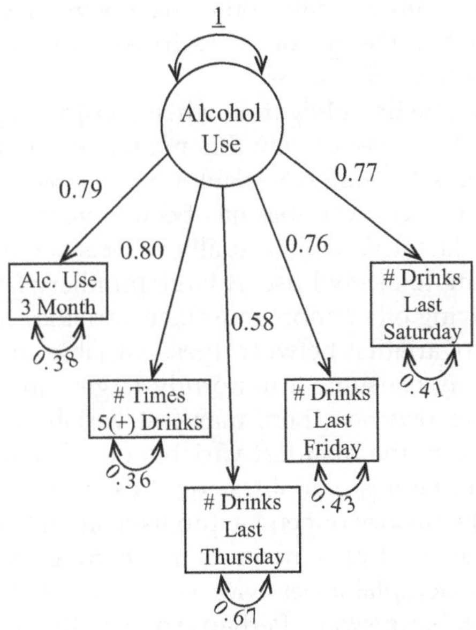


**FIGURE 8.2**
Standardized one-factor model.

over the last three months, suggesting that a smaller proportion of variation in last Friday's consumption may be explained than for typical consumption in a sitting over the last three months. We can calculate these two proportions from the diagram as $\frac{3.19^2}{3.19^2+7.35}=0.58$ and $\frac{2.40^2}{2.40^2+3.22}=0.64$, respectively. We could, if we wished, explore models that compare specific values of these unstandardized regression weights by comparing, for example, a model in which the unstandardized regression weights for these two variables were the same with the present model in which they were different. Model comparison will be taken up in Chapter 9.

The standardized single factor model shown in Figure 8.2 is appropriate when the sole interest is in understanding the relative unique contributions of the latent variable to each manifest variable, or in understanding the proportion of variability unaccounted for by the latent variable. Here, for example, the proportion of explained variation for the two variables mentioned earlier can be easily calculated as either the square of the factor loadings or as one minus the error term associated with the variable. Based on an inspection of these standardized regression weights, for example, it would appear that the past three-month reports of the number of times the student consumed five or more drinks in a sitting is the variable best explained by the latent variable. If, following the presentation in Chapter 7 in which we think of the latent variable as representing some "true score" version of a construct, we might conclude that the five-or-more-drinks variable is the most reliable measure of the construct of interest.

Finally, recall that the first obligation when considering a structural model is to be skeptical. We can also view this model as one in which the partial correlation between all manifest variables is zero, as discussed in Chapter 4. The fact that a missing variable has been labeled "Alcohol Use" is no guarantee that the latent variable actually reflects some psychological construct corresponding to alcohol use. A large number of other third-variable explanations involving other, more mundane variables could be responsible for the observed covariation between these variables. If, for example, men in the sample simply consume a uniformly larger amount of alcohol than women, covariances between these manifest variables will exist, but they will be entirely due to the manifest variable of sex and not to alcohol use per se. Many such manifest variables could exist, and to name just a few, these may be: family history or genetic predisposition to do these behaviors; a response set on the part of some participants to tell a researcher of some shocking socially unacceptable behavior; or, conversely, the desire to provide only socially acceptable answers. Perhaps some of the correlations related to last-week consumption are simply an artifact of some campus-wide event such as an athletic contest. Perhaps some covariation exists because of the internet format of the assessment. It could also be the case that no single third-variable explanation is responsible, but that the observed covariation is due to the joint effect of several of these alternative third-variable explanations. Of course, in this example, one may counterargue that yes, differences in consumption are due to the influence of other determinants, but that these variables are sequelae of alcohol use and not a confound. In this case, I believe many researchers in the area would agree with this assertion, but it bears recalling that the factor loadings seen here are not to be rarefied into Platonic ideal forms, even though they may be circles.

## Model Parsimony and Discrepancy between Predicted Covariance and Correlation Matrices

Use of the tracing rules to calculate the observed and predicted variances and covariances for this model reveals that the variances of the manifest variables are perfectly recovered by the model values, but that the predicted covariances and correlations in the two models are not perfectly recovered. For example, the correlation between the five-or-more-drinks variable and

Thursday number of drinks was $0.80 * 0.58 = 0.46$, while the observed correlation between these two variables was 0.52.

For models in which the compound paths involve only one arrow, it is possible to quickly calculate all the predicted variances and covariances or correlation using matrix algebra. This formula is introduced in order to show a recurring pattern in SEM models. Specifically, we can calculate the predicted variance covariance matrix by putting the arrow and sling information in the model into three matrices: $\Lambda$, a matrix of arrow information in the model; $\Phi$, a matrix of slings involving latent variables; and $\Theta_e$, a matrix of error slings.

The matrices are organized according to the following dimensions: $\Lambda$ has as many rows as there are manifest variables in the model and as many columns as there are latent variables; $\Phi$ has as many rows and columns as there are latent variables in the model; and $\Theta_e$ has as many rows and columns as there are manifest variables. The predicted variance/covariance matrix is then calculated as:

$$\Sigma = \Lambda\Phi\Lambda' + \Theta_e \tag{8.1}$$

In this example:

$$\Sigma = \begin{bmatrix} 0.79 \\ 0.80 \\ 0.58 \\ 0.76 \\ 0.77 \end{bmatrix} [1] \begin{bmatrix} 0.79 & 0.80 & 0.58 & 0.76 & 0.77 \end{bmatrix} + \begin{bmatrix} 038 & 0 & 0 & 0 & 0 \\ 0 & 0.36 & 0 & 0 & 0 \\ 0 & 0 & 0.67 & 0 & 0 \\ 0 & 0 & 0 & 043 & 0 \\ 0 & 0 & 0 & 0 & 0.41 \end{bmatrix} \tag{8.2}$$

Carrying out this multiplication results in the predicted correlation matrix at the bottom of Table 8.3. Analogous multiplication using the unstandardized solution values yields the values seen in the top of Table 8.3.

## Model Parsimony

Note that in this model there are fewer parameters (10) than unique elements of either variance/covariance matrix (15). Although not perfectly reproducing either average cross-products matrix, this single factor model may do a sufficiently satisfactory job of recovering average cross-products in the data. This idea of specifying a model that can explain patterns of covariation (and, in the case of the adjoined matrix, elevation) using relatively few parameters is behind the idea of specification of parsimonious SEM models

**TABLE 8.3**

Predicted Variance/Covariance and Correlation Matrices

| | **Alc. Use 3 Mo.** | **# Times 5(+) Drinks** | **Typical # Drinks Thurs.** | **Typical # Drinks Fri.** | **Typical # Drinks Sat.** |
|---|---|---|---|---|---|
| | | | Predicted Covariance Matrix | | |
| Alc.Use 3 Mo. | 8.95 | 2.09 | 3.74 | 6.83 | 7.57 |
| # Times 5(+) Drinks | 2.09 | 1.22 | 1.4 | 2.56 | 2.84 |
| Typical # Drinks Thursday | 3.74 | 1.4 | 7.58 | 4.59 | 5.09 |
| Typical # Drinks Friday | 6.83 | 2.56 | 4.59 | 14.62 | 9.29 |
| Typical # Drinks Saturday | 7.57 | 2.84 | 5.09 | 9.29 | 17.53 |
| | | | Predicted Correlation Matrix | | |
| Alc.Use 3 Mo. | 1 | 0.63 | 0.45 | 0.6 | 0.6 |
| # Times 5(+) Drinks | 0.63 | 1 | 0.46 | 0.61 | 0.61 |
| Typical # Drinks Thursday | 0.45 | 0.46 | 1 | 0.44 | 0.44 |
| Typical # Drinks Friday | 0.6 | 0.61 | 0.44 | 1 | 0.58 |
| Typical # Drinks Saturday | 0.6 | 0.61 | 0.44 | 0.58 | 1 |

to summarize data. Put another way, we may examine the discrepancies between the observed and predicted average cross-products matrix (comparison of the corresponding values of Table 8.2 with Table 8.3) and consider whether the observed deviations of the model to the data constitute random sampling variability or are, for practical reasons, ignorable. We return to the idea of parsimony, sampling variability, and fit when we introduce the ideas of fit measures for SEMs in Chapter 9, but it is first necessary to discuss how a particular model corresponds to a single set of parameter values and different strategies for doing so.

## Model Identification—Mathematical Identification

Chapters 3 and 4 highlighted the fact that no unique solution can logically exist if more parameters are included in the model than there are unique elements of the average cross-products matrix associated with the model. It is not always the case that having fewer parameters than elements of a variance/covariance matrix will necessarily have a single set of parameters. The comparison of the simple comparison of parameter number with the number of unique elements of the average cross-products matrix is termed "logical identification." Recall that one reason models with too many parameters are unacceptable is because an infinite number of sets of estimates fit the data equally well. (Think, for example, of a model in which a researcher attempts to estimate both an error path and error sling, for example.) Specification of a model that is guaranteed to have only one set of parameter estimates is termed "mathematical identification."

### Mathematical Identification in Latent Variable Scaling

Note, for example, that the models in Figures 8.1 and 8.2 both constrained the factor variance to be 1. When considering latent variables, it is necessary to make some assumptions regarding the numeric scale of these latent variables. Although setting variances to unity is acceptable in some situations, it would not be as straightforward if the latent variable is a dependent variable, as this would require the researcher to specify the variance of 1 in terms of the additional compound paths of the endogenous variable.

There are three ways to identify the measuring units associated with a latent variable: (1) setting the variance to unity (as done earlier); (2) fixing a loading to a specific numeric value (usually 1); and (3) constraining the sum of the squared factor loadings to 1. This last approach is not often discussed within the context of SEM, because it also requires specification of a nonlinear constraint, but it is featured in exploratory factor models under the term singular value decomposition (SVD).

### Factor Models Identified by Reference to a Manifest Variable

To see how different assumptions regarding the scale of a latent variable, compare Figure 8.3 with Figure 8.2. Figure 8.2 contains one fixed nonzero loading and a freely estimated latent variable variance. In general, this corresponds to the identification strategy of fixing one loading in the model to a nonzero value. Usually, the number 1 is used because doing so means that the latent variable is scaled in true score units of the associated manifest variable. By calculating the predicted correlation matrix associated with this model, you can verify that this model does just as good a job as the standardized one-factor model in Figure 8.2 in reproducing the observed correlation matrix associated with the data. First, note that the estimated error variances associated with the manifest variables are identical to those of Figure 8.2. Second, note that the estimated variance of the latent variable in Figure 8.3 now corresponds to the square of the loading in Figure 8.1. And third, note that the remaining loadings correspond to the values obtained in Figure 8.2, divided by the square root of the factor variance (i.e., the loading found for the indicator variable of # Times 5(+) drinks).

Note also that the parameter values in Figure 8.2 are not the only set of possible numbers if one assumes a factor variance of one. If one substitutes negative values of equal magnitude for the loadings of the standardized one-factor model, the same predicted correlations will also result. Conceptually, the latent variable in this case would be a variable of "Not Consuming

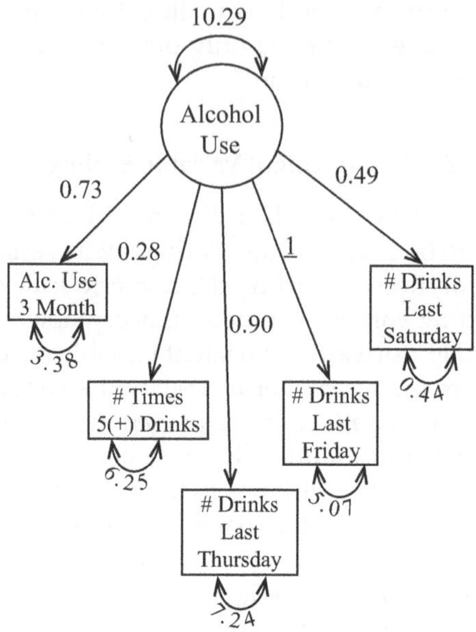


**FIGURE 8.3**
One-factor model with reference scale indicator (unstandardized).

Alcohol." Similarly, researchers who assess five types of intelligence may, on one hand, explain these patterns as due to a latent variable of "Intelligence" or, conversely, of negatively weighted loadings on a "Stupidity" factor. Those with a sense of humor may recall the old joke of lightbulbs as not being instruments that "fill the room with light," but instead "suck all the dark out of the room." There is no greater conceptual import of this phenomenon and, as we will discuss when explaining how SEM software produces parameter estimates, it is possible to use start values that provide the solution with the desired valence. Alternatively, one may instruct the software to search for only positive values, manually reverse the loadings (provided care is taken to appropriately reverse parameters associated with that factor when the latent variable is part of a larger model), or using Bayesian estimation to constrain the possible values of a loading to positive values.

In practice, it often makes little difference (in terms of overall model fit) which manifest variable is chosen as a reference variable. However, if the researcher chooses a poor indicator as the reference variable, it is possible that the resulting model may produce an improper solution (e.g., one in which parameter estimates account for negative error variances or correlations greater than one in absolute value). It is, however, the case that the Wald test associated with individual loadings in the model can be different depending on which manifest variable is chosen as a reference value (Gonzalez &

Griffin, 2001; Neale & Miller, 1997). For this reason, some individuals advocate formal model comparison procedures that test a given model against a localized null hypothesis of a zero loading for individual parameters.

## Identification Using Effects Coding

Confirmatory models are most often identified by constraining factors to unity (for exogenous factors if the researcher wants statistical significance tests of all loadings or wants to think of covariances between exogenous factors as representing correlations) or by fixing a loading to 1 (for endogenous factors). There are other times, however, when a researcher wants to express factor means and variances in terms of some general weighting across variables and not just in terms of one indicator variable. In situations where the mean of the factor is of conceptual interest (such as growth curve models), fixing the variance of a factor to one and the mean to zero may remove the conceptual interpretation of the scale from the measuring units of the manifest variables. Two models that accomplish this are effects coding and specification of a normalized basis.

The basic idea of the effects coding approach proposed by Little, Slegers, and Card (2006) is to scale the factor variance as a weighted function of the covariances of the indicators. This is accomplished by considering the average loading across the indicator variables and then specifying a nonlinear constraint on one of the manifest variables to reflect this weighting. For example, for five indicators,

$$\lambda_{11} = 5 - \lambda_{21} - \lambda_{31} - \lambda_{41} - \lambda_{51} \tag{8.3}$$

With a bit of math, it can be shown that the estimated factor variance under such a model constraint is a function of the covariance among the pairs of the observed variables and the factor loadings. For example, if there were only two variables in the model:

$$\psi_{11} = \frac{\text{cov}(y_1 y_2)}{\lambda_{11}\lambda_{21}} \tag{8.4}$$

More generally, if there are more than two variables, the factor variance will be the average of the covariances between each pair of indicators divided by the product of their respective loadings across all the indicator variables. If it is desired to also estimate the factor mean under such an identification strategy (as would occur when the model of interest is a growth curve), the constraint imposed on the intercepts is equal to the negative of the remaining intercepts. For example, for the five-indicator example considered here if $v$ is considered an estimated intercept:

$$v_1 = -(v_2 + v_3 + v_4 + v_5) \tag{8.5}$$

The freely estimated factor mean is then equal to the average of the observed manifest variable means weighted by their factor loadings. In this example (assuming $\alpha_1$ is the estimated factor mean):

$$\alpha_1 = \frac{\sum_{i=1}^{5} \bar{y}_i \lambda_i}{5} \tag{8.6}$$

Estimation of the factor mean as a weighted average of the manifest variable indicators relative to a particular measurement occasion has advantages in the context of growth modeling when the variables have similar metrics (such as being parallel measurements of the construct or sharing a common measurement scale such as that provided by a Likert format question). If the primary interest, however, is only in the parameterization of variance components, it is, of course, possible to identify the factor model without identifying the mean level of the latent variable (Figure 8.4).

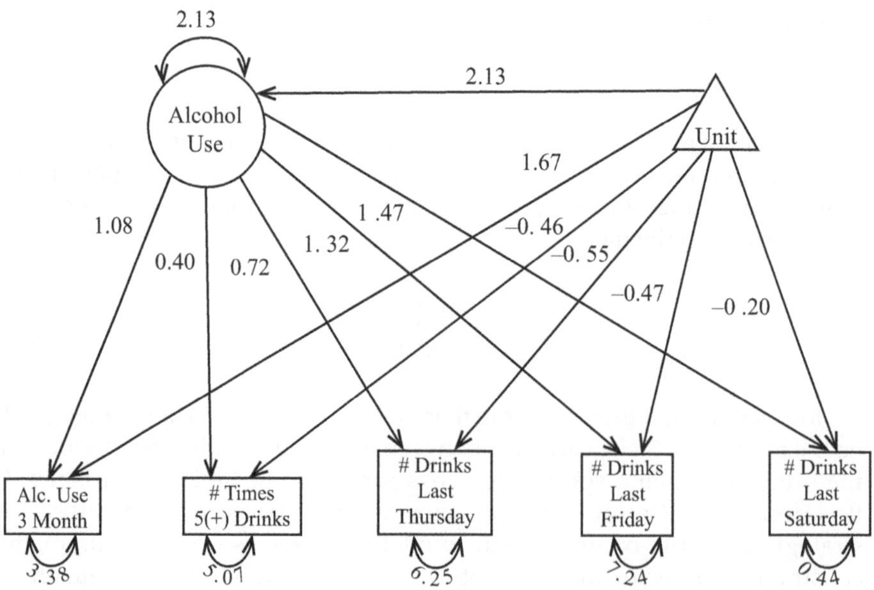


**FIGURE 8.4**
Single factor under effects coding identification.

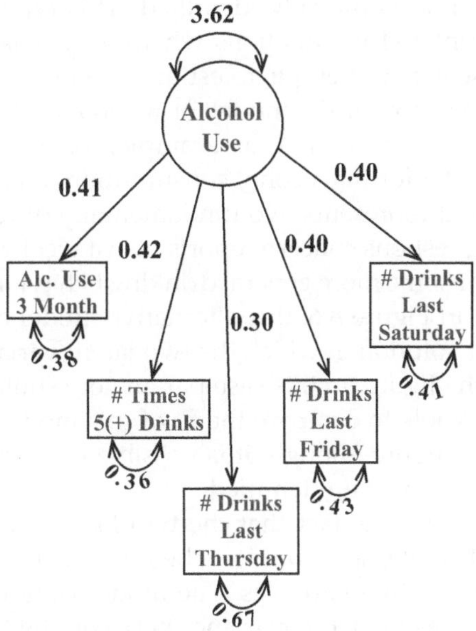


**FIGURE 8.5**
Single factor as normalized eigenvectors.

### Identification via Normalized Basis

A second way to identify a single factor model with no loadings or factor variances constrained to unity is to impose the restriction that the sum of the squared loadings equals the number 1 as shown in Figure 8.5. When we discuss exploratory factor models later, we refer to such scaled factor loadings as "eigenvectors" and to the factor variance as an "eigenvalue." The technical details of a singular value decomposition or, equivalently, a principal components analysis will be taken up later, but for the moment, it is instructive to examine the solution from such an identification strategy and to note, as for all previous models, that the predicted standardized solution associated with this metric is identical as well, as you can verify by calculating the predicted correlations from the standardized parameter values.

## Empirical Identification

Even if care has been taken to assure that a factor model is logically and mathematically identified, it is still possible that a single set of unique values cannot be estimated due to the numerical characteristics of the data at hand. Such models are termed not empirically identified. This type of problem occurs rarely and is encountered in situations where very small sample sizes are used, when the researcher attempts to estimate more parameters than what exist in the solution space for the model, or when a particularly unfortunate incorrect model has been chosen. As an example, consider the possibility that the single factor model for the alcohol consumption data might be criticized on the grounds that it represents two fundamentally different types of questions. Three of the questions concern reports of last week consumption, while the other two ask for some more general drinking behavior over the past three months. As shown in Figure 8.6, this alternative model could then be modeled as a two-factor solution in which the two latent variables are correlated. As can be seen in the figure, in this case parameter estimates were obtained, and it would be possible to compare the fit of this model relative to a single factor model. In this figure, though, it is possible to consider data scenarios that would make the model unidentified.

For example, consider the fact that the two factor loadings in the factor labeled "Typical Alcohol Use" would not be identified logically if the model consisted of only these two variables. The model is, however, identified in this case because the estimated covariance between the two latent variables means that the covariances of these two variables with the last week consumption variables are included in the model. If, however, it occurred in this

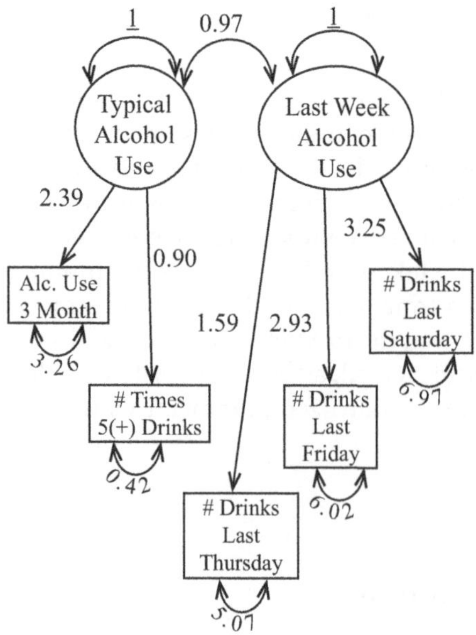


**FIGURE 8.6**
Correlated (oblique) two factor model.

particular data set that the estimated covariance between these two factors was exactly equal to zero, the two-indicator factor model would not be identified. Similarly, if the factor loading associated with one of the manifest variables of this factor happened to be exactly equal to zero (say # 5(+) Drinks), the model would also not be identified, due to the fact that the covariance of the remaining variable with the other three manifest variables would now be a function of the estimated correlation between the factors and the factor loading of the remaining variable (Alc. Use for the Past Three Months). Similarly, even an otherwise just-identified model may not be empirically identified if the exogenous variables (latent or manifest) are severely multicollinear. For moderately sized data sets and relatively simple models, empirical identification is not often a concern in practice. Nevertheless, as explained later in the chapter, the practice of using random start values during minimization often helps in identifying such cases when not initially evident from statistical software results.

### Local Under-Identification

Finally, the term "local under-identification" is sometimes used to refer to a model that, although logically identified as a whole, contains an unidentified subsection. These local under-identifications can be the result of logical, mathematical, or empirical under-identification.

## Iterative Parameter Estimates for Structural Models

Up to this point, we have presented only the parameter estimates and statistical significance levels of an SEM program without discussing how the software came to arrive at these estimates. This was done to focus first on the conceptual status of parameters in a path model and to explain how the tracing rules can be used to generate predicted values of the average cross-products matrix associated with a given score matrix. Hopefully, this has fostered some critical reasoning about path diagrams in general. Specifically, the principles of logical, mathematical, and empirical identification help to explain why latent variables with all possible covariances between error terms are not possible, why it is necessary to set the scale when estimating latent variables, and how reflecting factor loadings produce identical predicted average cross-products, for example.

It is important, however, to also understand how structural equation modeling software arrives at parameter estimates in order to better evaluate the trustworthiness of a given estimated model. Specifically, a given solution produced by an SEM program may not actually represent the optimal set of parameter values for a given model. The diagnostics for detecting and strategies for remediating this problem require a basic knowledge of the process used in SEM software to produce parameter estimates from data.

As an example, consider the path diagram for a three-indicator model representing only last week alcohol use across three days shown in Figure 8.7. Instead of analyzing the covariance matrix for these data, we will analyze the correlation matrix shown in Table 8.4 in order to use a simple example. This model might be appropriate, for example, if the researcher was interested in knowing the relative merits of these three manifest variables as indicators of a factor representing general alcohol consumption in the last week.

Considering this diagram as a model for a correlation matrix, we can note that it is a just-identified model in that six parameters are estimated to represent the six variances and covariances of the data. This model, like the other just-identified models in Chapter 4, can be solved using algebra. Such models are often referred to as "recursive" to distinguish them from other models

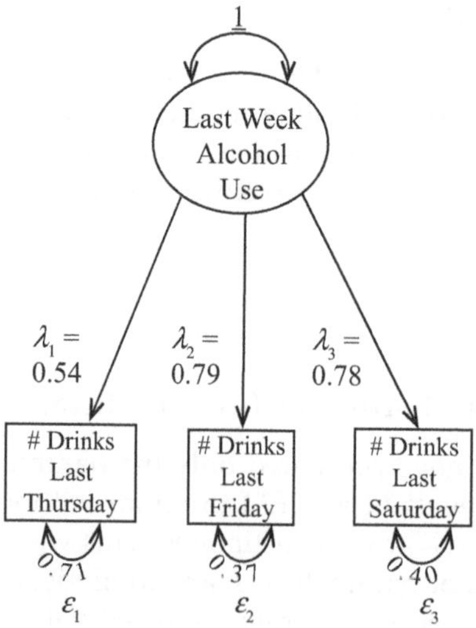


**FIGURE 8.7**
Three-indicator standardized solution.

**TABLE 8.4**
Correlation Matrix for Alcohol Consumption

| Variables | Thursday | Friday | Saturday |
|---|---|---|---|
| Thursday | 1 | 0.427 | 0.4168 |
| Friday | 0.4270 | 1 | 0.6157 |
| Saturday | 0.4168 | 0.6157 | 1 |

such as those involving feedback loops which cannot be solved algebraically. Doing so, however, would be tedious. One would first write out the predicted correlations for the data, solve one of the equations for $\lambda_1$, substitute it into the second equation and solve for $\lambda_2$, and then substitute these two definitions for $\lambda_1$ and $\lambda_2$ into the third equation, solving for $\lambda_3$. This numeric value would then be back-substituted into the prior two equations to then solve for $\lambda_2$ and then, by substitution again, $\lambda_1$. In addition to being an error-prone process to program, the solution process for any particular just-identified model is different. SEM software, by contrast, produces the solution to this problem in an iterative fashion by first considering the adequacy of initial guesses for the parameter values, and then successively improving these estimates until no further improvement can be found.

### Loss Functions

At a general level, a loss function is a measure of the cost associated with the discrepancy between candidate and true values and was introduced in contemporary statistics by Wald (1950). When applied to the question of parameter estimation, a loss function is a general measure of the adequacy of fit of a family of parameter estimates to the data at hand. In the case of a just-identified model which completely reproduces an average cross-products matrix, it is assumed that a loss function of zero is expected, indicating no discrepancy between the observed average cross-products and those predicted under the model.

### Least Squares Loss

Least squares loss is an intuitive and simple measure of this discrepancy. Least squares loss is simply the square of the difference between the observed and predicted average cross-products matrix from a given set of parameter values. In the previous data set, we can consider least squares estimation for only three factor loadings, as we can always determine a value for the error term that will reproduce the unit variance for the variable exactly.

In equation form, we could write our loss function as follows:

$$F_{LS} = (r_{12} - \lambda_1\lambda_2)^2 + (r_{13} - \lambda_1\lambda_3)^2 + (r_{23} - \lambda_2\lambda_3)^2, \tag{8.7}$$

where $r$ denotes the observed correlation and the product of corresponding $\lambda$'s involved in the predicted correlation under the factor model. The goal, then, is to find estimates of the parameters of the model for which this discrepancy is a minimum. It helps to think of the terrain of all possible values for two of the parameters of our model, $\lambda_1$, and $\lambda_2$. If we can, for the moment, assume that we knew $\lambda_3$, a plot of all possible values between 0.1 and 1 for $\lambda_1$, and $\lambda_2$ would look like that in Figure 8.8.

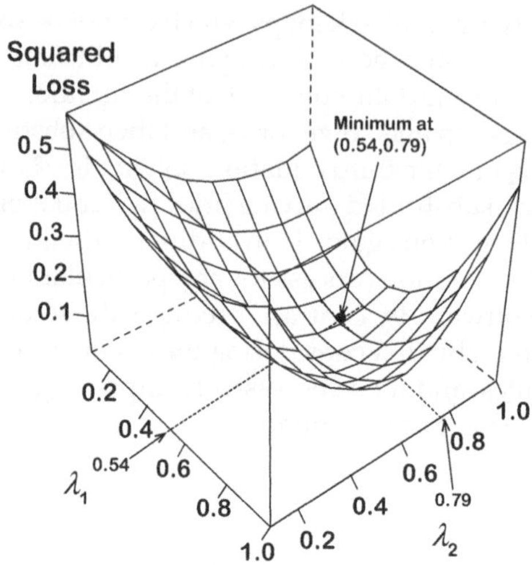


**FIGURE 8.8**
Example of well-defined search terrain.

## Start Values

As mentioned earlier, the process of solving this set of equations involves selecting an initial set of start values for the data. Software programs vary widely in terms of their sophistication in choosing initial values, with some programs using two-stage least squares estimates (2SLS), other programs using a set default number such as 1 or 0.5, and other programs generating random numbers (usually on the interval zero to 1, or −1 to 1) as the start values. Still others weight these random start values by the standard deviation of the variable(s) involved. When using a software package, it is important to know the strategy the software uses as a default.

In addition, all programs also have the option for the user to specify specific start values and, for reasons that will be discussed, it is sometimes helpful to consider values that might constitute reasonable numbers. For example, in these data, given that each loading is involved as part of the product for the correlation coefficients, we might consider the average square root of the correlations associated with a given loading as an initial guess for the parameter. (We could consider the average of Fisher $r$-to-$z$ transformed correlations, as a more precise estimate, but again, our goal here is to calculate a reasonable first guess.) Using the corresponding values from Table 8.4 gives us:

$$\begin{bmatrix} \lambda_1 \\ \lambda_2 \\ \lambda_3 \end{bmatrix} = \begin{bmatrix} \dfrac{\sqrt{0.4270}+\sqrt{0.4168}}{2} \\ \dfrac{\sqrt{0.4270}+\sqrt{0.6157}}{2} \\ \dfrac{\sqrt{0.4168}+\sqrt{0.6157}}{2} \end{bmatrix} = \begin{bmatrix} 0.6495 \\ 0.7191 \\ 0.7151 \end{bmatrix} \tag{8.8}$$

and the LS loss function for this answer is:

$$\begin{aligned} F_{LS} &= \left(0.4270-0.6495(0.7191)\right)^2 \\ &+\left(0.4168-0.6495(0.7151)\right)^2, \\ &+\left(0.6157-(0.7191)0.7151\right)^2 = 0.0142 \end{aligned} \tag{8.9}$$

which, while a small number, is not zero. The question then becomes: How can this initial estimate be changed so that we arrive at a better answer corresponding to the minimum value such as that shown in Figure 8.8?

### Newton's Method

Newton's method (also known as the Newton-Raphson method) is just such an iterative approach to finding parameter estimate values. What follows is a step-by-step description of how this works. Table 8.5 shows how this iterative approach arrives at values for least squares loss in the example model. Although the details involve a bit of calculus, I have tried to communicate the process in a way that explains the steps algebraically. The matrix formula for the calculation of a new (column) vector of parameter estimates is $\hat{\theta}^{\left(E^{i+1}\right)}$, given a column vector of initial estimates or "start values," $\hat{\theta}^{(i)}$. Making a new estimate involves calculating the first and second partial derivatives of the loss function. The idea of a partial derivative and how to calculate them will be explained shortly.

$$\hat{\theta}^{(i+1)} = \hat{\theta}^{(i)} - \left[\frac{\partial^2 F_{LS}}{\partial\theta\partial\theta}\right]^{-1}\left[\frac{\partial F}{\partial\theta}_{LS}\right] \tag{8.10}$$

**TABLE 8.5**

Newton-Raphson Iterative Solution Example

| Iteration | $\lambda_1$ | $\lambda_2$ | $\lambda_3$ | $F_{LS}$ |
|---|---|---|---|---|
| Initial | 0.64951 | 0.71907 | 0.71512 | 0.01418 |
| 1 | 0.48788 | 0.82393 | 0.81301 | 0.00396 |
| 2 | 0.54360 | 0.77534 | 0.76156 | 0.00068 |
| 3 | 0.53200 | 0.80063 | 0.78430 | 0.00015 |
| 4 | 0.54005 | 0.78969 | 0.77237 | 0.00003 |
| 5 | 0.53640 | 0.79555 | 0.77740 | 0.00001 |
| 6 | 0.53819 | 0.79314 | 0.77463 | $1.76 \times 10^{-6}$ |
| 7 | 0.53735 | 0.79451 | 0.77574 | $4.07 \times 10^{-7}$ |
| 8 | 0.53775 | 0.79398 | 0.77509 | $9.42 \times 10^{-8}$ |
| 9 | 0.53756 | 0.79431 | 0.77532 | $2 \times 10^{-8}$ |
| 10 | 0.53765 | 0.79419 | 0.77517 | $5 \times 10^{-9}$ |
| 11 | 0.53761 | 0.79427 | 0.77522 | $1.25 \times 10^{-9}$ |
| 12 | 0.53763 | 0.79425 | 0.77518 | $3.06 \times 10^{-10}$ |
| 13 | 0.53762 | 0.79427 | 0.77519 | $7.67 \times 10^{-11}$ |
| 14 | 0.53762 | 0.79426 | 0.77518 | $1.99 \times 10^{-11}$ |

## Partial Derivatives

For those of you who have not encountered Leibniz's notation or whose calculus skills are a little rusty, do not allow Eq. (8.10) to intimidate you. The right-most term in the equation is, conceptually, an estimate of the slope associated with the function minimizing the parameter estimate. In calculus, calculating a derivative is done by multiplying the leading coefficient of a function by its exponent and reducing the exponent by 1. For example, to calculate the partial derivative of the first loading, we will multiply out the loss function for $\lambda_1$ as $\theta_1$, treat $\lambda_1$ as a variable we wish to take the derivative of, and write:

$$\begin{aligned}\frac{\partial F_{LS}}{\partial \theta_1} &= [(r_{12}-\lambda_1\lambda_2)(r_{12}-\lambda_1\lambda_2)+(r_{13}-\lambda_1\lambda_3)(r_{13}-\lambda_1\lambda_3)]\frac{\partial}{\partial \theta_1} \\ &= [r_{12}^2-2\lambda_1\lambda_2 r_{12}+\lambda_1^2\lambda_2^2+r_{13}^2-2\lambda_1\lambda_3 r_{13}+\lambda_1^2\lambda_3^2]\frac{\partial}{\partial \theta_1} \end{aligned} \tag{8.11}$$

We can take the derivative of $\lambda_1$ the last two lines of Eq. (8.6), which will give us:

$$\begin{aligned}\frac{\partial F_{LS}}{\partial \theta_1} &= \left[r_{12}^2 - 2\lambda_1\lambda_2 r_{12} + \lambda_1^2\lambda_2^2 + r_{13}^2 - 2\lambda_1\lambda_3 r_{13} + \lambda_1^2\lambda_3^2\right]\frac{\partial}{\partial \theta_1} \\ &= \left[r_{12}^2 - 2\lambda_1\lambda_{2r_{12}} + \lambda_1^2\lambda_2^2 + r_{13}^2 - 2\lambda_1\lambda_3 r_{13} + \lambda_1^2\lambda_3^2\right]\frac{\partial}{\partial \theta_1} \\ &= -2\lambda_2 r_{12} + 2\lambda_1\lambda_2^2 - 2\lambda_3 r_{13} + 2\lambda_1\lambda_3^2\end{aligned} \quad (8.12)$$

Doing this for the three loadings gives us the following vector:

$$\frac{\partial F_{LS}}{\partial \theta} = \begin{bmatrix}\frac{\partial F_{LS}}{\partial \theta_1} \\ \frac{\partial F_{LS}}{\partial \theta_2} \\ \frac{\partial F_{LS}}{\partial \theta_3}\end{bmatrix} = \begin{bmatrix}-2\lambda_2 r_{12} + 2\lambda_1\lambda_2^2 - 2\lambda_3 r_{13} + 2\lambda_1\lambda_3^2 \\ -2\lambda_1 r_{12} + \lambda_1^2 2\lambda_2 - 2\lambda_3 r_{23} + 2\lambda_2\lambda_3^2 \\ -2\lambda_1 r_{13} + \lambda_1^2 2\lambda_3 - 2\lambda_2 r_{23} + \lambda_2^2 2\lambda_3\end{bmatrix} \quad (8.13)$$

The matrix of second derivatives, $\left[\frac{\partial^2 F_{LS}}{\partial\theta\partial\theta}\right]$, is a matrix of derivatives of derivatives and repeats the same process for all possible pairs of parameters based on the vector of first derivatives. The technical name for this matrix of second derivatives is the *Hessian*.

$$\left[\frac{\partial^2 F_{LS}}{\partial\theta\partial\theta}\right] = \begin{bmatrix}2\lambda_2^2 + 2\lambda_3^2 & -2r_{12} + 4\lambda_1\lambda_2 & -2r_{13} + 4\lambda_1\lambda_3 \\ -2r_{12} + 4\lambda_1\lambda_2 & 2\lambda_1^2 + 2\lambda_3^2 & -2r_{23} + 4\lambda_2\lambda_3 \\ -2r_{13} + 4\lambda_1\lambda_3 & -2r_{23} + 4\lambda_2\lambda_3 & 2\lambda_1^2 + 2\lambda_2^2\end{bmatrix} \quad (8.14)$$

To calculate our new set of parameter estimates, then, we calculate Eq. (8.5), substituting in the observed correlations and current parameter estimates.

$$\left[\frac{\partial F_{LS}}{\partial \theta}\right] = \begin{bmatrix}0.1258 \\ -0.0932 \\ -0.0849\end{bmatrix} \quad (8.15)$$

Then we calculate the estimated second derivatives,

$$\frac{\partial^2 F_{LS}}{\partial\theta\partial\theta} = \begin{bmatrix}2.0569 & 1.0141 & 1.0244 \\ 1.0141 & 1.8665 & 0.6265 \\ 1.0244 & 0.6265 & 1.8778\end{bmatrix}, \quad (8.16)$$

which enables calculation of the inverse. Recall from the matrix algebra discussion that the inverse is the adjoint divided by the determinant. If it is the case that the determinant of the second derivatives is zero, this suggests that the model may not be logically, mathematically, or empirically identified. Some software packages will produce statements of the particular parameters which appear to be functions of other parameters in the model when calculating the Hessian, and this information can be used to determine if a mistake has been made in model specification or, more rarely, empirical nonidentification. For this set of values, the Hessian is:

$$\left[\frac{\partial^2 F_{LS}}{\partial\theta\partial\theta}\right]^{-1} = \begin{bmatrix} 0.8161 & -0.3311 & -0.3348 \\ -0.3311 & 0.7376 & -0.0655 \\ -0.3348 & -0.0655 & 0.7270 \end{bmatrix}, \tag{8.17}$$

which then enables us to calculate the new set of parameter estimates for the data:

$$\theta^{(i+1)} = \begin{bmatrix} 0.6495 \\ 0.7191 \\ 0.7151 \end{bmatrix} - \begin{bmatrix} 0.8161 & -0.3311 & -0.3348 \\ -0.3311 & 0.7376 & -0.0655 \\ -0.3348 & -0.0655 & 0.7270 \end{bmatrix} \begin{bmatrix} 0.1258 \\ -0.0932 \\ -0.0849 \end{bmatrix} = \begin{bmatrix} 0.4879 \\ 0.8239 \\ 0.8130 \end{bmatrix}. \tag{8.18}$$

## Iteration to Convergence

If we then use this new set of factor loading estimates to generate a predicted correlation matrix, we find that the least squares loss is now 0.0040, a reduction from our initial loss function value of 0.0142, but still not equal to zero. We can, however, continue this process until the loss function is arbitrarily close to zero, as shown in Table 8.5. As you can see from the successive estimates for the factor loadings, estimated values for the factor loadings are changing very little toward the end. At this point, we say the model as converged and accept the parameter values as the least squares estimates.

## Remarks

This short example illustrates the basic idea behind successive iterative approximation to solutions of structural models. It will probably not surprise you to learn that this area of estimation is a research topic of its own and that several refinements of the approach are implemented in SEM software. B. S. Everitt's (1987) *Introduction to Optimization Methods and Their Application in Statistics* is a reasonably good introduction to some of the basic ideas involved in a variety of minimization approaches. Generally, these refinements include strategies for producing good starting values, for improving the strategy for updating estimates across iterations, and for reducing the computational burden associated with calculating derivatives of the loss function at each iteration.

Writing software that performs minimization is also an exercise requiring expertise in numerical analysis. This is necessary not only to ensure that the computing memory resources are used efficiently and using reasonable memory requirements, but also to minimize the effects of rounding error induced across multiple math operations. Practically speaking, this often involves rewriting the observed and predicted data as a single long vector and writing the math operations on an element-wise basis. One practical import of this for someone doing SEM is that performing minimization using direct matrix algebra, such as Excel, is inadvisable. A second point is that, as mentioned earlier, the inverse of the second derivatives may not be directly calculatable by the software program. This can sometimes happen if the start values chosen are problematic, which can be remedied by overriding the default start values. While some software programs helpfully point out the problematic parameters involved in such a dependency, other computer programs simply stop with a message that the Hessian cannot be inverted.

It is worth noting that, increasingly, SEM programs do not minimize the score matrix or an approximation to it as done here, but instead apply the loss function at the individual observation (row) level. Although in some respects this does not result in differences in estimation, it is easier to implement in the presence of missing data when full information maximum likelihood estimates are needed and has some technical advantages when the researcher wishes to specify interactions at the individual observation level.

## Implications, Problems, and Solution Strategies for Iterative Solutions

It bears mentioning that very often the default iterative estimation procedures in SEM software work well. However, there are situations where the computer program may fail to recover the best parameter estimates under the loss function. It is therefore important to be aware of estimation difficulties that can occur and diagnostics that may indicate a set of parameter estimates is not to be trusted. Some of these cautions are relatively straightforward and are discussed first, while others require a bit more attention.

### *Failure to Converge Using Default Iteration Limit*

Statistical software frequently specifies a default number of iterations for solving a given SEM. It may well be the case that the solution has not yet converged to a minimum value, but the printout or graphic display of the model may show parameter estimates that reflect the solution of the last iteration. It is important to inspect the results of your software to verify that the solution has been reported as converging. If convergence has not been achieved, the user must override the default by specifying a larger number of iterations. Although this is a straightforward issue, I have reviewed papers that reported solutions which have not converged.

### *False Convergence.*

Some software packages do not check to make sure that the Hessian (described earlier in the chapter) is of full rank, meaning that if the user has specified an unidentified model, the program may iterate indefinitely or, alternatively, arrive at a minimum and report parameter values, but which represent only one of many possible sets of equally well-fitting models. Although one remedy to this problem is to understand how to specify logically and mathematically identified models, the issue of empirical nonidentification can still occur in a model that seems reasonable. Use of SEM programs that check the rank of the Hessian or determine if it is invertible is another reassuring but not foolproof remedy. Alternatively, some SEM programs conduct multiple minimizations using a variety of randomly generated start values and as such may flag such estimation problems.

### *Failure to Converge Due to Terrain of Loss Function*

Even with increased iterations or a different minimization algorithm, a model may fail to converge. Diagnosing and fixing such problems vary depending on the particular software used. For some loss functions and structural models, the loss function may be essentially flat, meaning that parameter estimates may fluctuate wildly across iterations with no substantial improvement in fit and very slow progress toward the desired minimum of the fit function. Alternatively, the updated parameter estimates may result in iterations that "overshoot" the minimum parameter estimate, resulting in a failure to converge due to the use of inappropriately large adjustments. Remedies for such terrain problems can involve, again, use of better start values for parameters of the model, or the use of conjugate gradient minimization techniques, which, while more computer intensive, are less likely to encounter such difficulties during minimization.

### *Local Minima*

The topology of the loss function may also have minor "valleys" or "hills," thereby leading the algorithm to falsely declare that a minimum has been found when in fact a given solution is only a local minimum and not a global one. Although in principle this problem is yet another example of false convergence, inspection of the iteration history can often inform the researcher of a possibly torturous loss function topology. Some of the optimization approaches in SEM software are better than others in overcoming such difficulties. Proc Calis in SAS, for example, employs a minimization strategy depending on characteristics of the model to be fit. Alternatively, other software packages repeatedly minimize the loss function using several sets of random start values.

### *Concluding Remarks Regarding Minimization Problems*

Although the preceding discussion may cause some consternation as to whether one can ever really trust the output of a structural model, it again bears repeating that very often no minimization problems are encountered in fitting SEMs to data. Space limitations and the technical nature of minimization techniques preclude a systematic discussion of the relative merits of the different optimization approaches used by various packages, but a few of the optimization methods may help the user arrive at a satisfactory solution when difficulties are encountered. When these problems are encountered, the researcher may wish to seek technical assistance. That said, if one must research options on one's own or wishes to mention some technical possibilities during a statistical consultation, it is perhaps worth mentioning that Newton-Raphson, Levenberg-Marquardt, and trust-region techniques work fairly well in practice for a variety of structural models and often converge after only a few iterations for many models. All three approaches, though, require the computer to calculate the Hessian matrix in each iteration. If a given structural model is very large, quasi-Newton methods, especially those using the Broyden-Fletcher-Goldfarb-Shanno algorithm (often abbreviated as BFGS in reference materials) are often much less computer resource intensive and often yield satisfactory results.

## ML, GLS, and ADF Loss Functions

For SEMs consisting of continuous observed and latent variables, SEM programs generally default to maximum likelihood (ML) estimation and not least squares loss (LS). In addition, measures of model fit (discussed in Chapter 9) are also designed and optimized for maximum likelihood estimates. To understand this aspect of SEM, it helps to first compare LS and ML loss and to show an example of the differences between the predicted cross-products matrix under the two approaches. After this, mathematical definition of ML loss will be presented, as well as other loss functions commonly available.

For many statistical models, LS estimation is computationally convenient for determining the parameter values that will reproduce the average cross-product matrix exactly in just-identified linear models. Another way to say this is that LS loss can be used to identify the "zeros" in the multivariate quadratic equation, i.e., the numeric values for which the squared difference of the observed and predicted covariance matrices is zero. Within the context of linear models, least squares loss is designed to produce parameter estimates that have the smallest errors of prediction. You may also recall from your previous coursework in general linear models that LS and maximum likelihood estimates are identical if we can assume that errors of prediction are normally distributed (see, e.g., Kutner, Nachtsheim, Neter, & Li, 2005, pp. 15–32). For this reason, LS loss is also a maximum likelihood solution for this class of models.

Confirmatory factor models are different, however, in that they seek to achieve parsimony by invoking fewer parameters than unique elements of the average cross-products matrix. Least squares parameter estimates may not do the best job at estimating the most useful set of parameter estimates in this regard. Conceptually, maximum likelihood estimates are designed to recover parameter estimates that will reproduce the more important aspects of relationships in the data, and for this reason, maximum likelihood estimation is preferred.

In order to see how the predicted matrices of ML and LS estimates may differ, consider the correlation matrix shown in Table 8.6 for four hypothetical variables. This data set shows two marked relationships, the correlation between *V*1 and *V*2, and the somewhat smaller correlation between *V*3 and *V*4. There are also four minor correlations of *V*1 and *V*2 with *V*3 and *V*4.

**TABLE 8.6**

LS and ML Solutions for Hypothetical Data

| Variable | *V1* | *V2* | *V3* | *V4* |
|---|---|---|---|---|
| | | Hypothetical Data | | |
| *V*1 | 1 | 0.8 | 0.2 | 0.2 |
| *V*2 | 0.8 | 1 | 0.2 | 0.2 |
| *V*3 | 0.2 | 0.2 | 1 | 0.7 |
| *V*4 | 0.2 | 0.2 | 0.7 | 1 |
| | | Predicted LS Correlation Matrix | | |
| *V*1 | 1 | 0.67 | 0.33 | 0.33 |
| *V*2 | 0.67 | 1 | 0.33 | 0.33 |
| *V*3 | 0.33 | 0.33 | 1 | 0.17 |
| *V*4 | 0.33 | 0.33 | 0.17 | 1 |
| | | Predicted ML Correlation Matrix | | |
| *V*1 | 1 | 0.8 | 0.22 | 0.22 |
| *V*2 | 0.8 | 1 | 0.22 | 0.22 |
| *V*3 | 0.22 | 0.22 | 1 | 0.06 |
| *V*4 | 0.22 | 0.22 | 0.06 | 1 |

Consider the fit of a single factor model to such data. Although this is a poor model for the data, given that a two-factor structure is clearly present, the goal is to comment on how the ML model is somewhat better than the LS model. What are the most important relationships in the data that even an incorrect model should recover? One might argue that being able to reproduce the 0.8 correlation between $V1$ and $V2$ is the strongest relationship, as is reproducing the 0.7 correlation between $V3$ and $V4$. If forced to choose one, perhaps the 0.8 correlation is slightly more important, if only because the shared variance between $V1$ and $V2$ is 64%, while the shared variance between $V3$ and $V4$ is 49%. In any event, successful reproduction of the 0.2 correlations in the data is probably a minor concern unless many such correlations are present.

The predicted correlation matrices associated with the one-factor LS solution are shown in the second block of Table 8.6. The predicted correlation between $V1$ and $V2$ is 0.67, substantially lower than the observed correlation and representing a substantially lower proportion of shared variance. Similarly, the predicted correlation between $V3$ and $V4$ is only 0.17, which is quite different from the observed correlation of 0.70. The predicted correlation matrix for the ML solution is shown in the bottom block, where we see that the predicted correlation of 0.80 matches the observed correlation in the data, but that the correlation between $V3$ and $V4$ is not well modeled, being 0.06. The predicted smaller correlations of 0.2 are estimated at 0.22. In general, then, it can be argued that the predicted correlation matrix from the ML solution does a better job at reproducing the observed correlation matrix than the LS solution, even though both models are "wrong" in the sense that this reproduction is far from perfect.

## Matrix Expressions for Loss Functions

### *Maximum Likelihood (ML)*

Formally, the loss function for maximum likelihood can be expressed as follows for $\Sigma$, a matrix of predicted average cross-products; $S$, an observed matrix of average cross-products; and $m$, the number of manifest variables.

$$F_{\text{ML}} = \log|\Sigma| + \text{trace}\left(S\Sigma^{-1}\right) - \log|S| - m \tag{8.19}$$

The trace is the sum of diagonal elements. Geometrically, the determinant represents a multidimensional measure of the volume of a vector space. In the case of two orthogonal vectors, for example, the vector is the area of a rectangle. If an angle exists between the vectors indicating a correlation, the determinant is the area of a rhombus. The point being that, in conceptual terms, the ML fit function is a measure of the "volume" associated with a predicted cross-products matrix minus the volume of the observed cross-products matrix. The trace ($S\Sigma^{-1}$) component is the term that directly compares the fit of the two models. One can think of it roughly as representing $S / \Sigma$ from scalar algebra in that if $S = \Sigma$, the result will be an identity matrix. Accordingly, if $\Sigma$ completely recaptures $S$, the value of the trace will be $m$, the number of manifest variables in the data. Conceptually, again, the ML loss function represents a weighting strategy based on the likelihood of the predicted data under the estimated parameters of the model.

### *Least Squares Loss (LS)*

By comparison, consider the LS loss function in matrix form:

$$F_{LS} = \frac{1}{2}\text{trace}\left(S - \Sigma\right)^2 \tag{8.20}$$

This is a simpler matrix expression for the sum of the squares of the observed and predicted cross-products matrices. If $S$ is a correlation matrix and $\Sigma$ has ones on the diagonal, Eq. (8.10) is the sum of squares of the unique off-diagonals. In the more general case, if the diagonal elements of $\Sigma$ are not equal to $S$, Eq. (8.10) will be equal to the sum of squares of the unique off-diagonal elements plus one-half the sum of the squared differences between the corresponding diagonal elements of $S$ and $\Sigma$.

### *Generalized Least Squares (GLS)*

GLS is another loss function that is sometimes used and has the following loss function:

$$F_{GLS} = \frac{1}{2}\text{trace}\left(I - S^{-1}\Sigma\right)^2 \tag{8.21}$$

Notice in Eq. (8.21), the same general strategy of comparing $S$ to $S$ is done, but this time $S$ is in the "denominator" of the "fraction," meaning that we are determining how closely the model approximates the data. The GLS loss function is slightly different from the ML loss function in that it represents a weighting strategy of the data weighted by the variation and associations observed in the data. Historically, when computer resources were much more expensive than they are now, the GLS loss function was advocated because of its computational simplicity: Calculating matrix inverses were computationally expensive, and one only needed to do this once for the minimization algorithm in GLS. ML, by contrast, required the calculation of a matrix inverse in each iteration of the minimization function. It is somewhat comforting that results from the GLS loss function result in estimates asymptotically equivalent to ML. In general, use of the GLS loss function is not preferred because

GLS is not as robust to violations of multivariate normality as ML, especially when robust ML estimation is used and as mentioned earlier, fit statistics for SEMs are designed for use with maximum likelihood estimation.

### *Alternate Expression for Loss Functions and ADF*

Browne (1982, 1984) proposed a different expression for loss functions, leading to an alternative loss function which did not require the multivariate normality assumption of ML and GLS. A general form for the loss functions is:

$$(s-\sigma)^T W(s-\sigma) \tag{8.22}$$

where $s$ and $\sigma$ are column vectors containing the unique elements of $S$ and $\Sigma$. The central matrix, $W$, consists of a square matrix of weights. Browne showed that if $W$ is an identity matrix, then Eq. (8.22) represents LS estimation. If $W$ is a function of the elements of $S$, the observed variance/covariance matrix, the result is GLS, and if $W$ is a function of $\Sigma$, the result is ML estimation. As such, Eq. (8.22) represents a flexible way to think about loss functions in general.

Browne went on to show that if the weight function were instead a function of the fourth-order moments of the matrix corrected for the product of pairwise covariances, the result would be an asymptotically distribution-free estimate, termed ADF. In matrix form, these elements of W for ADF can be described as:

$$\Omega_{ij,kl} = \sigma_{ijkl} - \sigma_{ij}\sigma_{kl} \tag{8.23}$$

The rationale for the ADF estimator is that both ML and GLS capitalize on the assumption of multivariate normality, which permits much greater ease of estimation and inference. If this assumption is not warranted, however, then these loss functions are essentially using an incorrect weight matrix. Given that ADF does not make the requirement of multivariate normality and that the option is often available in statistical software, it may be tempting to use ADF rather than ML or its robust modifications. Unfortunately, the approach appears not to work well in practice, except in the case of extremely large sample sizes exceeding 1,000 to 2,000 observations (Curran, West, & Finch, 1996; Hu, Bentler, & Kano, 1992; Muthén & Kaplan, 1985). Some argument has been made that a bootstrap estimate of W can improve the performance of ADF (Yung & Bentler, 1994). Although the technique is largely unused, the approach to thinking about loss functions did have useful application in the area of SEMs containing categorical variables, where this approach was termed "WLS."

### *Weighted Least Squares (WLS)*

Sometimes data sets contain variables that are observed to have only discrete categories but that, conceptually, represent an underlying continuum, Olsson (1979) proposed WLS as a way to adjust the predicted covariance matrix for the categorical nature of the observed variables describing the approach. MacCallum, Zhang, Preacher, and Rucker (2002) is a readable explanation of the two-step approach outlined for the case of categorical variables. As might be expected, the WLS approach is rather computationally intensive, and especially so if several variables are considered categorical because it requires the inversion of $W$, but the approach tends to work well in practice.

### *Diagonally Weighted Least Squares (DWLS)*

Muthén (1993) proposed a "limited information" and more efficient approach to WLS known as diagonally weighted least squares (DWLS) estimation. This is sometimes also referred to as the WLSMV estimation. In brief, this approach uses a WLS estimator with polychoric correlations as input to create an asymptotic covariance matrix. The approach uses a calculus technique called Taylor expansion to avoid the computationally expensive full inversion of $W$. DWLS seems to perform well in practice and, in particular, seems to perform better than standard WLS for small samples (Rhemtulla, Brosseau-Liard, & Savalei, 2012). Unfortunately, the formulae for standard errors (as well as model fit statistics) are no longer appropriate in this case, but some software such as Mplus uses robust estimation adjustments (such as the "sandwich" estimator) to improve estimation of these values. As one might be able to tell from the earlier terms, some packages (such as Mplus and lavaan) refer to this approach (including the estimation adjustments) as WLSM (if only means are involved in the estimation) or WLSMV (if means and variances are used). The robust DWLS methods appear to work well with smaller samples and types of non-normal data (e.g., Rhemtulla et al., 2012).

### *Maximum Likelihood Robust (MLR)*

The goal of MLR is to perform ML estimation in which a standard error and $\chi^2$ are computed which are robust to non-normality. Standard errors for the solution are calculated using a sandwich estimator, a general technique which produces estimates in the presence of distributional violations of the model in the data. Like WLSMV, it is appropriate when the observed variables consist of at least some ordered categorical variables. WLSMV and MLR should, in general give very similar results. There is a small technical point between the approaches, however, in that the MLR approach assumes that the observed categorical variables were drawn from a normally distributed variable, while WLSMV assumes a logistic distribution.

## Assumption Checking

As mentioned in Chapter 5, it is not enough to specify an identified model that generates numerically plausible estimates, even if one has taken care to ensure that the estimated solution constitutes a global minimum and the observed cross-products matrix appears to correspond closely to the values predicted by the model. As Chapter 5 discussed, it is our job to critically examine whether the assumptions of the structural model appear justified and whether our conclusions could be an artifact of an incorrect understanding of the functional relationships between variables, the presence of a small number of influential observations, or heteroscedasticity of error.

Reconsider the five alcohol consumption variables in the correlated oblique factor model shown in Figure 8.6. If a researcher were to consider using these two latent variables of alcohol consumption as predictors of some other behavior such as grade in college classes, it is likely that neither of the regression weights would be significant given the high correlation between these two latent variables. A researcher may wish, instead, to consider a latent variable corresponding to last week alcohol consumption and then consider the other two alcohol variables, self-reports of past-three-month frequency of alcohol consumption, and number of times more than five drinks were consumed in a sitting, as related to this general factor, but perhaps sharing some additional covariance not due to this latent variable (as shown in Figure 8.9). The lines of such reasoning might be something like "Even though these two other variables have a different time window than last week, they might have some value in determining last week's consumption. They are probably also correlated due to other events, such as alcohol consumption before last week." If our primary interest in these variables is the degree to which a single factor explains a proportion of variability in the manifest variables, we might be most interested in the standardized solution for the data.

Although model fit will be taken up in the next chapter, stipulate for the moment that the fit of this model is quite good. It is, however, important that we examine whether the assumptions of this model are met using the procedures outlined in Chapter 5. These are readily available in many software packages such as Mplus, but it seems that such examination is often undiscussed in SEM books as well as research articles.

### Influential Observations

Although the sample size for this study is relatively large, examination of influential observations is still important, as this can inform researchers as to the degree to which parameter estimates in the model may be affected by a relatively small number of individuals. Examination of the leverage values from the total sample reveal a relatively large number of individuals who

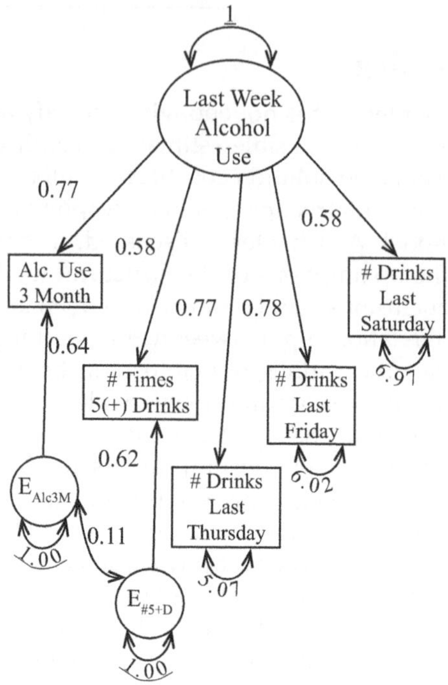


**FIGURE 8.9**
Single factor correlated errors.

may affect the parameter estimates of the model. Some of the alcohol use patterns among the influential observations can be described only as extreme, with students reporting consuming 23 and 24 drinks on each of the days, that they usually consume 10 drinks in a sitting, for example. These values appear plausible given that a very small number of college students do evidence such pathological levels of alcohol involvement.

Overall, when influence diagnostics are calculated for these data, we find that a rather substantial proportion of the data were influential by the general guideline of hat values, which exceed twice the number of parameters divided by the sample size (725 individuals, or about 28% of the data). Influential individuals are, as a group, characterized by rather high levels of alcohol consumption (consuming, on average 5.6, 6.4, and 3.7 drinks last Thursday, Friday, and Saturday, respectively). They reported that they usually consumed 7.4 drinks on any one occasion and drank five or more drinks in one sitting 1.8 times. By contrast, those who were not influential consumed only 2.3, 2.6, and 1.1 drinks across Thursday, Friday, and Saturday, usually consuming 4.0 drinks on any one occasion, and drinking five or more drinks 0.67 times. Table 8.7 shows the standardized and unstandardized factor

**TABLE 8.7**

Standardized and Unstandardized Factor Loadings Based on All Individuals ($N = 2{,}557$) and Excluding Influential Observations ($N = 1{,}832$)

| | Standardized | | Unstandardized | |
|---|---|---|---|---|
| | **All** | **Excl. Infl.** | **All** | **Excl. Infl.** |
| Thursday Drinks | 0.77 | 0.73 | 2.93 | 1.38 |
| Friday Drinks | 0.78 | 0.76 | 3.25 | 1.6 |
| Saturday Drinks | 0.58 | 0.34 | 1.59 | 0.32 |
| Alcohol Use 3-Month | 0.77 | 0.72 | 2.31 | 1.56 |
| # of 5+ Drinks | 0.78 | 0.83 | 0.87 | 0.34 |

loadings based on all individuals and based on the subset of the data excluding the influential observations. From this table we can see that although the estimates of the unstandardized values are quite different when influential observations are excluded, the estimates of the standardized loadings are quite similar, except for the case of Saturday consumption, given that the Alcohol Consumption factor explains far less variability in the Saturday drinking variable when influential observations are excluded.

In general, although these influential observations cause large departures from the standardized estimates of the factor model, their presence speaks to the likely fact that researchers studying similar populations with these items are likely to encounter similar influential observations where their effect on parameter estimates will be more marked. As a result, we would expect that smaller samples drawn from similar data with these items would not replicate well.

## Functional Form

Chapter 5 also discussed the importance of checking whether the functional form of the model is well captured by the model. To some extent this is a bit more problematic for technical reasons for latent variable models, but one useful diagnostic is to consider a LOESS regression of each manifest variable as a function of the factor score (an estimate of the unmeasured independent variable in the model). Figure 8.10 shows the estimated linear regression (the solid black line) and the LOESS curve (dashed line) for the complete data. As can be seen, the LOESS regression appears to show a marked departure from linearity for higher values of the Alcohol Use factor. In addition, the large number of individuals who consumed no drinks last Saturday also appear to pull the regression line down from the cloud of data points. Although it might seem reasonable to transform the Saturday drinks variable using some variance stabilizing transformation (perhaps suggested by a Box-Cox

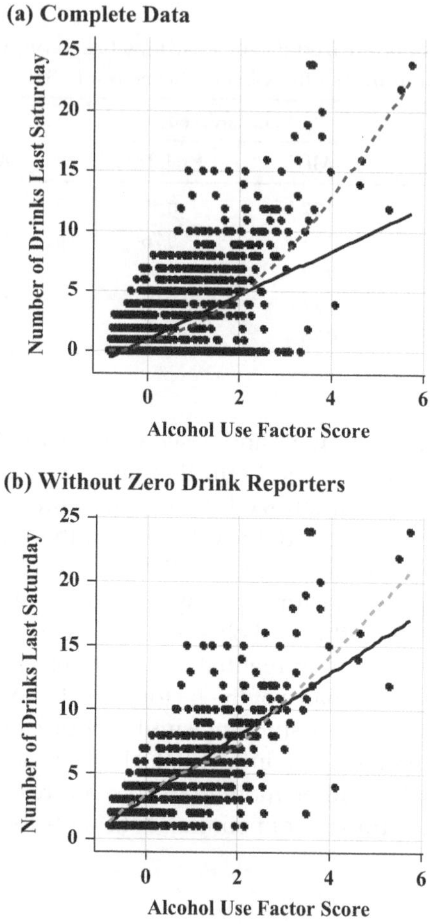


**FIGURE 8.10**
LOESS and linear regressions for Saturday drinks. (a) Complete data. (b) Without zero drink reporters.

transformation), this scatter plot suggests that it may instead be more reasonable to consider the data as being zero-inflated, meaning that the data likely consist of one population of individuals who did not consume alcohol and another population who does consume but to varying degrees. Such models are termed "mixture models" in that they involve one categorical latent variable (use versus nonuse) and a continuous variable for those who use, and the researcher is faced with the analytic choice of modeling such zero inflation at the level of each individual manifest variable or of modeling the latent variable as having these two parts. Although it is not possible to entertain such advanced models at this point, a general sense of whether the functional form of alcohol use is linear can be examined by simply removing individuals from the analysis who reported no Saturday drinks and rerunning the LOESS regression, as shown in Figure 8.10. The research questions then would be limited to that subset of the population who use alcohol. The general shape of the LOESS curve for this subset appears more or less similar to the linear form assumed by the factor model.

## Concluding Remarks

The goals of this chapter have been to survey identification strategies for SEMs, to survey the loss functions appropriate for an SEM, and to discuss assumption checking procedures that apply to SEMs. Assumption checking is not limited to general linear models such as regression and analysis of variance and is an important part of model specification and interpretation. Cautions regarding overinterpretation of the latent variable in a model have also been raised. The question, however, of the degree to which a given SEM represents a good fit to the data or is appreciably better than competing models has not been addressed and is taken up in the next chapter.

# 9

# *Model Fit and Comparison*

If the data do not fit the theory, they must be disposed of.

**–Anonymous**

To this point we have discussed only model identification, parameter estimation, and the statistical significance of individual parameters using the Wald test. In SEM studies, however, we are often interested in the degree to which an SEM represents a good fit to one of the three average cross-product matrices introduced in Chapter 2. In some contexts, we wish to make comparisons between competing models as to which provides the best fit to the available data in terms of the trade-off of model parsimony and successful recapture of variation and covariation in the data (as well as level effects if the average SSCP is used). We now take up the topic of general model fit and model comparison procedures. This chapter begins with a brief refresher of the normal and chi-square distributions, given their importance in SEM theory; proceeds with a discussion of stand-alone fit measures; and, using Marsh, Balla, and McDonald's terminology, reviews Type I and Type II comparative fit measures. Formal model comparisons for nested models are then presented as well as information-theoretic model comparisons for non-nested models.

## Normal and $\chi^2$ Distribution Review

Some confusion sometimes exists when the topic of $\chi^2$ distributions are mentioned in discussions of SEMs. Some individuals believe that there must be some connection between contingency table analysis and SEMs because of the use of the term $\chi^2$. There is none, and it bears recalling that Pearson's $\chi^2$ test is only an approximation of a $\chi^2$ distribution, and a rather conservative one at that (von Eye, Rovine, & Wood, 1990). Reviewing some of the basic distributional theory used in parametric statistics provides a way to understand the important elements of the $\chi^2$ distribution and how they are used to assess model fit in SEM.


DOI: 10.1201/9781003595267-9

## $\chi^2$ with 1 Degree of Freedom

Consider data distributed according to the standard normal distribution (i.e., with population mean $\mu = 0$ and standard deviation $\sigma = 1$). If observations are drawn infinitely many times from a standard normal distribution, the resulting distribution is termed a $\chi^2$ distribution with one degree of freedom. Degrees of freedom are usually denoted using the Greek letter $\nu$. Figure 9.1 shows these two distributions. As noted in the figure, the resulting $\chi^2$ distribution has mean 1 and variance 2 or, more generally, $\mu = \nu$ and $\sigma^2 = 2v$.

Notice it is not possible for values drawn from a $\chi^2$ distribution to ever be negative, since negative values from the normal distribution are squared. Further, the probability of a $\chi^2$ value being exactly equal to zero is technically zero, since observation of a value drawn from the normal distribution being precisely equal to zero is nil.

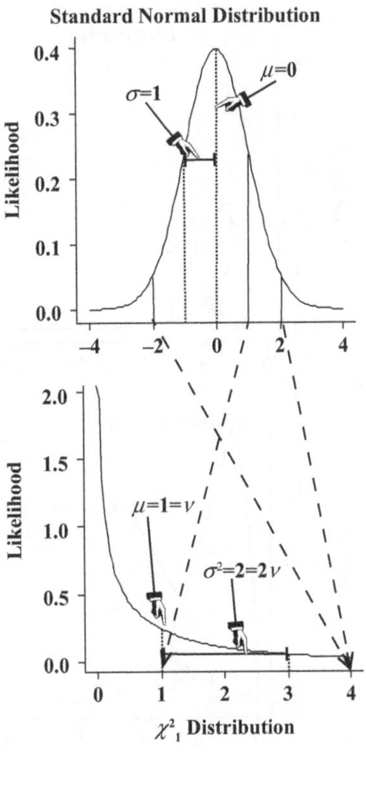


FIGURE 9.1
Standard normal and $\chi_1^2$ distributions.

## Additivity of $\chi^2$

$\chi^2$ distributions are, in turn, additive, meaning that if the process is repeated again, this time drawing an observation from the $\chi^2$ distribution and then drawing another and adding the two to produce a new distribution, the result is a $\chi^2$ distribution with two degrees of freedom ($v = 2$). The resulting distribution is shown in the top figure in Figure 9.2. The process can again be repeated with this distribution, producing a $\chi^2$ distribution with $v = 4$. Alternatively, one could also draw an observation from a $\chi^2$ distribution with $v = 1$ and an observation from a $\chi^2$ distribution with $v = 2$, add them, resulting in a $\chi^2$ with $v = 3$.

As the number of degrees of freedom increase, the mean of the distribution shifts to the right and the variance increases. The distribution of scores also becomes less and less skewed as the degrees of freedom increase. The skewness of the $\chi^2$ distribution is equal to $\sqrt{\frac{8}{v}}$. It can also be noted that the mode of the distribution is equal to the larger of $(0, v-2)$.

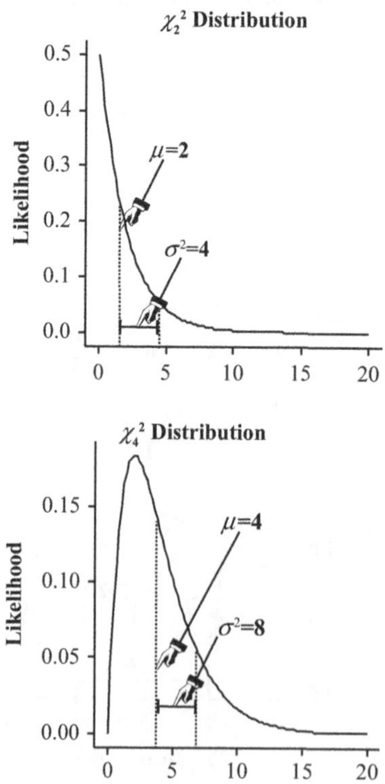


**FIGURE 9.2**
$\chi_2^2$ and $\chi_4^2$ distributions.

## Non-Central $\chi^2$ Distribution

The non-central $\chi^2$ distribution is relevant in some approaches to assessment of statistical power and in some assessments of model fit. It represents a rather straightforward extension of the central $\chi^2$ presented earlier, except that before squaring the variates, a constant is added to the variable. This constant can vary across the different $\chi^2$ variates being added. The sum of the squared means is referred to as the non-centrality parameter and is usually denoted by $\delta$. In equation form, assuming the constant is denoted by $\mu$ and $k$ variates are being added, the non-centrality parameter then is:

$$\delta = \sum_{i=1}^{k} \mu_i^2 \tag{9.1}$$

For example, Figure 9.3 shows a central $\chi_3^2$ and a non-central $\chi_3^2$ distribution with $\delta = 3$ (i.e., each of the draws from the $\chi_3^2$ could have one added to it). To determine the critical value associated with the statistical significance of the central $\chi_3^2$ distribution, we find the area under the curve associated below where 95% of the area falls. For this distribution, this corresponds to a critical value of 7.81 and, if we observe a test statistic greater than this amount, we will reject the null hypothesis. In many cases, the alternate hypothesis is associated with a non-central $\chi^2$ distribution. The shaded area in Figure 9.3 shows this proportion of area under the curve, and this quantity is known as the statistical power of the alternate model. For this non-central distribution, for example, the shaded proportion in Figure 9.3 represents 27% of the area under the curve.

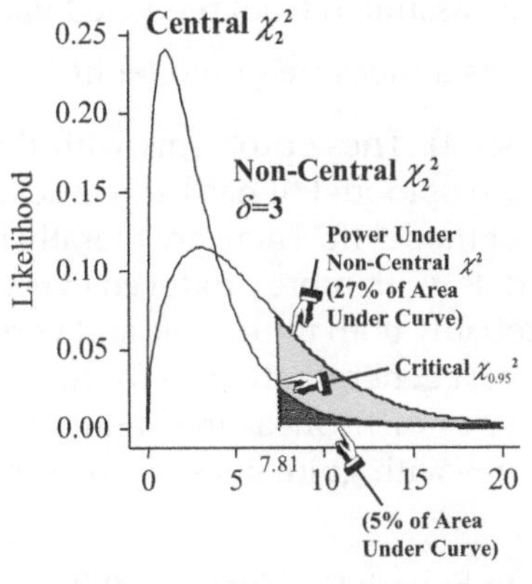


**FIGURE 9.3**
Central and non-central $(\delta = 3)$ $\chi^2$ distributions.

## $\chi^2$ Model Fit for SEMs

The description of the $\chi^2$ distribution is sufficient for us to introduce one stand-alone measure of fit, the $\chi^2$ test statistic, as one way to assess the fit of an SEM. Specifically, if the data of $N$ observations were in fact generated by the proposed structural model and the assumption of multivariate normality is met, then the maximum likelihood fit function FF, when multiplied by $N-1$, produces a test statistic distributed as a $\chi^2$ distribution with degrees of freedom equal to the difference between the number of unique elements of the average cross-products matrix and the number of free parameters in the model. Accordingly, if the probability level associated with this distribution is less than $\alpha$, we would reject the hypothesis that these data came from this model.

One important thing to note, then, about the use of the $\chi^2$ test is that it is a test of *poor* fit. In failing to reject the model, one has not necessarily found support for the proposed model. Although the $\chi^2$ test would seem a relatively straightforward tool, for small samples, it will probably not produce a statistically significant result due to low power. It is also the case that large samples will also lead to a rejection of the proposed model even if there is a relatively high correspondence between the observed and predicted average cross-products matrices. This problem is largely due to the fact that the condition for the model's appropriate use has not been met, namely that of multivariate normality. Nonlinear relationships between variables, influential observations, as well as non-normality of error scores and other exogenous variables can also produce a statistically significant $\chi^2$ statistic. For example, in the alcohol factor model of Figure 8.9 in Chapter 8 on page 148, the $\chi^2$ statistic was significant if based on the entire sample ($\chi^2(4) = 96.65, p < 0.001$), but it was no longer significant if influential observations were removed from the data set $\left(\chi^2(4) = 3.10, p = 0.54\right)$. For example, in order to adjust the deviation of the model $\chi^2$ from its (assumed true) predicted value, we divide the $\chi^2$ by its degrees of freedom as a measure of model fit (e.g., $\frac{96.65}{4}$ or $\frac{3.10}{4}$ for the two models just discussed). These problems with the use of the $\chi^2$ statistic as a stand-alone measure of model fit has led to a search for other criteria for judging the adequacy of model fit. There are literally hundreds of possible fit measures for SEM models, with more produced every year. A systematic survey of all of them is probably both unrealistic and very quickly outdated. The discussion here presents a general rationale for model fit in SEMs, describes some of the general types of fit measures, and surveys some of the most frequently used measures with guidelines for their interpretation.

### Standardized Root Mean Square Residual (SRMSR)

Examination of the differences between the predicted and observed average cross-products matrix would seem a natural place to evaluate whether associations between individual associations between manifest variables are well or poorly modeled. This has led to the motivation of some fit measure that examines all deviations between the predicted and observed covariances as a measure of model fit. Given that discrepancies between observed and predicted covariances are a function of the loss function and contain positive and negative values, squaring these discrepancies, summing, and taking the square root seems to be a reasonable adjustment similar to the computation of a standard deviation. This produces the covariance-based SRMR:

$$\text{SRMR} = \sqrt{\left(\frac{1}{k}\right)\sum_{i=1}^{k}\left(\frac{\left(s_{ij} - \hat{\sigma}_{ij}\right)}{s_{ii}s_{jj}}\right)^2} \tag{9.2}$$

where:

- $s_{ij}$ = observed covariance between variables $i$ and $j$
- $\hat{\sigma}_{ij}$ = model-implied covariance between variables $i$ and $j$
- $s_{ii}$ and $s_{jj}$ = observed variances of variables $i$ and $j$

Given, however, that different measuring units of manifest variables are present in a given data set, examination of such an average root deviation does not have a ready conceptual interpretation because it varies from study to study depending on scale and number of variables. Accordingly, the average squared residual from the standardized version of the model and using correlations is used for the SRMSR as shown in Eq. (9.3).

$$\text{SRMR} = \sqrt{\left(\frac{1}{k}\right)\sum_{i=1}^{k} - \left(r_{ij}^{\{\text{obs}\}} - r_{ij}^{\{\text{model}\}}\right)^2} \tag{9.3}$$

where:

- $k$ = the number of non-redundant elements in the covariance or correlation matrix
- $r_{ij}^{\text{obs}}$ = the observed correlation between variable $i$ and $j$
- $r_{ij}^{\text{model}}$ = the model-implied correlation between variables $i$ and $j$

It is worth noting that the initial work in SEM calculated the average residual based on the covariance matrix and referred to the corresponding fit measure as the MSR. Given the fact that the SRMSR is the most widely used version of root mean residual, others have shortened it to RMSR (e.g., Mulaik, 2009a) or even RMR, with an explicit mention of whether the correlation matrix is analyzed (e.g., Marsh, Balla, & McDonald, 1988). As a result, it is appropriately cautious to check the specific formula being used in your software package.

Specific values denoting good and poor fit for the SRMSR are, however, somewhat vague. Often authors seem to be content to use values less than 0.05 as indicative of acceptable fit (e.g., Byrne, 2012) or less than 0.08 (Hu & Bentler, 1999). Hu and Bentler originally suggested that, taken together, an SRMSR value larger than 0.09 and a CFI less than 0.96 were indicative of poor fit, but Fan and Sivo (2005) criticized this recommendation as resulting from the specific types of models considered by Hu and Bentler.

### Root Mean Square Error of Approximation (RMSEA)

The RMSEA (Steiger & Lind, 1980; also reported in Steiger, 1990) is one such stand-alone measure of model fit based on $\delta$, the non-centrality parameter of the $\chi^2$ distribution. If we use $\chi^2_H$ and $df_H$ to denote the $\chi^2$ statistic and degrees of freedom from the model, the estimate of the non-centrality parameter, $\delta_H$, is given by the formula:

$$\delta_H = \max\left(\left(\chi^2_H \; df_H\right), 0\right) \tag{9.4}$$

This definition follows from the fact mentioned earlier that the mean of a $\chi^2$ distribution is equal to its degrees of freedom. This estimate is then rescaled to take into account sample size:

$$\delta_H = \frac{\max\left(\left(\chi^2_H \; df_H\right), 0\right)}{N-1} \tag{9.5}$$

The RMSEA is then calculated as the ratio of the non-centrality parameter, $\delta_H$ to the sample size:

$$\text{RMSEA} = \sqrt{\frac{\delta_H}{df_H}} \tag{9.6}$$

Lower values of the RMSEA are indicative of better fit. Methodologists differ in their recommendations as to what values are acceptably low, however. While many argue that values 0.01 or less are indicative of excellent fit, Steiger (1989) and Browne and Mels (1990) suggested that values less than 0.05 were indicative of good fit. MacCallum, Browne, and Sugawara (1996) characterize observed values between 0.05 and 0.08 as indicating mediocre fit. Values close to 0.10 are taken as indicative of only fair or poor fit. Chen, Curran, Bollen, Kirby, and Paxton (2008) argue against the general adoption of the 0.05 (or any other cutoff) and propose that evaluation of fit be done on the basis of the particular model, its degrees of freedom, and the sample size used.

The RMSEA has the further advantage that it includes a confidence interval to assess the precision of this estimate. Additionally, if one accepts the 0.05 cutoff value as a close fit, many programs also produce the statistic PCLOSE, and a test of the alternative one-sided hypothesis is that the RMSEA is greater than 0.05. If PCLOSE is not statistically significant (i.e., greater than 0.05), this is taken as support of the conclusion that the fit of the model is close. If PCLOSE is less than 0.05, it is taken as evidence that the RMSEA is substantially greater than 0.05. Even though a particular RMSEA may be impressively low, a large confidence interval around this estimate would caution a researcher about the precision of this estimate. To some extent the factors affecting the confidence interval of the RMSEA are the same as those affecting any parameter estimate: If the sample size used is small and the number of estimated parameters in the model is large, the confidence interval will be correspondingly large. Given the fact that a low-sample-size study can have a substantial, but nonsignificant $\chi^2$ statistic, the estimated RMSEA may appear substantial but with a large confidence interval. Greater sampling error also results for models with small degrees of freedom, and Kenny, Kaniskan, and McCoach (2014) argue that the RMSEA should not be examined if the model under consideration has low degrees of freedom. Kenny et al. (2014) proposed that researchers control for these issues by first assuring that the RMSEA associated with the independence model (described later) is poor (greater than 0.158) before proceeding to a calculation and interpretation of the RMSEA for a proposed model.

## $\chi^2$ Difference Tests

The discussions of the stand-alone $\chi^2$ statistic and the RMSEA reveal some of the general frustration with the search for a single stand-alone measure of acceptable model fit. In many situations, however, researchers may be focused on comparing competing models or, alternatively, to designing fit measures involving increment in fit of a given model over some baseline model. Because the $\chi^2$ distribution is additive, this suggests that a researcher can compare a model in which a parameter is freely estimated to a model in which that parameter is fixed to a specific value (such as 0 or 1 or equality with other parameters in the model). Such models are said to be nested, and the comparison of two nested models can be thought of in path notation as the navigation from a more restricted model to a less restricted model by the addition of slings or arrows to arrive at a more complex model. (Equality constraints within a more restricted model can be relaxed by adding additional arrows or slings that do not have the equality requirement to the diagram for the more elaborate model.) As Bollen and Long (1993) note, even though individual $\chi^2$ statistics may be elevated due to non-normality of the data, this caution does not apply to $\chi^2$ difference tests given the additivity of $\chi^2$ statistics discussed earlier. Satorra and Bentler (2001) discussed this property and proposed a scaled modification of the $\chi^2$ difference test which adjusts for the non-normality of each stand-alone $\chi^2$ statistic. This modified $\chi^2$ test can be calculated using the lavtestLRT function or the Difftest option in lavaan and Mplus, respectively.

The expressions of some of the models one wishes to compare may need to be recast slightly to see how two models are nested. For example, one could write a single factor model with unit variance as a two-factor model in which factor variances and the covariance between the two are constrained to one. This covariance could then be freed in a subsequent model and, in this way, we can think of the single factor model as nested within the two-factor oblique model such as that shown in Figure 8.6. In turn, one could also consider the two-factor orthogonal model as nested within the two-factor oblique model of Figure 8.6. Note, for example, that the single factor model is not nested within an orthogonal version of Figure 8.6, as these two models have the same degrees of freedom, and one cannot specify one model from the other by adding paths or variances.

By extension, it is possible to also test the statistical significance of any given parameter in a model by comparing it to a model in which the parameter is fixed to zero. So, for example, in order to test whether a single factor of alcohol use accounts for the college drinking data or whether a (highly) correlated factor (with no correlated errors between the manifest variables) is a better model, we find the following: $\chi^2(5) = 107.70$. For the correlated two-factor model, we have $\chi^2(4) = 96.64$. $\chi^2_{\text{diff}}(1) = 107.70 - 96.64 = 11.06; p = 0.0009$, which we would take as statistically significant evidence that the correlated factor model is superior to the single factor model. Note, however, that the $\chi^2$ for the two-factor correlated factor model is equal to that obtained for the single-factor, correlated error model and there is no way to compare this model to the single factor model with correlated errors given that the two models are not nested.

## Independence Null

It is tempting then, to consider some kind of model that makes a statement about the average cross-products matrix of the data, but that could accommodate any kind of SEM. The independence model, a structural model in which only variances of the manifest variables are used, fulfills this requirement. For SEMs that are based on the average adjoined SSCP matrix, this

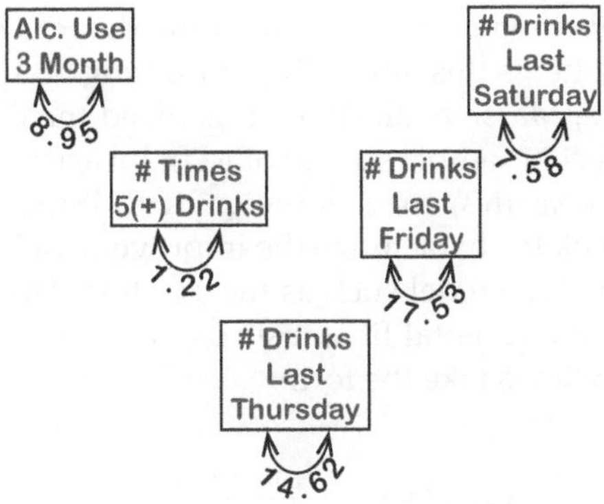


**FIGURE 9.4**
Independence null model for freshman drinking data.

model freely estimates the means of the structural model, given that the remaining diagram constitutes a statement about independent variances of manifest variables.

For the freshman drinking variables in Figure 8.6, for example, Figure 9.4 represents the independence null model.

Although the $\chi^2_{\text{diff}}$ test informs the researcher that the proposed model is preferable to a model of uncorrelated variables, practically speaking, the independence null has little conceptual value. It only informs the researcher as to whether any relationships exist between any of the variables at all, as in the case of small-sample-size studies relative to the number of variables in the model or when associations between variables are extremely weak. As such, it can be thought of as the "worst plausible model," a benchmark for evaluating how well an SEM model accounts for the observed relationships among variables. But knowing that a model fits *better* than a model of complete independence does not, on its own, guarantee that the model fits the data *well*. To assess the adequacy of fit more precisely, SEM researchers rely on a variety of model fit indices.

## Type I and Type II Fit Functions

These indices differ in what aspect of the model they evaluate. Some – known as *Type I fit indices* – assess the *absolute discrepancy* between the observed data and the model's predictions. Others – referred to as *Type II fit indices* – evaluate the *relative improvement* in fit over the independence model. Together, these two types of indices offer complementary perspectives: Type I indices answer the question "Is this model consistent with the data?" while Type II indices ask, "How much better is this model than assuming no structure at all?"

At a general level, Marsh et al. (1988) organized their discussion and presentation of the effect of sample size on SEM fit measures as being composed of stand-alone fit measures, such as the $\chi^2$ or RMSEA, and incremental fit measures, which seek to characterize the improvement in fit of a model relative to some alternative model such as the $\chi^2_{\text{diff}}$ test. They further organized their discussion of incremental fit statistics into Type 1 and Type 2 fit functions. Type 1 fit functions take the form:

$$\text{Type1} = \left| \frac{F_{\text{Target}} - F_{\text{Base}}}{\max\left(F_{\text{Target}}, F_{\text{Base}}\right)} \right| \tag{9.7}$$

where $F_{\text{Target}}$ represents the value of a stand-alone fit function for a model, and $F_{\text{Base}}$ is the value for the Base or Null model, such as the independence null just discussed. Type 2 fit functions, by contrast, take the following general form:

$$\text{Type2} = \frac{\left|F_{\text{Target}} - F_{\text{Base}}\right|}{\left|F_{\text{Base}} - \exp\left(F_{\text{True}}\right)\right|} \tag{9.8}$$

where $\exp\left(F_{\text{True}}\right)$ represents the expected value of the stand-alone fit function if the target model is true.

## Tucker-Lewis Index (TLI), aka NNFI

The TLI (also known as the Non-Normed Fit Index, or NNFI) is an example of a Type 2 fit measure and is widely used. It is given by the formula:

$$\text{TLI} = \frac{\dfrac{\chi^2_{\text{Base}}}{df_{\text{Base}}} \; \dfrac{\chi^2_{\text{Target}}}{df_{\text{Target}}}}{\dfrac{\chi^2_{\text{Base}}}{df_{\text{Base}}} \; 1} \tag{9.9}$$

The value of 1 appears in the denominator because, as mentioned in the review of the $\chi^2$ distribution, the expected value of a variable from a $\chi^2$ distribution is its degrees of freedom. Historically, the TLI was first proposed by Tucker and Lewis for use in exploratory factor analysis. The mathematical rationale for the index is roughly patterned on the notion of a reliability index from analysis of variance and can be thought of analogous to an intra-class correlation coefficient (see Mulaik, 2009a, pp. 330–333, for the mathematical details involved). The NNFI is not a computational formula for an intra-class correlation. It is only an estimate, and even that by analogy. Bentler (1990, p. 239) describes conditions under which the NNFI can take on negative values or be larger than 1, for example. In their extensive survey of the behavior of SEM fit indices as a function of sample size, Marsh et al. (1988) noted that the TLI appeared to be the least affected by sample size but appeared to have larger variance in sampling studies. In order to remedy discomfort with reporting TLI values greater than 1, McDonald proposed simply setting the TLI to 1 if it was initially estimated as being larger than 1. This modification performs, as you would expect, largely the same as the unadjusted TLI. It is important, however, to check which formula your particular SEM software uses for this reason.

In terms of cutoff values to indicate acceptable fit, it is important to recall that TLI values are dependent on the average strength of associations present in the data. It is quite easy to encounter TLI values that exceed 0.90 for extremely highly correlated behaviors. For example, in the alcohol use factor shown in Figure 8.6, the TLI was 0.96 for the complete data and 1.00 if influential observations were removed. If, however, a researcher analyzes data from an experiment in which several orthogonal predictor variables were included in the path diagram, the TLI should be expected to be lower because, for a sub-matrix of the covariance matrix, the independence model correctly characterizes the data. What counts as acceptable fit for the TLI appears to vary as a function of specific area of psychology. For example, in many clinical and cognitive neuroscience settings, values greater than 0.95 are expected. By contrast, in developmental, educational, personality, or social psychology settings, values greater than 0.90 appear acceptable. This is not to say that one area or another is more rigorous, but it reflects that research varies in terms of whether it is high stakes or is investigating complex psychological processes or deals with more complex longitudinal data.

## Normed Fit Index (NFI) and Comparative Fit Index (CFI)

The NFI, by contrast, is an example of a Type 1 fit index, and is given by the formula:

$$\mathrm{NFI} = \frac{\chi_{\mathrm{Base}}^{2} \chi_{\mathrm{Target}}^{2}}{\chi_{\mathrm{Base}}^{2}} \tag{9.10}$$

The NFI has the property that it cannot exceed 1 or be less than zero. Marsh et al. (1988) found it more affected by sample size than the TLI. Specifically, it tends to underestimate its population value in smaller samples, and significantly so for samples smaller than 800.

To correct for this, McDonald and Marsh (1990) and Bentler (1990) each independently proposed that fit be scaled relate to the non-centrality parameter estimates of the respective target and base models. This fit measure was termed the CFI and is defined as follows for the estimated $\delta$ of the base and target models:

$$\mathrm{CFI} = \frac{\delta_{\mathrm{Base}} - \delta_{\mathrm{Target}}}{\delta_{\mathrm{Base}}} \tag{9.11}$$

In order to see how this is related to the TLI in terms of $\chi^2$ statistics, we can note that the estimated $\delta$ is equal to the $\chi^2$ minus its degrees of freedom. The CFI can then be written as:

$$\mathrm{CFI} = \frac{\chi^2_{\mathrm{Base}} - df_{\mathrm{Base}}\left(\chi^2_{\mathrm{Target}} - df_{\mathrm{Target}}\right)}{\chi^2_{\mathrm{Base}} - df_{\mathrm{Base}}} \tag{9.12}$$

Similar to McDonald's adjustment of the TLI, in order to keep this fit measure bounded on the interval 0 to 1, the index is set to zero if estimated at a negative value and as 1 if greater than 1. Mulaik (2009a) reports that the CFI appears to show much smaller bias than the NFI in small samples ($N = 50$), and this bias is generally negligible at moderate to large sample sizes. It should also be noted that if the CFI is less than 1, it is always less than the TLI. As with the TLI, values generally larger than 0.90 indicate acceptable fit, and values larger than 0.95 are taken to indicate good fit.

## Information-Theoretic Fit Values

The $\chi^2$ difference test was a way to directly compare the fit of nested models, but in many situations researchers are interested in comparing the fit of non-nested models. These general models are based on the econometric work of Akaike (presented in 1971 but published later in 1973). Information criteria are based on the log-likelihood function with some penalty on model complexity, where complexity is operationalized as the number of parameters in the model. These criteria are not measures of model fit. A researcher can fit several models and select as the best a given model that does not fit the data very well.

### Akaike Information Criterion (AIC)

The AIC is defined as:

$$AIC = 2LL - 2p \tag{9.13}$$

where *LL* is the log-likelihood estimate of the model and *p* is the number of parameters in the model. There is no "cutoff value" of acceptable fit for the AIC, as the log-likelihood value is idiosyncratic to the model. Researchers wishing to use the AIC simply select that model with the smallest AIC value.

### Variants of the AIC

Minor differences appear across software packages in their operationalization of the AIC. For example, Mplus uses the formula:

$$AIC = 2LL + 2q \tag{9.14}$$

where $q$ is the number of free parameters in the model. This particular version of the AIC has the (minor) advantage that it can take on only positive values. Yet other formulae have been used, including one involving $\chi^2$:

$$AIC = \chi^2 + k(k+1)2df \tag{9.15}$$

for analysis of $k$ variables, which has the advantage that it is based on the model $\chi^2$ and its degrees of freedom $df$ and allows you to see what the AIC would be in a completely saturated model. If a model is completely saturated, it has zero $df$, and a $\chi^2$ of zero, leaving the AIC to be equal to $k(k+1)$. In general, these three variants of the AIC should all produce the same ordering of models, but it is again important to know and report which value is being used.

### Bayesian Information Criterion (BIC)

The BIC represents another approach to a penalty function for model complexity. Introduced by Schwarz (1978), it takes the form:

$$BIC = \chi^2 + p(\ln(N)) \tag{9.16}$$

where, as before, $p$ represents the number of parameters in the model, and $\chi^2$ is the $\chi^2$ associated with the model. $N$ represents the sample size of the data. As with the AIC, model selection is based on the smallest BIC value. Mulaik (2009a, p. 361) notes that the penalty function of the BIC will tend to prefer models with larger numbers of parameters as the sample size increases.

Despite this limitation, it appears that the BIC seems to function somewhat better in Monte Carlo simulations than the AIC, suggesting that the penalty function that takes into account the sample size involved is more successful in discriminating between models (e.g., Lin, Huang, & Weng, 2017).

### Summary of Information Criteria

It is somewhat difficult to recommend one or the other information criterion, as they appear to some extent to vary in performance depending on the types of SEMs being compared. Vrieze (2012) points out that, in terms of asymptotic theory, the BIC is a consistent estimator for model comparison if the true model is among the alternatives being considered, whereas the AIC is not a consistent estimator. If, however, the true model is not among those considered by the researcher, the AIC is more efficient than the BIC, i.e., it will (asymptotically) choose the model that minimizes the mean squared error of estimation in the data. AIC can also sometimes identify a more complex model that is better in prediction contexts (Burnham & Anderson, 2004). As these observations are asymptotic theory results, however, they may not provide the clear choice a researcher would like when looking at the data from one study.

### Bayes Factor

In reviewing the types of fit comparisons, one can readily see that researchers began by gross examination of the deviation of the predicted, from the observed cross-products matrix, to attempts to make parametric theory estimates of model fit, to measures that indicate "closeness of fit" or at least not appreciable "lack of fit" to the data and, for the case of non-nested models, to use information theoretic values to compare competing model fits using penalties for model complexity. When comparing models, though, the question remains whether one model substantially improves on another model.

Bayes factors are one such approach. As its name implies, the Bayes factor derives from Bayesian, as opposed to frequentist inference and, as such, features most easily in Bayesian SEMs as opposed to the frequentist SEMs presented in this book. An approximation of the Bayes factor is described here, however. Formally, the Bayes factor can be defined as:

$$BF = \frac{\Pr(\text{Data} \mid \text{Model}_1)}{\Pr(\text{Data} \mid \text{Model}_2)} \tag{9.17}$$

where Pr denotes the probability estimate; Data, the observed data; and $\text{Model}_1$ and $\text{Model}_2$, two competing SEMs. Mathematically, these probabilities are estimated from a Bayesian analysis of the data, i.e.:

$$\mathrm{BF} = \frac{\int \Pr(\theta_1 \text{ Model}_1)\left(\Pr(\text{ Data}\theta_1 \text{ Model}_1)\right)d\theta_1}{\int \Pr(\theta_2 \text{ Model}_2)\left(\Pr(\text{ Data}\theta_2 \text{ Model}_2)\right)d\theta_2} \tag{9.18}$$

where $\theta_1$ and $\theta_2$ denote the vectors of parameters in $\text{Model}_1$ and $\text{Model}_2$ respectively.Kass and Raftery (1995) proposed the following cut-off values for judging whether the difference between two models was "substantial," as shown in Table 9.1.

Kass and Wasserman (1995) propose an approximation for the Bayes factor using the BIC when the Bayesian estimator is used:

$$\mathrm{BF} = \frac{e^{0.5\mathrm{BIC}_{H1}}}{e^{0.5\mathrm{BIC}_{H0}}} \tag{9.19}$$

Strictly speaking, calculating a Bayes factor requires the use of a Bayesian SEM. However, some researchers have used Kass and Raftery's guidelines as a way of assessing the difference between BIC values from frequentist SEMs (e.g., Lin et al., 2017; Reddy, Anderson, Liebschutz, & Stein, 2013). Note that calculating this value in a computer program will run into numerical accuracy problems given that BIC values are frequently large numbers. Taking the log of the Bayes factor as follows and then exponentiating it is advisable.

$$log(\mathrm{BF}) = 0.5(\mathrm{BIC}_{H0}) - 0.5(\mathrm{BIC}_{H1}) \tag{9.20}$$

## Fit Statistic Overview

Many fit statistics have been covered so far, and a brief summary of the interpretation and guidelines of their use is appropriate, as summarized in Table 9.2.

**TABLE 9.1**

General Guidelines for Bayes Factor Model Comparison

| Bayes Factor | Conclusion |
|---|---|
| 1 to 3 | Not worth more than a bare mention |
| 3 to 20 | Positive |
| 20 to 150 | Strong |
| 150 | Very strong |

**TABLE 9.2**

Overview of Fit Statistics

| Fit Statistic | Good Fit Defined | Comments |
|---|---|---|
| $\chi^2$ | Lower, hopefully nonsignificant | Affected by non-normality, which is amplified in large samples |
| $\chi^2$ difference test | Lower, hopefully nonsignificant | Requires nested models. Less affected by non-normality. Satorra and Bentler's adjusted approach is better. |
| TLI (a.k.a. NNFI) | values > 0.90 or 0.95 | Can exceed 1. |
| NFI | values > 0.90 or 0.95 | More affected by sample size than TLI. |

## Statistical Power

Up to this point we have not discussed the design of SEM studies. As with any proposed analysis, however, consideration of power is an important part of experimental design. The assessment of statistical power is more complicated in SEMs than in models such as regression or analysis of variance in that interest in the latter focuses on the statistical significance of a given parameter or subset of parameters. In SEM, researchers may also be interested in the overall fit of the model, successful identification of the correct model under nested alternatives, and likely fit values of the "true" model under minor model misspecifications. Two approaches to assessing power are discussed here: (1) assessment of power in comparing nested models, which can involve the special case of assessing statistical significance of a single parameter; and (2) Monte Carlo approaches to power, which consider overall model fit and the power associated with all parameters of the model.

### Statistical Power for Nested Models

To test for the statistical power of an individual parameter or to test the power of a target model relative to some base model, you will need to first specify parameter estimates that can be used to generate the predicted cross-products matrix associated with some target model. The method described here was first proposed by Satorra and Saris (1985). In keeping with the pattern of "examples first, then theory" let's consider the case of a researcher wishing to know what the power of a model to correctly detect an oblique factor model rather than a single factor model for a given data set. (One could, alternatively, consider the statistical power associated with a given regression weight, but the logic of the analysis is the same.)

Step 1: *Generate Predicted Covariance Matrix Under Alternate Hypothesis.* Frequently, factor loadings are not known with a great degree of certainty, and so it is often helpful to use the tracing rules to generate values for the loadings and error terms associated with the standardized score matrix. Specifically, one might set error variances to be one minus the internal consistency estimate for a manifest variable, and loadings to the square root of that quantity. For example, to test the power of an SEM to correctly detect a correlated two-factor solution instead of a single factor solution, we might consider a path diagram with factor loadings equal to 0.7 for all five variables and error variances equal to $1-0.7^2$. Factor variances are fixed at one, and the proposed correlation between the two factors could be 0.8. (One can obviously experiment with other values of factor loadings and degree of correlation between the factors.) As a check that this has been done correctly, the researcher may want to fit this predicted covariance matrix to the alternate hypothesis to verify that the $\chi^2$ for the model is zero.

Step 2: *Use Predicted Covariance Matrix Associated with the Simpler (Incorrect) Model to Find Non-Centrality Parameter.* The predicted covariance matrix from Step 1 is now fit to the data from the base (or reduced) model. This predicted covariance matrix will specify the sample size the researcher is interested in assessing the power for. The $\chi^2$ statistic from this model minus its degrees of freedom constitutes the non-centrality parameter for the analysis. For the data example given here with a sample size of 100, the non-centrality parameter from the SEM program is 4.964. This is the non-centrality parameter associated with a $\chi^2$ with 1 degree of freedom (because the $\chi^2$ difference test had only 1 degree of freedom). If, however, a sample size of 200 is considered, the non-centrality parameter is 9.927.

Step 3: *Find the Critical $\chi^2$ Value for the Analysis.* The critical value for the $\chi^2$ statistic is found for the degrees of freedom associated with the base model. This value can be found in any tables of the $\chi^2$ distribution or, more simply, by using the quantile function for the $\chi^2$ in many common statistics packages. For example, in *R*, the line of code for our example would be:

```
qchisq (p = 0.05, df = 1, lower. tail = FALSE)
```

or the formula for a given cell in Excel would be:

```
=CHISQ.INV.RT (0.05,1)
```

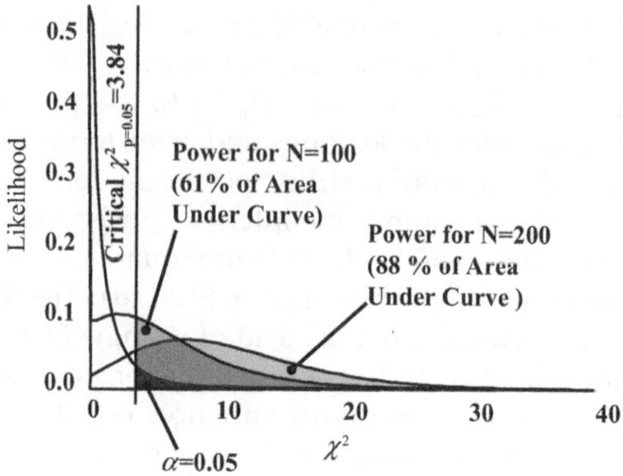


**FIGURE 9.5**
Null hypothesis and power for $N = 100$ and $N = 200$.

in SAS, the line of code would be:

```
crit_Chi=cinv(0.95,1);
```

The critical value for a single *df* $\chi^2$ at $a = 0.05$ is 3.84.

Step 4: *Calculate the Power for the Desired Scenario.* Using the critical value found in Step 3 and the non-centrality parameter determined in Step 2, and the degrees of freedom associated with the base model, the probability of rejecting the base model is calculating using either a table of non-central $\chi^2$ or, again more conveniently, using statistical functions from available statistical software. For example, if one is interested in calculating these probabilities using *R*, the following code would be used (Figure 9.5):

```
pchisq(q=3.841459, df=1, ncp=4.964, lower.tail =
  FALSE, log.p = FALSE) #N=100
pchisq(q=3.841459, df=1, ncp=9.927, lower.tail =
  FALSE, log.p = FALSE) #N=0
```

## Power Analysis via Monte Carlo Simulation

In some situations, the researcher may be interested in investigating the behavior of model misspecification, missing data, or distributional violations of an SEM. In these cases, it is somewhat easier to examine the results of several SEMs generated from a model. One straightforward way to do this is to use the Monte Carlo generator of Mplus (Muthén & Muthén, 2002), although other programs such as the SIMSEM module in *R* are also available.

Under a Monte Carlo approach, a true model is specified for the data, and this model is used to generate a large number of simulated data sets (e.g., 1,000). A random number is often specified to start the simulations to ensure identical results when rerunning the program. This alternative model is then analyzed using either the true model or some alternative, such as the base model considered in the previous section. Programs such as Mplus also allow the user to specify various degrees of missing data or missing data patterns in the data as well.

Although Muthén and Muthén (2002) is a much longer systematic presentation, it helps to summarize the information that such Monte Carlo simulations provide. First, the simulations provide an estimate of the distribution of model fit statistics under the second model or in the presence of missing data. The standard deviation of parameter estimates across simulations can be compared to the estimated standard error to investigate possible bias in standard errors for parameters of interest. The coverage, or proportion of times that the confidence interval associated with the simulation included the true value, is also shown as indicative of bias in estimation. Finally, the proportion of times that parameters of interest were statistically significant is also provided. As a result, although the method initially provided for assessing power of an individual parameter provides a parametric estimate of power for a given parameter, the Monte Carlo approach appears to have many advantages, not the least of which is the simultaneous estimation of power for all parameters in the model and the ability to examine the effect of missing data and assumption violations on parameter estimates and model fit.

For example, the Mplus program for examination of the aforementioned scenario is:

```
TITLE: Short example of power analysis using Monte Carlo
 Simulations
MONTECARLO:
NAMES = x1-x5;
NOBSERVATIONS = 200;
NREPS = 1000;
SEED = 53487;
MODEL POPULATION:
[x1-x5*0];
f1 by x1*0.7 x2*0.7;
f2 by x3*0.7 x4*0.7 x5*0.7;
f1@1 f2@1;
f1 with f2*0.8;
x1-x5*0.51;
MODEL:
[x1-x5*0];
f1 by x1*0.7 x2*0.7 x3*0.7 x4*0.7 x5*0.7;
f1@1;
x1-x5*0.51;
```

**TABLE 9.3**

Selected Output from Mplus Power Analysis

| Population | Average | Std. Dev. | S.E. Average | MSE | 95% Coverage | % Significant |
|---|---|---|---|---|---|---|
| F1 BY | | | | | | |
| X1 | 0.61 | 0.08 | 0.07 | 0.01 | 0.74 | 1 |
| X2 | 0.62 | 0.07 | 0.07 | 0.01 | 0.77 | 1 |
| X3 | 0.68 | 0.07 | 0.07 | 0.01 | 0.93 | 1 |
| X4 | 0.68 | 0.07 | 0.07 | 0.01 | 0.94 | 1 |
| X5 | 0.68 | 0.07 | 0.07 | 0.01 | 0.94 | 1 |

To perform the same analysis as described above, we would need to save the results of simulations under the single and two-factor models. The $\chi^2$ difference statistic can be calculated by subtracting the $\chi^2$ for the two-factor model from the single factor model. When this is done for the aforementioned model, assuming a sample size of 200, we observe that the power for the $\chi^2$ difference statistic is equal to 0.88, the same as that arrived at using the parametric approach. In addition to the relatively simpler approach to calculating power for simulations, it is also possible to calculate the power associated with other measures. The average RMSEA for the single factor model was 0.08, for example, while the average RMSEA for the two-factor model was 0.03.

Other parts of the Mplus printout may be of interest in their own right. Table 9.3 shows one part of the Mplus output for the previously mentioned scenario with a sample size of 200. From this analysis, we can see that the effect of mis-specifying the model as a single factor model results in a marked downward bias in the estimated factor loadings for X1 and X2, but this bias is less pronounced for X3, X4, and X5. Letting $\theta$ denote an estimated parameter, the bias in estimation for X1 can be calculated, for example, as:

$$\text{Percent Bias} = 100\left(\frac{\theta_{\text{Base}}\theta_{\text{Alt}}}{\theta_{\text{Alt}}}\right) = 100\frac{0.610.7}{0.7} = 12.86 \tag{9.21}$$

This means that the mis-specified model would result in an approximately 13% lower score under the base model which assumed a single factor. The power (% Significant) of the individual parameters is close to 1. The standard deviation of estimated parameters across all replications, 0.08, is close to the average standard error of the model. The coverage, or percentage of times that the 95% confidence interval included the true population value, was markedly lower for X1 and X2 (0.74 and 0.77), but largely unaffected for X3, X4, and X5.

The preceding is an admittedly short introduction to the general topic of Monte Carlo simulation investigations of power and bias. The approach can, however, be used to investigate comparison of non-nested models, various types of missing data scenarios, and alternatives to the multivariate normality assumed by the SEM model. For further information, the reader is referred to Muthén and Muthén (2002).

## Summary

It is helpful here to recall Popper's observation, mentioned in Chapter 1, that all scientific models are false in that they must be oversimplifications in order to be useful. This is especially true in the case of comparisons of SEMs. Large sample sizes will almost certainly provide statistical significance of poorly fitting stand-alone fit measures such as $\chi^2$ due to the violation of assumed normality. As such, the test provides us with accurate conclusions (i.e., that the conditional normality assumption is not tenable), but the structural form of the SEM may still be a very good explanation of the phenomenon under investigation. Statistical models, however, live in an ecological competition of "survival of the fittest." At times these models can be nested in terms of their complexity, which permits use of the $\chi^2$ difference test. Other stand-alone fit indices such as the CFI and TLI are incremental fit indices relative to the base independence model, but other base models, such as a Rigdon's (1998) suggestion that a random intercept factor model be used as a base model, are also possible. Acceptable fit values for such base models are, however, difficult to determine and not appropriate to all SEMs. (For example, SEMs that include experimental independent variables would not be appropriately modeled with an equal-correlation random intercept factor.) In other settings, model comparisons rely on a more general assessment of model complexity from information theory. The Bayes factor appears to be an interesting gauge of the "practical benefit" of one model over another, though improvements in such measures are currently under development.

# 10

# *Measurement Models*

Number is different from quantity.

**– Gregory Bateson (1979)**

## Introduction

To this point we have introduced latent variable models by saying that multiple manifest variables are all measures of a construct. The particular way or degree that these manifest variables represent the construct has not been discussed, however. Two types of measurement models are possible. If researchers are interested only in the variance components of variables and the quality of the manifest variables as measures of the latent variable, the researcher is interested in using latent variable models that are based on the deviation score form of the variables. Manifest variable measurements are taken to be assessed at the interval level of measurement, meaning that the mean level of the variables is arbitrary and that pointage differences at various levels of a given manifest variable represent the same quantity. When research interests include statements about mean level effects as well, the corresponding measurement model can involve ratio level of measurement, meaning that the manifest variables are thought to have a meaningful zero point in addition to equal measurement units along the scale. Alternatively, a researcher may assume a common interval-level scale (as occurs in some growth models) underlies the data, or that the scale points represent only an ordinal or ranked level of measurement, which could be modeled as an SEM by assuming some reference or link function.

This chapter will concentrate on measurement models for variance components, as many SEM models are based on the covariance matrix, and much of the terminology of test theory is developed for this case. Measurement models that include the mean will be discussed in the chapter on growth (Chapter 13), and measurement models for ordinal scales will be discussed in the chapter dealing with latent response variables (Chapter 17).

Basically, measurement model choices for data deal with the questions of whether: (1) the measurement units associated with some underlying latent variable are identical or vary across the manifest variables; (2) whether one or


DOI: 10.1201/9781003595267-10

more latent variables best explain patterns of variation and covariation in the data; and (3) whether the amount of measurement error is the same across manifest variables. The measurement models are presented in increasing order of their complexity, from strictly parallel measurement to multifactor measurement models. As an example, we consider corresponding measurement models for a study involving five manifest variables. We then present the results of a simulation study highlighting the roles of measurement quality and sample size in determination of the appropriate measurement model.

## Types of Measurement Models

### Strict Measurement Invariance

The easiest measurement model assumes that the true score measuring units of the manifest variables and the precision of measurement are identical across the variables. As such, we may think of the manifest variables as rules graded in the same units (e.g., centimeters) with the same tick marks. Because means are not specified in the model, the strict measurement invariance model is agnostic as to whether the means of the variables are the same. It could be, by analogy, that some of our "rulers" are broken off at the left. We discuss the issue of mean equivalence in Chapter 13 on growth. Within a single factor model, this assumption amounts to a constraint of equality in estimated error variances and factor loadings (or, equivalently, that the factor loadings are constrained to unity and the factor variance freely estimated).

For example, in Figure 10.1 we can see the estimated model for a sample of five variables of randomly generated data, which will be described in more detail later. The strictly parallel model for these data contains but two freely estimated parameters ($\lambda_1$ and $\varepsilon$), and so it is quite parsimonious. Conceptually, this model assumes the scale of each manifest variable has the same measurement units (i.e., an increase of one point of the latent variable results in the same increase in each manifest variable) and the same degree of measurement precision (i.e., the error variance is the same). This simple measurement model is referred to as "strict invariance" by factor analysts, as "interchangeable parallel forms" in test theory, and as the assumption of "compound symmetry" in repeated measures analysis of variance (as discussed in Chapter 13 on growth) and some types of multilevel models.

As a practice example, you might think of strict invariance as measurements of objects being made by any of the three rulers shown in Figure 10.2. In that figure the three measurement devices (which we can think of as variables in a study), have equally spaced units and, given that the tick marks are the same, equal precision of measurement. As you can see, the numbers

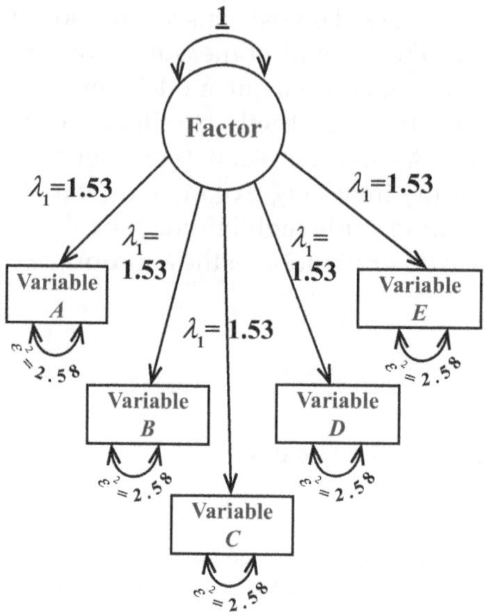


**FIGURE 10.1**
Strictly parallel measurement.

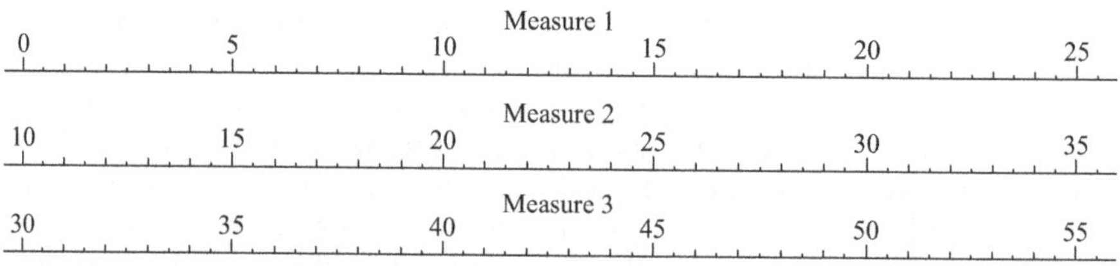


**FIGURE 10.2**
Physical measurement devices showing strict invariance.

assigned by using the rulers vary, because we are making measurements without regard to mean level.

If this model represents the data well, it is possible, using the tracing rules, to calculate the proportion of variability in each manifest variable accounted for by latent variable $\frac{1.53^2}{1.53^2+2.58}=\frac{2.34}{4.92}=0.48$ coefficient associated with the data. Alternatively, we can calculate the standardized factor loading as $\sqrt{0.48}=0.69$. Although it might not seem reasonable to believe that this model is conceptually realistic, it can be an effectively parsimonious model for research situations when the research involves relatively few observations. As discussed in Chapter 13, the strictly parallel measurement model is known in models such as repeated measures analysis of variance as the assumption of compound symmetry. Researchers using repeated measures analysis of variance make the strictly parallel measurement model assumption of compound symmetry but use the Huynh-Feldt and Greenhouse-Geisser corrections to repeated measures for smaller and larger departures from compound symmetry, respectively.

## Tau Equivalent Measures

The tau equivalent measurement relaxes the assumption of equal accuracy of measurement of the strictly parallel measurement model by allowing the error variances to vary across the manifest variables. For $m$ manifest variables, such a model will then have $m+1$ parameters. For the random data, such a model is shown in Figure 10.3.

As can be seen from the figure, the measuring units associated with the manifest variable are the same across measures, but under this model, the precisions vary, with intra-class correlation coefficients ranging from 0.38 to 0.52. In keeping with the ruler analogy mentioned in the previous section, the measuring units of the "rulers" represented by the manifest variable are in the same scale, but the tick-marks associated with Variable E are not as closely spaced as those of Variable *A* as shown in Figure 10.4.

In addition to providing a more forgiving structural model that preserves the measurement scale interpretation of the manifest variables, tau

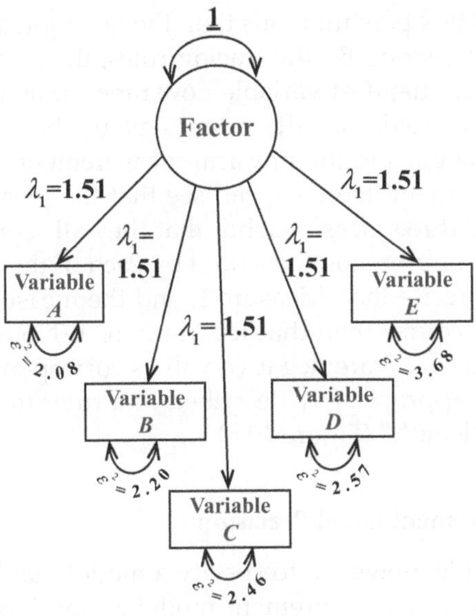


**FIGURE 10.3**
Tau equivalent measurement.

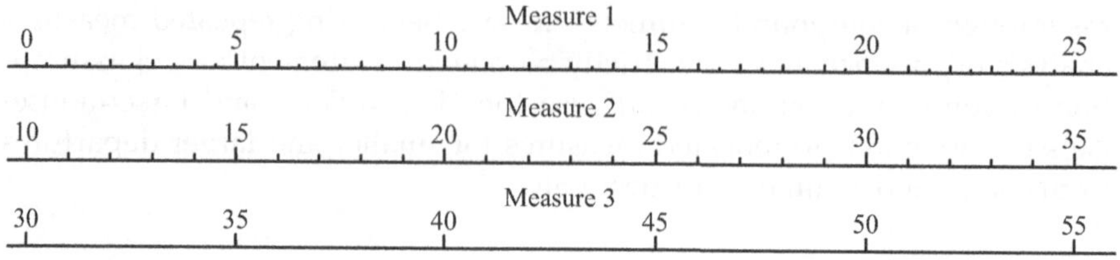


**FIGURE 10.4**
Tau equivalent measurement ruler version.

equivalent measurement models have also been useful in the development of short-form versions of assessment instruments. If the tau equivalent measurement model fits well, it provides support for the contention that the scale units of the short form are essentially the same as that of a longer, more intensive, or expensive assessment.

## Congeneric Measurement

Yet another way to relax the measurement assumption is to assume that the true score measuring units of the manifest variables are expressed in different measurement units (i.e., the factor loadings are different numbers) and the degree of measurement precision varies across the measures as well. This model is termed the congeneric measurement model and is shown in Figure 10.5. This model is less parsimonious than the tau equivalent model, having only 5 degrees of freedom. By the tracing rules, the intra-class correlations associated with each manifest variable now range from 0.46 (Variable $A$) to 0.51 (Variable $C$). To again use the ruler analogy, the congeneric measurement model is analogous to the physical measurements one could make in congeneric measurement. Here we can see that the measurement units are different across the three measures, but that they all represent some idea of length. The degree of precision measured by each of the instruments varies: Measure 3 is less precise than Measure 1, and the measurement spacing for Measure 2 is more course than that for Measure 1. Note also that the measuring units of Measure 3 are negative values corresponding to length, i.e., Measure 3 is more appropriately described as a ruler that measures "shortness" rather than "length" (Figure 10.6).

## Congeneric Measurement Equal Precision

It is logically possible, however, to specify a model that is more parsimonious than the congeneric measurement model as specified, but that has the same degree of parsimony as the tau equivalent model. This model has been

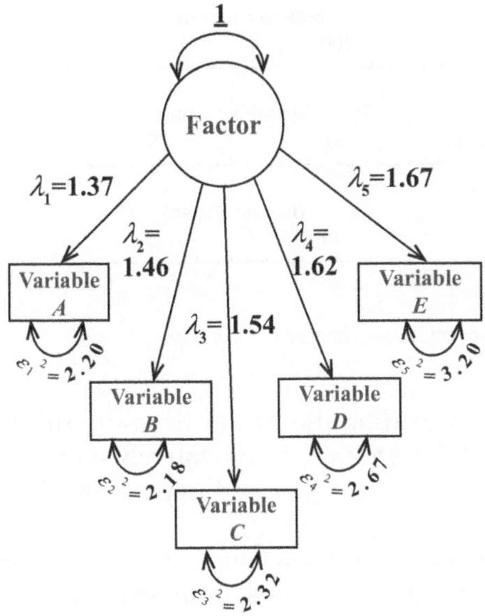


**FIGURE 10.5**
Congeneric measurement.

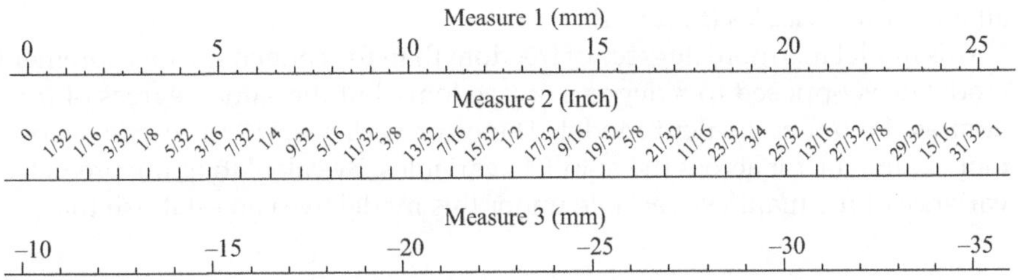


**FIGURE 10.6**
Congeneric measurement ruler version.

proposed in Wood, Wiedermann, and Steinley (2023). To understand this model conceptually, consider the case of measurements made with a ruler, which are made in different units (say centimeters and inches) as was true for the congeneric measurement model. Measure 1 measures in 1/10 of a millimeter, Measure 2 measures in millimeters, and Measure 3 measures in negative millimeter units. Yet, in all three cases, the precision of these measurements is equal – just equal in terms of the spacing of the measuring units employed (Figure 10.7).

Alternatively, measurement errors using these instruments may be thought to be inaccurate not because of the particular spacing of tick marks, but due

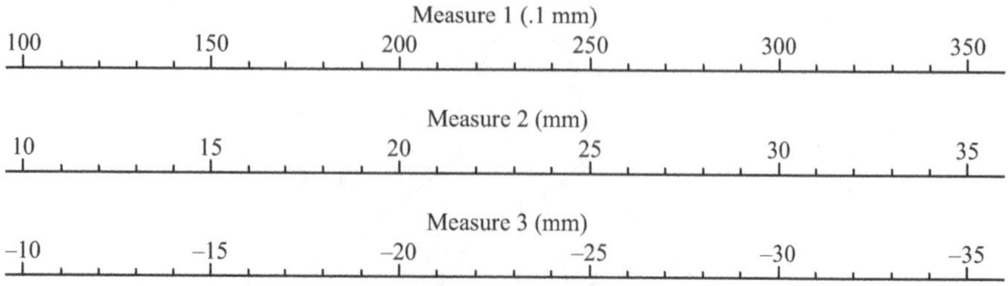


**FIGURE 10.7**
Congeneric measurement equal precision ruler version.

to the imprecision of reading the values from the ruler. As such, the measurements would all be accurate to a relatively equal extent; however, the numeric value of the inaccuracy would depend on the measuring units of the scale.

To implement an equal precision congeneric model, we borrow the alternative specification for error variance discussed in the chapter on reliability models (Chapter 7) and specify both a path and variance for the error term associated with a manifest variable, but this time constraining the path from the error to manifest variable to be equal to its respective factor loading, and specifying that the variance of the error variable be constrained equal across all manifest variables (Figure 10.8).

This model has more degrees of freedom than the congeneric measurement model (9 as opposed to 5 degrees of freedom), but the same degrees of freedom as the tau equivalent model. The error variance, $\varepsilon^2$, represents a common scaling factor across the manifest variables. By calculating the predicted variance of the manifest variable under this model, we can establish that:

$$\text{ICC} = \frac{1}{1+\varepsilon^2} = \frac{\lambda^2}{\lambda^2 + \lambda^2\varepsilon^2} \tag{10.1}$$

By this math, we can see that this covariance model assumes that the intra-class correlation for all variables is equal to 0.48.

## Random Intercept Measurement Model

A model in which both the tau equivalent and the congeneric factor models are present, called the Random Intercept Factor model (Maydeu-Olivares & Coffman, 2006), is also discussed in Wood, Steinley, and Jackson (2015). Figure 10.9 shows the random intercept model for the example data set.

Conceptually, the random intercept model may be appropriate if, for example, it is believed that the general nature of change over time follows some type of slope-intercept model discussed in Chapter 13 on growth and

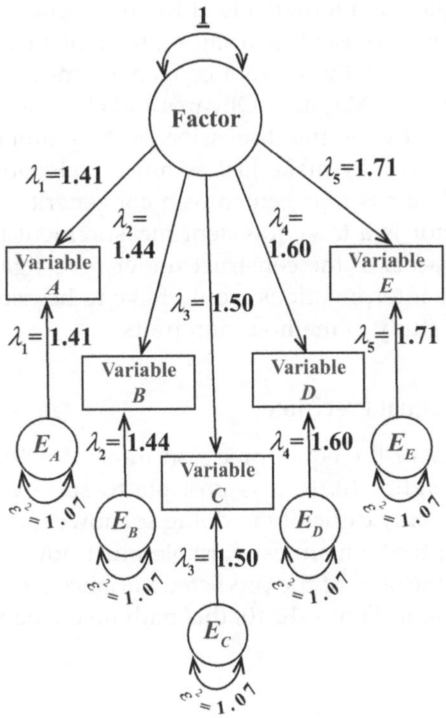


**FIGURE 10.8**
Congeneric measurement with equal precision.

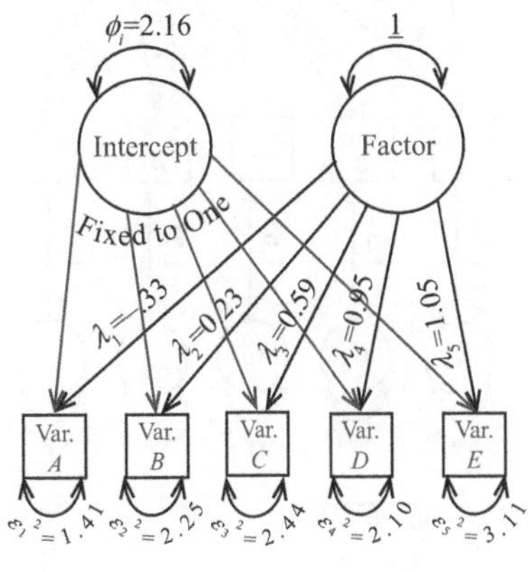


**FIGURE 10.9**
Random intercept measurement model.

change, for example; or, alternatively, if the researcher believes that a common method factor is present as a tau equivalent measurement model in addition to the conceptually relevant congeneric measurement model. This is the interpretation that Maydieu-Olivares and Coffman gave to the random intercept model. Following the discussion of the naming problem in latent variables, however, it may make just as much conceptual sense to believe that the method factor is represented as a congeneric measurement model while the trait factor is a tau-equivalent measurement model, or that both factors represent facets of the construct under investigation. The adjudication between these interpretations would have to be determined within the context of a larger study of methods and traits.

## Random Intercept Equal Precision

Following the logic of the equal precision model for the random intercept model shown in Figure 10.10, it is possible to specify an equal precision model as well. The only difficulty in doing so, however, comes from the fact that the error path to the manifest variable must now be constrained to be the square root of the sum of the predicted variance based on the two latent variables. As shown in Figure 10.10, this path takes the form of a nonlinear constraint.

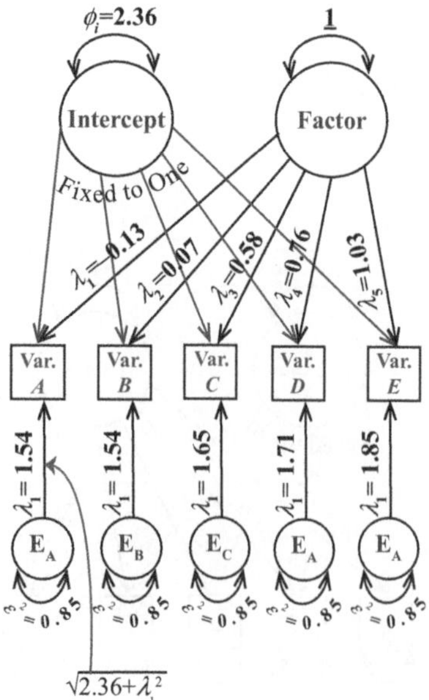


**FIGURE 10.10**
Equal precision random intercept model.

As before, the scaling factor can be used to calculate the proportion of variance due to the latent variables as 0.54, meaning that the proportion of variability due to measurement error is 0.46. As mentioned earlier for the random intercept model, the choice of error structure present in the data depends on the conceptual interpretation given to the random intercept and congeneric factors. A researcher who believes that the random intercept factor represents only method variance could constrain the error paths to be equal to their respective congeneric factor loadings, for example.

## Two-Factor Triangular Model

Finally, two congeneric latent variables may best explain patterns of variation and covariation in the data. For such a model to be mathematically identified, it becomes necessary to impose some additional constraint on the model. One could, for example, consider that if one attempted to fit a two-factor solution in which all variables loaded on all variables, minimization of such a solution could be "confused" in that there is no basis for choosing one loading – say, a loading on the first manifest variable – that should be assigned to the first or the second factor. As a result, one approach is to constrain a loading on one of the latent variables to zero, as shown in Figure 10.11.

This approach is often referred to as a "triangular decomposition," as the pattern of freely estimated loadings form a triangle when successive factors constrain additional loadings to zero (e.g., a third factor might have zero loadings on both variables $A$ and $B$). To designate the factor loadings of the

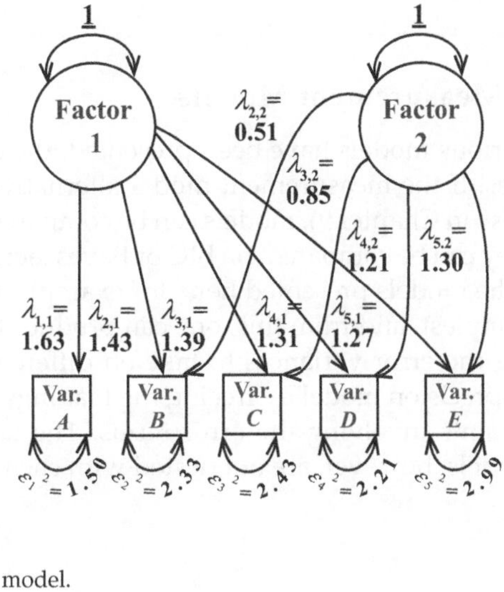


**FIGURE 10.11**
Triangular two factor model.

model, two subscripts are now shown for each loading, with the row indicating the number of the variable and the column indicating the number associated with the factor.

The term "Cholesky factorization" is often used to describe such a triangular decomposition strategy, although this is most appropriate when the researcher extracts as many factors as there are variables. When a reduced dimensionality is considered in the model, the term "reduced Cholesky factorization" is more appropriate (e.g., Neale & Cardon, 1992, p. 192). Although a triangular two-factor model may explain variability better than the previous alternatives in this chapter, it is less than obvious that it should be adopted as a final model. In one respect, such a model is less parsimonious than the random intercept factor model mentioned earlier in that it contains many more free parameters. Second, as noted by Loehlin (2004, p. 73), the resulting solution can be different, depending on which loading is set to zero on the second factor. (Although the article also features some formulas for translating between Cholesky, reverse-ordered Cholesky, simplex, and correlated factor models, it is easiest to directly estimate these from the data directly.) For reduced Cholesky factorizations, the fit of the model may be quite different as well depending on this choice. For these reasons, although a two-factor triangular decomposition may be informative regarding the number of latent variables present in the data, other models, such as the bifactor, oblique, or orthogonal factor models, may be explored as more interpretable alternatives. The unrotated factor solution from exploratory factor models may also be a useful starting point for exploring alternative triangular models or other possible parsimonious structures for the data.

## Comparison of Measurement Models

Now that these various models have been presented, the question of which model is best arises. If the measurement model alternatives are nested (see "$\chi^2$ Difference Tests" in Chapter 9), models can be compared via $\chi^2$ difference tests, and if not, they can be compared via BIC or Bayes factors, also described in Chapter 9. For the models presented here, for example, the strictly invariant model is the simplest, and from this, one can produce the tau equivalent model by allowing the error variances to take on different values or could produce the equal precision model by freeing the loadings and imposing the appropriate constraints involving the error terms. The tau equivalent and equal precision models, however, are not nested with each other.

## Can a Model Be Chosen if None Is "True" or if the Data Are Less Than Ideal?

The last chapter discussed ways to assess the power of a structural model to detect whether a given parameter was statistically significant. Although it is important to assess whether a given study is well powered to test such hypotheses, the question of whether a given data set can correctly adjudicate between candidate measurement models has some important conceptual differences.

In addition, the size of the data set being analyzed and the ability to correctly identify a particular model as well fitting are a function of the amount of measurement error in the manifest variables, the number of variables assessed, and the sample size. Recall from Chapter 1's comments on philosophical assumptions of the SEM that no one "true" model underlies the data.

Selection of the "best" structural model amounts to identifying that model which describes systematic patterns of variation and covariation in the data while ignoring those idiosyncrasies of the data, which can be, with the advantage of parsimony, safely ignored. If a given data set consists of relatively few observations or consists of data with a high degree of measurement error, a simpler model, such as the strict invariance model, may be all that we can reasonably "say" about variation and covariation in the data. Just as the "dots" used in newspaper photographs are useful representations when looked at superficially but cannot stand closer inspection with a magnifying glass, small data sets or coarse assessments may provide only a suggestive general patterning of relationships. Our task, then, as methodologists is to "right-size" out statistical models by not adopting overly elaborate models, which would have low statistical power and improper solutions, or overly simple statistical models, which fail to characterize important aspects of the model. To some extent, however, no such "Goldilocks" model that is "just right" for the data may exist. Our task is to consider that model that best captures the data, to subject it to all the critical tests we believe a reasonable skeptic could advance, and to present a prospective model that has survived our attempts to destroy it. Conversely, however, larger data sets or variables that are measured very precisely may result in rejection of even highly complex models. This rejection may provide insights regarding the non-normality of the error distribution, the nonlinear relationship of latent variables to their manifest variable indicators, or other nonlinear or interaction effects in the data. As researchers can increasingly make more precise measurements, and as it becomes increasingly easy to gather large amounts of complex, multivariate data, the question of how to address model misfit will only become more pronounced.

### Ignore the Misfit, Fix the Data, Fix the Model, or Discard Both?

How should we respond to these documented aspects of failure in our model? Are these model elaborations conceptually critical to understanding the phenomenon under investigation? For example, suppose that a researcher observed a curvilinear relationship between a latent variable and its manifest variable indicator. The researcher as a result has several options: (1) the researcher may note this lack of fit and argue that these aspects of the data are relatively minor characteristics relative to the aspects of the data that are summarized by the model; (2) the manifest variable may be redefined so that the relationship appears largely linear; (3) the manifest variable scores may be grouped into discrete categories and analyzed as an ordered categorical data using latent response variables; (4) the nonlinear relationship may be modeled as a nonlinear structural model such as Parafac (see, e.g., Nesselroade & Cattell, 1988); (5) the researcher may conclude that the data are uninformative of the phenomenon; or (6) the researcher may conclude that structural modeling is inappropriate for the data.

The choice between these alternatives must rest on the current state of theory in the area. Perhaps model misfit represents "decimal dust," which can be ignored in practice. Using an otherwise useful model which is nevertheless "wrong" has analogies to the historical fact that astronomers were quite content to use Newtonian laws of gravitation, despite the fact that they did not successfully account for the orbit of Mercury until the advent of relativity theory. The deviation of Mercury's orbit from predicted values was simply not a large enough phenomenon to refute the existing Newtonian theory.

If the researcher secures a small sample, it is reasonable to expect that simpler models will provide a satisfactory fit to the data. As sample size increases, however, we can expect that the correct model will be identified, but that other intermediary models may be "best." Similarly, if a researcher using a given sample size were to increase the precision of measurement, the odds of successfully correctly identifying the correct model increase.

## A Monte Carlo Study of Right-Sizing Measurement

The nature of the real-world trade-off between accurate measurement of constructs, sample size, and number of constructs that can be successfully addressed in a single study is somewhat poorly understood. Methodologists have told me, for example, that as long as the correct statistical model is one of those entertained by the researcher, fit indices such as the BIC will always select the correct model, and that the only reason this does not work in practice is because our data do not completely meet the assumptions of the model. Very productive researchers have told me that it is unnecessary to have a high degree of measurement error in large-scale studies because the sheer size of the sample will cause the "signal to shine through" due to the power afforded by the sample size. Yet others have done power studies of single parameters of an SEM and argue that the power of general model comparisons must be an even more powerful aggregate of the power of the single parameter estimates. These are common beliefs in structural modeling. I will argue here, however, that the first position is not true and that beliefs in the redeeming qualities of large samples and the power of general model comparisons based on individual parameters are overly optimistic.

Monte Carlo studies can be useful in understanding "right-sizing" of statistical models and some of the practical problems involved in model comparison. Consider, for example, a case of data generated by the random intercept model with equal precision. In the example, error variance values are chosen so that the reliability of each variable was the same. Suppose that the researcher is interested in identifying the correct structural model and that the intra-class correlation associated with each variable was 0.70 (meaning that one might, for example, expect single factor loadings to approximate 0.84). Assume further that the true model generating the data is that shown in Figure 10.12.

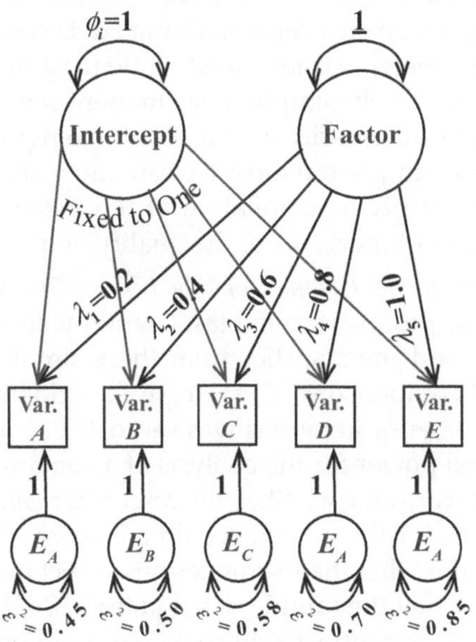


**FIGURE 10.12**
Example random intercept model

Assume further that we choose to use the BIC as our fit measure for adjudicating between models, given that the seven models described earlier are not all nested with each other. From a model comparison, note that this model contains two factors rather than the single factor assumed by the strict, tau, and congeneric models, and that, for the last variable, a roughly equal proportion of variance is due to the two latent variables. In addition, because the error variances have been chosen so that the explained variation due to the latent variables is constant across the variables, we can recognize that the "true" model is a random intercept model with equal precision of measurement.

Although the power to detect individual parameters of a model may be quite large, the question of which model will be selected among alternatives using a given fit statistic is a different inferential question. Exploring the proportion of times that the different models are selected based on a fit statistic from a Monte Carlo simulation can help address this question.

*Comments*: In the real world there is no "true" model, but an SEM can identify the model that best summarizes the data and, together with the diagnostic tests proposed, can be part of a procedure of identification of a model that reasonably summarizes important aspects of the data.

In general, we would expect that small samples or imprecise measurement are more likely to be well summarized by simpler models, and this is simply a characteristic of model fitting. At times it is tempting for some to tout the superior measurement of small data sets that were very labor intensive to gather, as in transcribed semi-structured interview formats. The fact that relatively simple measurement models summarize such data well is not necessarily a sign of the strength of such assessment strategies but is simply the most we can know about the data given its small sample nature. Interpretation of the reliability of measurement must still guide comparisons of the quality of measurement.

*Power Analyses for Model Comparisons for ICC* = 0.70: Monte Carlo analysis similar to those proposed in the last chapter were estimated to address this question, and the statistics from these simulations were saved to enable model comparisons. To increase the odds of successful convergence, the number of iterations was set to 10,000 instead of the default 1,000. Statistical power for the analysis of a scenario in which the ICC is 0.70 must first consider whether the researcher can expect that the estimated solutions for the models would converge. The first two columns of Table 10.1 show that the researcher can expect to have all models converge provided that the sample size is at least 300. Extremely small samples, such as 50, by contrast, will converge about only half the time, and a sample size of more than 400 will improve the convergence rate. Closer examination of the failures to converge, not reported in Table 10.1, reveals

**TABLE 10.1**

Convergence Rates for Sample Size and $N = 300$ Studies as a Function of ICC

| Sample Size | Proportion Converged | ICC | Proportion Converged |
|---|---|---|---|
| 50 | 0.54 | 0.2 | 0.71 |
| 100 | 0.63 | 0.3 | 0.68 |
| 200 | 0.76 | 0.4 | 0.65 |
| 300 | 0.86 | 0.5 | 0.69 |
| 400 | 0.92 | 0.6 | 0.72 |
| 500 | 0.95 | 0.7 | 0.86 |
| 600 | 0.97 | 0.8 | 0.98 |
| 1,000 | 1 | 0.9 | 1 |
| 2,000 | 1 | | |
| 3000 | 1 | | |

that these failures were, with two exceptions, in attempts to estimate the two-factor model to the data. Convergence failures associated with over-complex models have been a commonly noted problem (Savalei & Kolenikov, 2008).

BIC statistics from each model were then compared and the model was selected if it had the lowest BIC from among the seven candidate models considered. The proportion of times that each of the seven models was selected in those replications that converged across all models is shown in Figure 10.13.

As can be seen in the figure, the proportion of times that the correct model is selected increases as a function of sample size, but it is far from certain that a researcher will correctly select the random intercept precision model if the sample size is only 300. If one counts either random intercept model as a successful model identification, the proportion of times either model is selected is 73.7%. If, by contrast, the sample size is increased to 400, the percentage of converged solutions across all models increases to 92%, and the percentage of times that the random intercept precision model is correctly chosen increases to 63.4%. The percentage of times that the two-factor model is (incorrectly) chosen increases as well, however, at 36.0%.

If the researcher decides to adopt a sample size of 300, it is also reasonable to consider how critical the estimated ICC is for model selection. Figure 10.14 shows how model selection can vary for this sample size across a variety of scenarios of ICC.

It is possible that use of another model comparison statistic might improve the ability of the researcher to correctly identify the alternative model when the null hypothesis is false. As noted before, several of the measurement

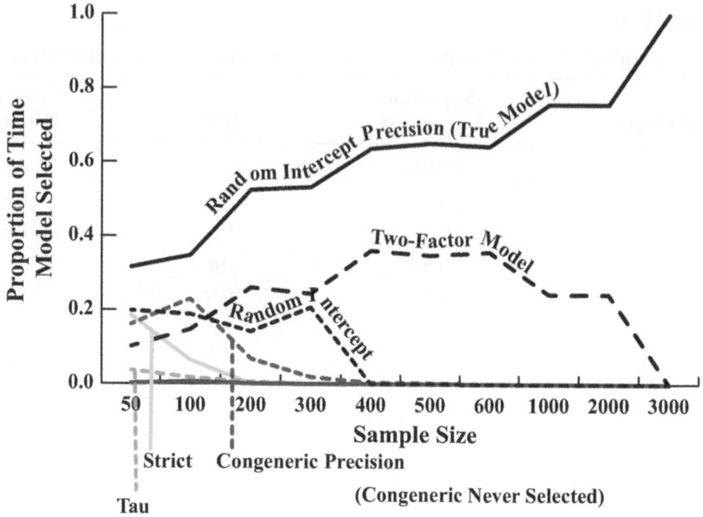


**FIGURE 10.13**
Proportion of models selected based on minimum BIC for ICC of 0.70 as a function of sample size.

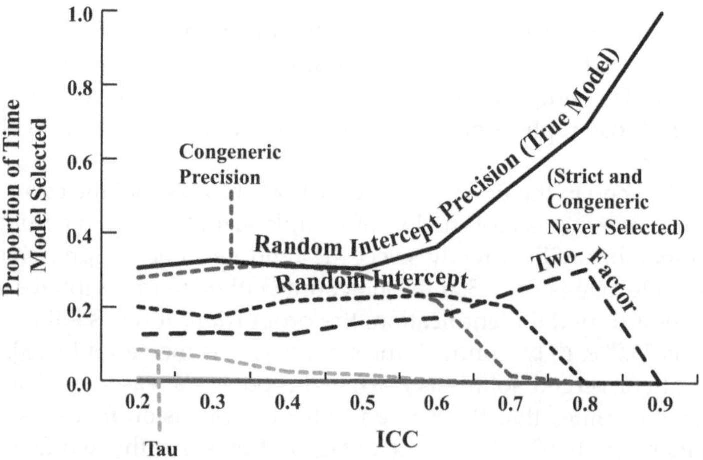


**FIGURE 10.14**
Proportion of models selected based on minimum BIC for sample size of 300 as a function of ICC.

models are nested within each other, although some are not (for example, the tau equivalent measurement model is not nested with the congeneric precision model). To explore this, $\chi^2$ difference tests were conducted on the simulation data making nested comparisons of the strict, congeneric precision, random intercept precision, random intercept, and two-factor models. If, during model selection, it was found that the congeneric precision model fit the data best, these models were compared to the congeneric model. If the congeneric model was selected, this model was then compared to the tau equivalent model. Similarly, if the random intercept model was found to be the best-fitting model from those considered so far, it was also compared to the tau equivalent and congeneric models (Figure 10.15).

$\chi^2$ difference tests for models with an ICC = 0.70 are presented in Figure 10.16. As shown, the power to correctly detect the correct model is 81.4% if one includes both random intercept models, a number slightly higher than the 73.6% obtained using the BIC in Figure 10.16. It seems that, for smaller samples, the $\chi^2$ difference test does a better job of discriminating the random intercept precision model from the congeneric precision, tau, and strict models. The pattern remains, however, that the ability of the data to discriminate between the random intercept and two-factor models remains at around 0.8 for sample sizes ranging from 1,000 to 3,000.

Similarly, one may examine the power of the $\chi^2$ difference test under the various ICC scenarios for a sample size of 300 shown in Figure 10.16. This figure shows that the ability to correctly identify the true model is slightly high tor the ICC = 0.70 condition than it was in Figure 10.15.

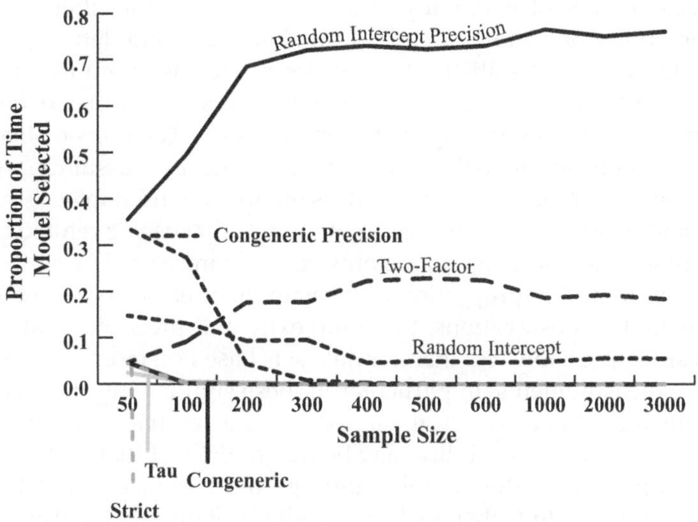


**FIGURE 10.15**
Model comparisons as a function of sample size using $\chi^2$ difference tests for ICC = 0.70.

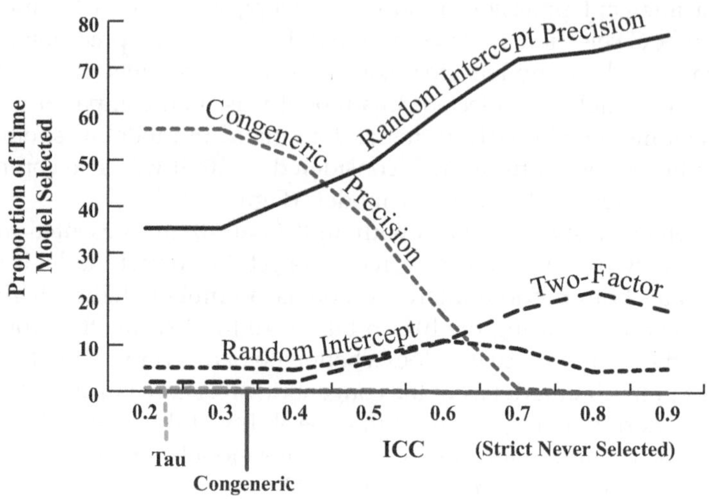


**FIGURE 10.16**
Model comparisons using as a function of ICC for $\chi^2$ difference tests for $N = 300$.

## Summary

The preceding exercise is intended to serve multiple purposes. First, it demonstrates that the question of whether a particular model can explain a nonzero amount of variability in a data set is quite a different question from the ability of a model to adjudicate between candidate explanations. Second, the use of fit statistics to evaluate models is by no means perfect, and even for large samples, it is possible that a researcher may overestimate the complexity of relationships between variables. There is no silver bullet for judging between models even if the data meet all assumptions of the model comparison procedure. Second, as mentioned in Chapter 1, there is no "true" model, but rather only classes of models that can be refuted based on the existing data. For data that are only coarsely measured (as in the case of variables with a high proportion of measurement error), or data consisting of relatively few observations, the complexity and precision of appropriate models are appreciably less than for data sets based on more precise or more extensive measurements. In particular in these simulations, the fact that the two-factor model persistently remained as a substantial alternative regardless of sample size or reliability may be due to the fact that, in this case, the adjudication between the random intercept and two-factor models rests on the decision to conclude that the factor loadings of the factor fall on a straight line. In one sense, this decision is based on a sample size of five, the number of manifest variables in the study.

One may argue that the random intercept, if identified as best-fitting in a particular real-world data set, is merely a more parsimonious version of a two-factor model and that other two-factor models exist involving estimation of fewer parameters than the triangular factors of the two-factor model. Although most texts at this point move to a discussion of simple structure as the only parsimonious models to be explored, in Chapter 12 we discuss some of the alternative patterning possibilities.

# 11

# Matrix Notation Models

To this point we have relied on path diagrams and individual equations to illustrate different types of models. When the number of variables and model complexity increase, however, it can be both more compact and useful to represent the model in matrix form. Matrix models use a notation system of Greek letters to indicate the path, variance, or covariance indicated in a model. These matrices can then be patterned to convey which parameters of the model are fixed and which are free. In addition, matrices written in this form can be readily used to calculate the average cross-products matrix of interest or, if parameter names are used, to produce a matrix of predicted average cross-products in terms of the parameters of the model. A number of matrix expressions for types of structural models exist, and these vary in terms of the sparseness and number of matrices involved and, in the case of older matrix models, they may also vary in terms of the types of models which may be estimated under the matrix formulation.

## ASA′ Model

The simplest and oldest way to produce a predicted matrix for a given data set is simply to construct a matrix of arrows, or paths, and an accompanying symmetric matrix of slings, or variances/covariances. The limitation of this approach, though, is that it is applicable only to structural models in which no consecutive arrows are involved in a given compound path. This approach does work for a variety of models, including ANOVA and regression as well as orthogonal and oblique factor models.

### The Arrow Matrix

In the case of, say, a two-factor model involving six variables, we might write our arrow matrix with six rows and eight (six manifest plus two latent) variables. If no path exists from one of the latent variables to the manifest, we enter a zero in that matrix cell. To see an example of this, consider the diagram in Figure 8.6, shown again here as Figure 11.1. The arrow matrix for this example consists of the loadings from the two latent variables (the first

 DOI: 10.1201/9781003595267-11

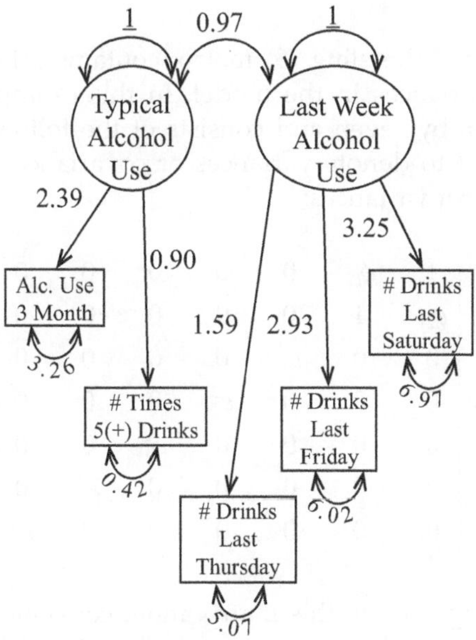


**FIGURE 11.1**
Correlated two-factor model for alcohol use variables.

two columns) and the (implicit) arrows from the conditional variance (i.e., error variance) associated with each manifest variable (the last five columns). Using $\lambda_{ij}$ to denote row $i$ and column $j$ of this matrix:

$$A = \begin{bmatrix} \lambda_{11} & 0 & 1 & 0 & 0 & 0 & 0 \\ \lambda_{12} & 0 & 0 & 1 & 0 & 0 & 0 \\ 0 & \lambda_{32} & 0 & 0 & 1 & 0 & 0 \\ 0 & \lambda_{42} & 0 & 0 & 0 & 1 & 0 \\ 0 & \lambda_{52} & 0 & 0 & 0 & 0 & 1 \end{bmatrix} \tag{11.1}$$

More generally, we might write this matrix out in terms of submatrices $\Lambda$ of factor loadings and an identity matrix $I$ as:

$$A = \left[\Lambda \middle| I\right] \tag{11.2}$$

with the vertical bar indicating that the two matrices are concatenated horizontally.

## The Sling Matrix

As mentioned above, the sling ($S$) matrix contains information about the variances and covariances in the model. In this example, the $S$ matrix is of dimension seven by seven and consists of the following free and fixed parameters, using $\phi$ to denote variances or covariances of latent variables and $\varepsilon_1^2$ to denote error variances:

$$S = \begin{bmatrix} 1 & \phi_{12} & 0 & 0 & 0 & 0 & 0 \\ \phi_{21} & 1 & 0 & 0 & 0 & 0 & 0 \\ 0 & 0 & \varepsilon_1^2 & 0 & 0 & 0 & 0 \\ 0 & 0 & 0 & \varepsilon_2^2 & 0 & 0 & 0 \\ 0 & 0 & 0 & 0 & \varepsilon_3^2 & 0 & 0 \\ 0 & 0 & 0 & 0 & 0 & \varepsilon_4^2 & 0 \\ 0 & 0 & 0 & 0 & 0 & 0 & \varepsilon_5^2 \end{bmatrix} \tag{11.3}$$

As before, we might write this matrix more conveniently in terms of its submatrices: $\Phi$, the covariance matrix of the latent variables; $\Theta_\delta$, the covariance matrix of error variables; and suitably sized null matrices to keep the matrix square:

$$S = \begin{bmatrix} \Phi & 0_{jxi} \\ 0_{ixj} & \Theta_\delta \end{bmatrix} \tag{11.4}$$

## Predicted Cross-Products

The predicted cross-products matrix (in this case a covariance matrix) is readily obtained by the formula $\Sigma = ASA'$, which, for this example, would look like:

$$\Sigma = \begin{bmatrix} \lambda_{11} & 0 & 1 & 0 & 0 & 0 & 0 \\ \lambda_{12} & 0 & 0 & 1 & 0 & 0 & 0 \\ 0 & \lambda_{32} & 0 & 0 & 1 & 0 & 0 \\ 0 & \lambda_{42} & 0 & 0 & 0 & 1 & 0 \\ 0 & \lambda_{52} & 0 & 0 & 0 & 0 & 1 \end{bmatrix} \begin{bmatrix} 1 & \phi_{12} & 0 & 0 & 0 & 0 & 0 \\ \phi_{21} & 1 & 0 & 0 & 0 & 0 & 0 \\ 0 & 0 & \varepsilon_1^2 & 0 & 0 & 0 & 0 \\ 0 & 0 & 0 & \varepsilon_2^2 & 0 & 0 & 0 \\ 0 & 0 & 0 & 0 & \varepsilon_3^2 & 0 & 0 \\ 0 & 0 & 0 & 0 & 0 & \varepsilon_4^2 & 0 \\ 0 & 0 & 0 & 0 & 0 & 0 & \varepsilon_5^2 \end{bmatrix}$$

$$\begin{bmatrix} \lambda_{11} & \lambda_{12} & 0 & 0 & 0 \\ 0 & 0 & \lambda_{32} & \lambda_{42} & \lambda_{52} \\ 1 & 0 & 0 & 0 & 0 \\ 0 & 1 & 0 & 0 & 0 \\ 0 & 0 & 1 & 0 & 0 \\ 0 & 0 & 0 & 1 & 0 \\ 0 & 0 & 0 & 0 & 1 \end{bmatrix} \tag{11.5}$$

When multiplied out, this gives:

$$\begin{bmatrix} \lambda_{11}^2+\varepsilon_1^2 & \lambda_{11}\lambda_{12} & \phi_{12}\lambda_{11}\lambda_{32} & \phi_{12}\lambda_{11}\lambda_{42} & \phi_{12}\lambda_{11}\lambda_{52} \\ \lambda_{12}\lambda_{11} & \lambda_{12}^2+\varepsilon_2^2 & \phi_{12}\lambda_{12}\lambda_{32} & \phi_{12}\lambda_{12}\lambda_{42} & \phi_{12}\lambda_{12}\lambda_{52} \\ \phi_{21}\lambda_{32}\lambda_{11} & \phi_{21}\lambda_{32}\lambda_{12} & \lambda_{32}^2+{\varepsilon_3}^2 & \lambda_{32}\lambda_{42} & \lambda_{32}\lambda_{52} \\ \phi_{21}\lambda_{42}\lambda_{11} & \phi_{21}\lambda_{42}\lambda_{12} & \lambda_{42}\lambda_{32} & \lambda_{42}^2+\varepsilon_4^2 & \lambda_{42}\lambda_{52} \\ \phi_{21}\lambda_{52}\lambda_{11} & \phi_{21}\lambda_{52}\lambda_{12} & \lambda_{52}\lambda_{32} & \lambda_{52}\lambda_{42} & \lambda_{52}^2+\varepsilon_5^2 \end{bmatrix} \tag{11.6}$$

If we wanted instead to estimate a four-predictor regression model predicting the first variable, past three-month alcohol use, from the remaining variables, we might think of the predictor variables as being perfect measures of individual latent variables. Using $b_i$ to denote the regression weights, $\sigma^2$ to denote the conditional criterion variance, $\sigma_i^2$ for the predictor variance, and $\sigma_{ij}$ to denote predictor covariances, we could write the $A$ and $S$ matrices as:

$$A = \begin{bmatrix} 1 & b_1 & b_2 & b_3 & b_4 \\ 0 & 1 & 0 & 0 & 0 \\ 0 & 0 & 1 & 0 & 0 \\ 0 & 0 & 0 & 1 & 0 \\ 0 & 0 & 0 & 0 & 1 \end{bmatrix} \tag{11.7}$$

and

$$S = \begin{bmatrix} \sigma^2 & 0 & 0 & 0 & 0 \\ 0 & \sigma_1^2 & \sigma_{12} & \sigma_{13} & \sigma_{14} \\ 0 & \sigma_{12} & \sigma_2^2 & \sigma_{23} & \sigma_{24} \\ 0 & \sigma_{13} & \sigma_{23} & \sigma_3^2 & \sigma_{34} \\ 0 & \sigma_{14} & \sigma_{24} & \sigma_{34} & \sigma_4^2 \end{bmatrix}, \tag{11.8}$$

which would give us a large, square matrix of predicted cross-products containing the following equation for the variance of the criterion:

$$\begin{aligned}\sigma_Y^2 &= \sigma^2 + \sigma_1^2 b_1^2 + 2\sigma_{12} b_1 b_2 + 2\sigma_{13} b_1 b_3 + 2\sigma_{14} b_1 b_4 + \sigma_2^2 b_2^2 \\ &+ 2\sigma_{23} b_2 b_3 + 2\sigma_{24} b_2 b_4 + \sigma_3^2 b_3^2 + 2\sigma_{34} b_3 b_4 + \sigma_4^2 b_4^2\end{aligned} \tag{11.9}$$

and four covariances between the predictors and the criterion of:

$$C_{X,Y} = \begin{bmatrix} \sigma_1^2 b_1 + \sigma_{12} b_2 + \sigma_{13} b_3 + \sigma_{14} b_4 \\ \sigma_{12} b_1 + \sigma_2^2 b_2 + \sigma_{23} b_3 + \sigma_{24} b_4 \\ \sigma_{13} b_1 + \sigma_{23} b_2 + \sigma_3^2 b_3 + \sigma_{34} b_4 \\ \sigma_{14} b_1 + \sigma_{24} b_2 + \sigma_{34} b_3 + \sigma_4^2 b_4 \end{bmatrix} \tag{11.10}$$

The section corresponding to the observed variances and covariances of predictors remains unchanged.

Although it is useful to know that one can, with two multiplications, obtain the entire predicted cross-products matrix for a large class of models, sadly, the technique does not work for path diagrams that have compound paths involving trips in one direction of more than one arrow. The matrix expression of SEMs has been the object of a few papers, with different patterns suggesting the generality of systems of linear equations. Two such systems, the Reticular Action Model (RAM) approach proposed by McArdle and McDonald (1984) and the familiar LISREL notation provide a useful introduction to such models. The former approach is explicitly used in many SEM programs such as OpenMx and, by extension, Onyx, while the latter, used in the computer program LISREL, is often used in presentations describing structural models.

## RAM Model

As before, we will consider matrices $A$ and $S$ to consist of path coefficients and a variance/covariance matrix. The only difference now, however, is that we will consider $A$ to be a square matrix containing the path information involved in all variables, latent and manifest. This allows, for example, for specifying an arrow from one latent variable to another in addition to the loadings. The matrices $A$ and $S$ will be used to generate a cross-products matrix for all variables in the model, latent and manifest. To generate the predicted cross-products matrix for just the manifest variables under the model, the data are subsetted using a filter matrix $F$, described in the next section.

The model can be calculated directly using matrix algebra and, as it does not involve many steps, it is presented here. Those uninterested in the particulars can skip the next section.

**RAM Model via Matrix Algebra**

To begin, consider a vector $v$ for each individual that consists of an individual's scores, both latent and manifest. For the oblique factor model, this would look like the following:

$$v = \begin{bmatrix} F_1 \\ F_2 \\ V_1 \\ V_2 \\ V_3 \\ V_4 \\ V_5 \end{bmatrix} \tag{11.11}$$

Consider also the $A$ matrix of all possible arrows. In our example:

$$A = \begin{bmatrix} \lambda_{11} & 0 & 1 & 0 & 0 & 0 & 0 \\ \lambda_{12} & 0 & 0 & 1 & 0 & 0 & 0 \\ 0 & \lambda_{32} & 0 & 0 & 1 & 0 & 0 \\ 0 & \lambda_{42} & 0 & 0 & 0 & 1 & 0 \\ 0 & \lambda_{52} & 0 & 0 & 0 & 0 & 1 \end{bmatrix} \tag{11.12}$$

Let's now write out a factor of "outside causes" to the system we want to analyze. This will be identical to Eq. (11.11), except we will replace any endogenous variables with their conditional variables.

$$u = \begin{bmatrix} F_1 \\ F_2 \\ e_1 \\ e_2 \\ e_3 \\ e_4 \\ e_5 \end{bmatrix} \tag{11.13}$$

Let's now define the variables of our system as being composed of the relationships between variables ($A$) multiplied by the relevant variables ($v$), plus variables not explained by our system ($u$). In equation form, this would be:

$$v = Av + u \tag{11.14}$$

or, for this example:

$$\begin{bmatrix} F_1 \\ F_2 \\ V_1 \\ V_2 \\ V_3 \\ V_4 \\ V_5 \end{bmatrix} = \begin{bmatrix} 0 & 0 & 0 & 0 & 0 & 0 & 0 \\ 0 & 0 & 0 & 0 & 0 & 0 & 0 \\ \lambda_{11} & 0 & 0 & 0 & 0 & 0 & 0 \\ \lambda_{12} & 0 & 0 & 0 & 0 & 0 & 0 \\ 0 & \lambda_{32} & 0 & 0 & 0 & 0 & 0 \\ 0 & \lambda_{42} & 0 & 0 & 0 & 0 & 0 \\ 0 & \lambda_{52} & 0 & 0 & 0 & 0 & 0 \end{bmatrix} \begin{bmatrix} F_1 \\ F_2 \\ V_1 \\ V_2 \\ V_3 \\ V_4 \\ V_5 \end{bmatrix} + \begin{bmatrix} F_1 \\ F_2 \\ e_1 \\ e_2 \\ e_3 \\ e_4 \\ e_5 \end{bmatrix} \tag{11.15}$$

This can then be solved for $v$:

$$v - Av = u \tag{11.16}$$

$$(\mathrm{I} - A)v = u \tag{11.17}$$

$$v = (I - A)^{-1} u \tag{11.18}$$

We are now able to calculate the cross-products matrix for our model. We do this by calculating the sum of squares and cross-products matrix as premultiplication by the transpose:

$$vv' = (I - A)^{-1} uu' (I - A)^{-1'} \tag{11.19}$$

$uu'$ is a variance/covariance matrix of all variables, manifest and latent, which can be termed $S$.

$$vv' = (I - A)^{-1} S (I - A)^{-1'} \tag{11.20}$$

The difficulty would seem to be to calculate $(\mathrm{I} - A)^{-1}$. McArdle and McDonald (1984), however, showed that

$$(I - A)^{-1} = I + A + AA + AAA\ldots \tag{11.21}$$

and that, for recursive systems (e.g., those with no feedback loops), eventually the product of $A$ matrices will eventually become a zero matrix. The matrices $AA$, $AAA$, etc., have the added benefit of representing the magnitude of the compound arrows involving two, three, and more paths in the diagram (i.e., the "indirect effects" of the model in addition to the "direct" effects estimated by $A$; see the "Mediational Models" section in Chapter 4).

For our example, $AA$ is zero, which leaves us

$$\begin{bmatrix} 1 & 0 & 0 & 0 & 0 & 0 & 0 \\ 0 & 1 & 0 & 0 & 0 & 0 & 0 \\ 0 & 0 & 1 & 0 & 0 & 0 & 0 \\ 0 & 0 & 0 & 1 & 0 & 0 & 0 \\ 0 & 0 & 0 & 0 & 1 & 0 & 0 \\ 0 & 0 & 0 & 0 & 0 & 1 & 0 \\ 0 & 0 & 0 & 0 & 0 & 0 & 1 \end{bmatrix} + \begin{bmatrix} 0 & 0 & 0 & 0 & 0 & 0 & 0 \\ 0 & & 0 & 0 & 0 & 0 & 0 \\ \lambda_{11} & 0 & 0 & 0 & 0 & 0 & 0 \\ \lambda_{12} & 0 & 0 & 0 & 0 & 0 & 0 \\ 0 & \lambda_{32} & 0 & 0 & 0 & 0 & 0 \\ 0 & \lambda_{42} & 0 & 0 & 0 & 0 & 0 \\ 0 & \lambda_{52} & 0 & 0 & 0 & 0 & 0 \end{bmatrix}$$
$$= \begin{bmatrix} 1 & 0 & 0 & 0 & 0 & 0 & 0 \\ 0 & 1 & 0 & 0 & 0 & 0 & 0 \\ \lambda_{11} & 0 & 1 & 0 & 0 & 0 & 0 \\ \lambda_{12} & 0 & 0 & 1 & 0 & 0 & 0 \\ 0 & \lambda_{32} & 0 & 0 & 1 & 0 & 0 \\ 0 & \lambda_{42} & 0 & 0 & 0 & 1 & 0 \\ 0 & \lambda_{52} & 0 & 0 & 0 & 0 & 1 \end{bmatrix} \tag{11.22}$$

The predicted covariance matrix of all variables is:

$$\begin{bmatrix} 1 & 0 & 0 & 0 & 0 & 0 & 0 \\ 0 & 1 & 0 & 0 & 0 & 0 & 0 \\ \lambda_{11} & 0 & 1 & 0 & 0 & 0 & 0 \\ \lambda_{12} & 0 & 0 & 1 & 0 & 0 & 0 \\ 0 & \lambda_{32} & 0 & 0 & 1 & 0 & 0 \\ 0 & \lambda_{42} & 0 & 0 & 0 & 1 & 0 \\ 0 & \lambda_{52} & 0 & 0 & 0 & 0 & 1 \end{bmatrix} \begin{bmatrix} 1 & \phi_{12} & 0 & 0 & 0 & 0 & 0 \\ \phi_{12} & 1 & 0 & 0 & 0 & 0 & 0 \\ 0 & 0 & e_1^2 & 0 & 0 & 0 & 0 \\ 0 & 0 & 0 & e_2^2 & 0 & 0 & 0 \\ 0 & 0 & 0 & 0 & e_3^2 & 0 & 0 \\ 0 & 0 & 0 & 0 & 0 & e_4^2 & 0 \\ 0 & 0 & 0 & 0 & 0 & 0 & e_5^2 \end{bmatrix}$$
$$\begin{bmatrix} 1 & 0 & \lambda_{11} & \lambda_{12} & 0 & 0 & 0 \\ 0 & 1 & 0 & 0 & \lambda_{32} & \lambda_{42} & \lambda_{52} \\ 0 & 0 & 1 & 0 & 0 & 0 & 0 \\ 0 & 0 & 0 & 1 & 0 & 0 & 0 \\ 0 & 0 & 0 & 0 & 1 & 0 & 0 \\ 0 & 0 & 0 & 0 & 0 & 1 & 0 \\ 0 & 0 & 0 & 0 & 0 & 0 & 1 \end{bmatrix} \tag{11.23}$$

Which is:

$$\begin{bmatrix} 1 & \phi_{12} & \lambda_{11} & \lambda_{12} & \phi_{12}\lambda_{32} & \phi_{12}\lambda_{42} & \phi_{12}\lambda_{52} \\ \phi_{12} & 1 & \phi_{12}\lambda_{11} & \phi_{12}\lambda_{12} & \lambda_{32} & \lambda_{42} & \lambda_{52} \\ \lambda_{11} & \phi_{12}\lambda_{11} & \lambda_{11}^2+e_1^2 & \lambda_{11}\lambda_{12} & \phi_{12}\lambda_{11}\lambda_{32} & \phi_{12}\lambda_{11}\lambda_{42} & \phi_{12}\lambda_{11}\lambda_{52} \\ \lambda_{12} & \phi_{12}\lambda_{12} & \lambda_{12}\lambda_{11} & \lambda_{12}^2+e_2^2 & \phi_{12}\lambda_{12}\lambda_{32} & \phi_{12}\lambda_{12}\lambda_{42} & \phi_{12}\lambda_{12}\lambda_{52} \\ \phi_{12}\lambda_{32} & \lambda_{32} & \phi_{12}\lambda_{32}\lambda_{11} & \phi_{12}\lambda_{32}\lambda_{12} & \lambda_{32}^2+e_3^2 & \lambda_{32}\lambda_{42} & \lambda_{32}\lambda_{52} \\ \phi_{12}\lambda_{42} & \lambda_{42} & \phi_{12}\lambda_{42}\lambda_{11} & \phi_{12}\lambda_{42}\lambda_{12} & \lambda_{42}\lambda_{32} & \lambda_{42}^2+e_4^2 & \lambda_{42}\lambda_{52} \\ \phi_{12}\lambda_{52} & \lambda_{52} & \phi_{12}\lambda_{52}\lambda_{11} & \phi_{12}\lambda_{52}\lambda_{12} & \lambda_{52}\lambda_{32} & \lambda_{52}\lambda_{42} & \lambda_{52}^2+e_5^2 \end{bmatrix} \tag{11.24}$$

The submatrix of the third through fifth rows and columns of Eq. (11.24) is equal to the values from the ASA′ model. To subset this matrix so that it contains only the manifest variables, a filter ($F$) matrix is used.

The filter matrix is a rectangular matrix with rows equal to the number of manifest variables and columns equal to the number of variables in the system. Elements of the $F$ matrix are zero unless a given variable is a manifest variable, in which case the entry corresponding to the row and column of that variable is 1. For our example:

$$F = \begin{bmatrix} 0 & 0 & 1 & 0 & 0 & 0 & 0 \\ 0 & 0 & 0 & 1 & 0 & 0 & 0 \\ 0 & 0 & 0 & 0 & 1 & 0 & 0 \\ 0 & 0 & 0 & 0 & 0 & 1 & 0 \\ 0 & 0 & 0 & 0 & 0 & 0 & 1 \end{bmatrix} \tag{11.25}$$

## RAM Model

The complete form of the RAM model is, in one equation:

$$\Sigma = F(I-A)^{-1} S(I-A)^{-1\prime} F' \tag{11.26}$$

Which, in our example, gives us Eq. (11.27):

$$\Sigma = \begin{bmatrix} \lambda_{11}^2+e_1^2 & \lambda_{11}\lambda_{12} & \phi_{12}\lambda_{11}\lambda_{32} & \phi_{12}\lambda_{11}\lambda_{42} & \phi_{12}\lambda_{11}\lambda_{52} \\ \lambda_{12}\lambda_{11} & \lambda_{12}^2+e_2^2 & \phi_{12}\lambda_{12}\lambda_{32} & \phi_{12}\lambda_{12}\lambda_{42} & \phi_{12}\lambda_{12}\lambda_{52} \\ \phi_{12}\lambda_{32}\lambda_{11} & \phi_{12}\lambda_{32}\lambda_{12} & \lambda_{32}^2+e_3^2 & \lambda_{32}\lambda_{42} & \lambda_{32}\lambda_{52} \\ \phi_{12}\lambda_{42}\lambda_{11} & \phi_{12}\lambda_{42}\lambda_{12} & \lambda_{42}\lambda_{32} & \lambda_{42}^2+e_4^2 & \lambda_{42}\lambda_{52} \\ \phi_{12}\lambda_{52}\lambda_{11} & \phi_{12}\lambda_{52}\lambda_{12} & \lambda_{52}\lambda_{32} & \lambda_{52}\lambda_{42} & \lambda_{52}^2+e_5^2 \end{bmatrix} \tag{11.27}$$

## Direct and Indirect Effects

Models with direct and indirect effects have compound paths greater than length one. Suppose that the researcher is interested in calculating the predicted variance/covariance matrix under a model in which the alcohol consumption in the past three months ($F1$) caused last week's consumption ($F2$). The arrow matrix for such a model would then be

$$A = \begin{bmatrix} 0 & 0 & 0 & 0 & 0 & 0 & 0 \\ \gamma_1 & 0 & 0 & 0 & 0 & 0 & 0 \\ \lambda_{11} & 0 & 0 & 0 & 0 & 0 & 0 \\ \lambda_{12} & 0 & 0 & 0 & 0 & 0 & 0 \\ 0 & \lambda_{52} & 0 & 0 & 0 & 0 & 0 \\ 0 & \lambda_{42} & 0 & 0 & 0 & 0 & 0 \\ 0 & \lambda_{52} & 0 & 0 & 0 & 0 & 0 \end{bmatrix} \tag{11.28}$$

$AA$ would then be:

$$AA = \begin{bmatrix} 0 & 0 & 0 & 0 & 0 & 0 & 0 \\ 0 & 0 & 0 & 0 & 0 & 0 & 0 \\ 0 & 0 & 0 & 0 & 0 & 0 & 0 \\ 0 & 0 & 0 & 0 & 0 & 0 & 0 \\ \gamma_1 & 0 & 0 & 0 & 0 & 0 & 0 \\ \gamma_1\lambda_{42} & 0 & 0 & 0 & 0 & 0 & 0 \\ \gamma_1\lambda_{52} & 0 & 0 & 0 & 0 & 0 & 0 \end{bmatrix} \tag{11.29}$$

Which would make $(I - A)^{-1}$:

$$\begin{bmatrix} 1 & 0 & 0 & 0 & 0 & 0 & 0 \\ \gamma_1 & 1 & 0 & 0 & 0 & 0 & 0 \\ \lambda_{11} & 0 & 1 & 0 & 0 & 0 & 0 \\ \lambda_{12} & 0 & 0 & 1 & 0 & 0 & 0 \\ \gamma_1 & 1 & 0 & 0 & 1 & 0 & 0 \\ \gamma_1\lambda_{42} & \lambda_{42} & 0 & 0 & 0 & 1 & 0 \\ \gamma_1\lambda_{52} & \lambda_{52} & 0 & 0 & 0 & 0 & 1 \end{bmatrix} \tag{11.30}$$

Which would make the $F(I - A)^{-1} S(I - A)^{-1'} F'$ matrix:

$$\begin{bmatrix} \lambda_{11}^2 + e_1^2 & \lambda_{11}\lambda_{12} & \gamma_1\lambda_{11} & \gamma_1\lambda_{11}\lambda_{42} & \gamma_1\lambda_{11}\lambda_{52} \\ \lambda_{12}\lambda_{11} & \lambda_{12}^2 + e_2^2 & \gamma_1\lambda_{12} & \gamma_1\lambda_{12}\lambda_{42} & \gamma_1\lambda_{12}\lambda_{52} \\ \gamma_1\lambda_{11} & \gamma_1\lambda_{12} & \gamma_1^2 + \Psi_2 + e_3^2 & \lambda_{42}\gamma_1^2 + \lambda_{42}\Psi_2 & \lambda_{52}\gamma_1^2 + \lambda_{52}\Psi_2 \\ \gamma_1\lambda_{42}\lambda_{11} & \gamma_1\lambda_{42}\lambda_{12} & \lambda_{42}\gamma_1^2 + \lambda_{42}\Psi_2 & \gamma_1^2\lambda_{42}^2 + \Psi_2\lambda_{42}^2 + e_4^2 & \lambda_{42}\lambda_{52}\gamma_1^2 + \lambda_{42}\lambda_{52}\Psi_2 \\ \gamma_1\lambda_{52}\lambda_{11} & \gamma_1\lambda_{52}\lambda_{12} & \lambda_{52}\gamma_1^2 + \lambda_{52}\Psi_2 & \lambda_{52}\lambda_{42}\gamma_1^2 + \lambda_{52}\lambda_{42}\Psi_2 & \gamma_1^2\lambda_{52}^2 + \Psi_2\lambda_{52}^2 + e_5^2 \end{bmatrix} \tag{11.31}$$

## LISREL Notation

LISREL notation uses separate matrices to convey information about different types of arrows and slings in an SEM. These are given slightly different names depending on whether the arrows and slings are associated with exogenous or endogenous variables. There are two LISREL notation systems. The older notation system, often called the "full LISREL notation," dates to the original program and is sometimes used in technical presentations of SEMs. The newer system, referred to as the "All-Y" model, is simpler in that it involves fewer matrices and is often the notation system used in computer programs such as Mplus or lavaan.

Discussion of these models begins with an explanation of a given observation vector of variables in terms of the latent variables and covariates of an SEM. Discussion then usually turns to calculations of likelihoods under the SEM or to the way that the model can be estimated by means of appropriate average cross-products.

### "Full LISREL" Notation

The full LISREL notation model explains an SEM via two equations. An equation is written for each observation in terms of the corresponding latents as:

$$y = \tau_y + \Lambda_y\eta + \varepsilon \text{ and } x = \tau_x + \Lambda_x\xi + \delta \tag{11.32}$$

where $y$ and $x$ represent vectors of manifest variables, which are the observed scores for an observation corresponding to indicators of endogenous and exogenous factors, respectively. Often the letter $m$ is used to denote how many manifest variables are under consideration, but this subscript will be dropped to make the presentation simpler. $\tau_y$ and $\tau_x$ denote the manifest variable intercepts, $\Lambda_y$ and $\Lambda_x$ denote matrices of factor loadings, $\eta$ and $\xi$ represent the unobserved factor score(s) for the individual on the endogenous and exogenous constructs, and $\varepsilon$ and $\delta$ represent vectors of errors of measurement for the endogenous and exogenous variables. Relationships between the latent variables are modeled as

$$\eta = \alpha + B\eta + \Gamma\xi + \zeta \tag{11.33}$$

where $\alpha$ represents the intercepts of the latent endogenous variables, $B$ is a matrix of path (arrows) between endogenous variables, $\Gamma$ is a matrix of paths from exogenous latent variables to endogenous latent variables, and $\zeta$ is a vector of residual scores associated with the endogenous factors.

These equations provide the notation for all paths in the model. Four matrices then explain all the sling information in the model. $\theta_\delta$ and $\delta_\varepsilon$ represent the variance/covariance matrix of the error variables associated with the exogenous and endogenous latent variables in the model, and $\Phi$ and $\Psi$ represent the variance/covariance matrices associated with the exogenous and endogenous latent variables, respectively. A heuristic path diagram is shown at the top of Figure 11.2 for the arrow and sling information of the model.

Notice, however, that there is one limitation of the model shown in the equations and the top of Figure 11.2: No paths are specifiable running from manifest variables such as covariates to the latent variables. If your model contains a mix of manifest and latent exogenous variables and the software uses this older LISREL notation, it becomes necessary to create an error free "latent variable" corresponding to this variable so that it can be estimated with these equations. For this reason, software using the newer All-Y notation, which specifically includes the possibility of covariates, is preferable.

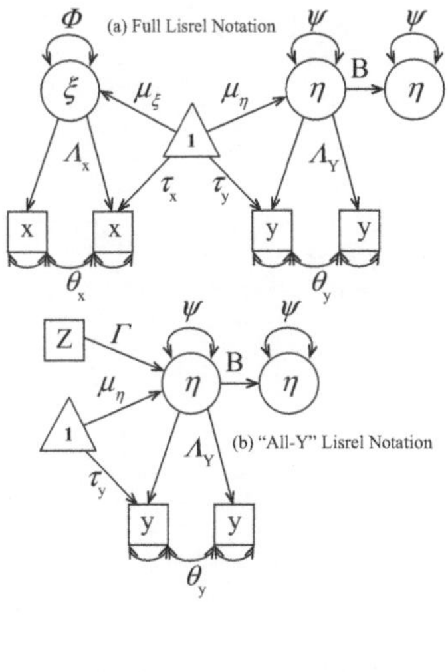


**FIGURE 11.2**
Heuristic path diagram for (a) full LISREL and (b) All-Y LISREL notations.

## All-Y Notation

The All-Y notation explains relationships between variables by means of a model of manifest variables associated with endogenous variables,

$$y = \tau_y + \Lambda_y \eta + \varepsilon, \tag{11.34}$$

and models relationships between latents as

$$\eta = \alpha + B\eta + \Gamma Z + \zeta \tag{11.35}$$

where the same matrices are used as in Eq. (11.33), except that Γ represents a matrix of path coefficients from covariates to the vector of latent variable scores, and Z represents a vector of covariate values. Latent variable scores that were exogenous under the full LISREL model are simply given no predictor variables. Generally speaking, path coefficients from one covariate to another do not figure in SEM presentations. A heuristic path diagram for the "All-Y" LISREL model is shown on the bottom of Figure 11.2.

Although other matrix models for SEM have been proposed, the ASA′ model was introduced for its simplicity, the RAM model for its generality, and the LISREL models for their use in technical discussions of SEM and some software specification issues. Although we have discussed various measurement models for individual factors, the discussion now turns to a variety of relationships between latent variables.

# 12

# *Parsimonious Factor Models*

> [T]he grand aim of all science ... is to cover the greatest possible number of empirical facts by logical deductions from the smallest possible number of hypotheses or axioms.
>
> **– Einstein, quoted in Nash (1963)**

> Science may be described as the art of systematic oversimplification – the art of discerning what we may with advantage omit.
>
> **– Karl Popper (1992)**

As mentioned in Chapter 10, one of the first questions concerning a latent variable analysis is whether the characteristics of the data may be parsimoniously summarized by one or more latent variables. In the discussion of triangular factorization, even if a two-factor triangular factorization fits well, the interpretation of the resulting factor loadings may not have a clear conceptual interpretation, and the estimated factor loadings can vary depending on which loadings are zeroed out across subsequent factors. This chapter discusses a number of parsimonious multifactor models. In some cases, a particular parsimonious model may correspond to a theory that the researcher wishes to propose. Other parsimonious models may constitute part of a reasonable counterargument to an originally proposed model. In the case of models that can constitute counterarguments, exploratory factor models such as those discussed in Chapter 15 can reframe (i.e., rotate) a latent variable model to a form corresponding to such an alternative explanation.

## Reasons for Drinking Example

Suppose, for existence, that a researcher is interested in looking at the three sets of items shown in Table 12.1, which are a subset of items taken from a measure of drinking motives (Cooper, 1994). We may well wonder if these three sets of items reflect distinct subsets as indicated by the column headings, or if they are all simply general indicators of an individual's willingness to give some reason for drinking.

DOI: 10.1201/9781003595267-12

**TABLE 12.1**

Reasons for Drinking. "I Drink ..."

| A Priori Content Area | | |
|---|---|---|
| **Social** | **Conformity** | **Enhancement** |
| Because it makes gatherings fun | To get in with a group | Because it gives a pleasant feeling |
| Because it improves parties | To be liked | Because I like the feeling |
| Because it helps enjoy the party | So I will not be kidded | To get high |
| To be sociable | Because friends pressure me | Because it is fun |
| To celebrate special occasions | So I will not feel left out | Because it is exciting |

## Orthogonal Factors

One possibility is to consider these three subsets of items as representing distinct uncorrelated factors. This is somewhat analogous to the independence null hypothesis mentioned in Chapter 8, except that the uncorrelated sections of the covariance matrix now involve blocks of items rather than each individual item. Although such a model is unrealistic in many situations, if true, it would be a very parsimonious model. A rough schematic of the path diagram for orthogonal factors for these 15 items is shown in Figure 12.1. This diagram also demonstrates that all manifest variables demonstrate "simple structure" in that each manifest variable has a loading on only one latent variable.

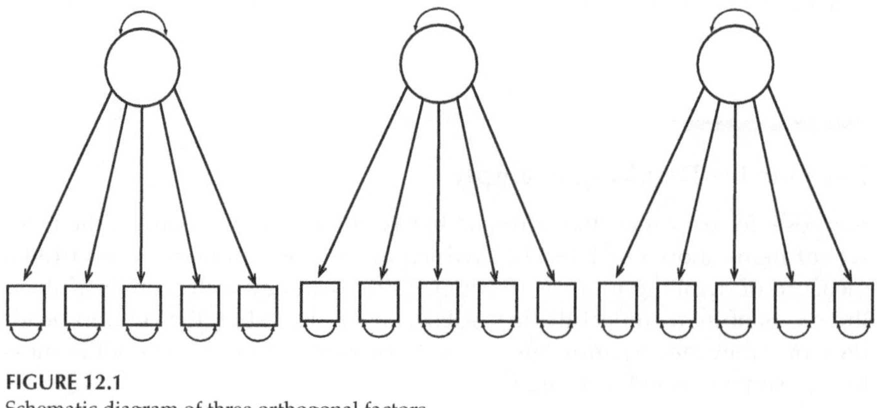

**FIGURE 12.1**
Schematic diagram of three orthogonal factors.

### Relations to Exploratory Factor Analysis

Despite the relatively unlikely nature of the orthogonal factor model, summarizing groups of variables by means of uncorrelated subsets of items is the starting point for many exploratory factor analytic techniques. Although we will discuss these more in depth later, it bears remembering that this is the general pattern that is first examined in some types of exploratory factor models.

Estimated standardized loadings and error terms are shown in the two left-most columns of Table 12.2 for 2,385 first-time college students described

**TABLE 12.2**

Orthogonal and Correlated Standardized Confirmatory Factor Loadings

| | Orthogonal | | Oblique | |
|---|---|---|---|---|
| **Item: "I drink …"** | **Standardized Loadings Λ** | **Error diag($\theta_\varepsilon$)** | **Standardized Loadings Λ** | **Error diag($\theta_\varepsilon$)** |
| Social | | | | |
| Because it makes gatherings fun | 0.94 | 0.12 | 0.93 | 0.13 |
| Because it improves parties | 0.92 | 0.15 | 0.92 | 0.15 |
| Because it helps enjoy the party | 0.83 | 0.31 | 0.84 | 0.3 |
| To be sociable | 0.69 | 0.52 | 0.69 | 0.52 |
| To celebrate special occasions | 0.66 | 0.57 | 0.66 | 0.56 |
| Conformity | | | | |
| To get in with a group | 0.91 | 0.17 | 0.91 | 0.17 |
| To be liked | 0.91 | 0.17 | 0.91 | 0.17 |
| So I will not be kidded | 0.81 | 0.34 | 0.81 | 0.34 |
| Because friends pressure me | 0.73 | 0.47 | 0.73 | 0.47 |
| So will not feel left out | 0.67 | 0.55 | 0.67 | 0.55 |
| Enhancement | | | | |
| Because it gives a pleasant feeling | 0.88 | 0.23 | 0.91 | 0.17 |
| Because I like the feeling | 0.85 | 0.29 | 0.91 | 0.17 |
| To get high | 0.7 | 0.5 | 0.81 | 0.34 |
| Because it is fun | 0.82 | 0.32 | 0.73 | 0.47 |
| Because it is exciting | 0.76 | 0.42 | 0.67 | 0.55 |

**Factor Correlations**

| | **Social** | **Conformity** | **Enhancement** |
|---|---|---|---|
| Social | 1 | 0.19 | 0.78 |
| Conformity | | 1 | 0.19 |
| Enhancement | | | 1 |

in Chapter 8 (Sher & Rutledge, 2007). The loadings appear to show that the individual items are relatively good indicators of the three factors. In some situations, grouping items into subsets such that each item has a loading on only one factor may make conceptual sense, such as when an attempt is made to group items into distinct topics; in other situations, it may not be a good starting point. For example, in a classroom test the items may reflect a general level of understanding, but some items are additionally related because they deal with a particular topic or chapter. If the items are longitudinal assessments of the same variable and the researcher is interested in identifying growth processes, a general factor of growth where the factors load on all items may be more appropriate.

## Correlated (Oblique) Factors

It is often the case that a parsimonious explanation involving multiple factors involves latent variables that are correlated. For this example, these are shown in the two right-most columns of Table 12.2. Examination of the standardized loadings for the orthogonal and correlated factor models reveals few differences at the level of the estimated standardized loadings. The fit of these two models, however, is quite different. For example, the estimated TLI of the orthogonal model is 0.86, while the TLI for the correlated factor model is 0.95. The three estimated correlations between the factors are shown at the bottom of Table 12.2. Further, the orthogonal model is nested within the correlated factor model, allowing us to calculate a $\chi^2_3$ difference test of 1,851.39, which is highly statistically significant. The path diagram corresponding to the correlated factor model is shown in Figure 12.2. The general aim of exploratory factor model rotations, discussed in Chapter 16, is to parsimoniously summarize relationships within clusters of items with such a pattern of factor loadings on only one factor with estimated correlations between factors.

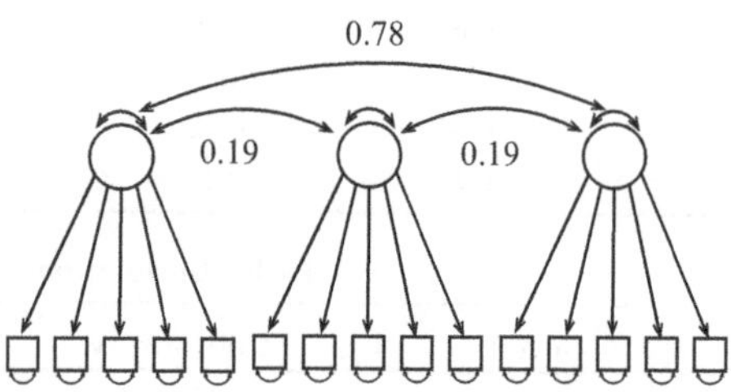


**FIGURE 12.2**
Schematic of correlated three-factor model.

## Bifactor Models

In bifactor models a single general latent variable is assumed to exist, with additional subsets of items explaining covariation between subsets of the items. In some testing situations, for example, the examinee is asked to read examples of text and answer multiple questions about each text. The smaller factors then model additional covariation between questions specific to a given text, while the general factor represents a general measure of achievement.

A rationale for use of the bifactor model in other studies can be made as well. In the example of reasons for drinking, a general factor may exist corresponding to whether people have many or few reasons for drinking, with subfactors being related to the content areas of the three factors mentioned so far. Recalling our discussion of the naming problem for latent variables, however, this is not the only explanation, as other inter-individual differences may account for the general factor, such as the sex of the participant (if men are more likely to endorse these reasons than women), personality (if, for example, individuals who are more extroverted as more likely to endorse the items), method factors associated with the common format of the items, or response bias in willingness to endorse the items. It should be noted that the bifactor model, by virtue of the several loadings estimated for the general factor, contains many more estimated parameters than the correlated factor model, a property that has led some to criticize the bifactor model as being overly general (e.g., Bonifay, Lane, & Reise, 2016). A schematic bifactor solution for the drinking motives data is shown in Figure 12.3. It should be noted that this model assumes that model factors are uncorrelated. This is a requirement for mathematical identification given the structure of the model, which contains only three factors; however, some correlations between the subfactors could be estimated in models involving more subfactors. EFA rotation techniques discussed in Chapter 15, such as Bi-Geomin and BI-CF-Quartimax, attempt to summarize factor models to such patterns of loadings.

The estimated factor loadings for the model are shown in Table 12.3. The column labeled general corresponds to the general factor, while the second column refers to factor loadings specific to that factor. As shown in the table, if a general factor accounts for most of the motives, it is composed mostly of items dealing with social and enhancement motives, with the conformity motives representing a relatively unrelated set of items. If the BIC values from the Bifactor (BIC = 66,488.65) and Correlated Factor models (BIC = 68,911.30) are compared, it appears that the Bifactor model fits the data better, possibly due to the fact that the Enhancement reasons which are related to the Social reasons have a different factor structure than the unique Enhancement items.

It should be noted that it is also possible to specify more parsimonious bifactor models such as those with general tau-equivalent factors which

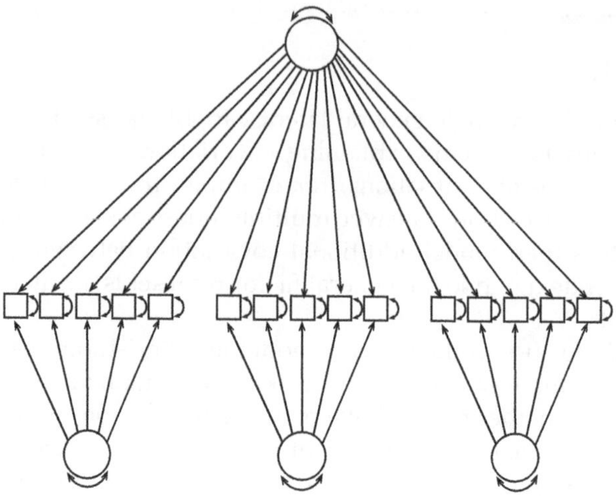

**FIGURE 12.3**
Schematic of bifactor factor loadings.

**TABLE 12.3**
Bifactor Confirmatory Factor Loadings

| | Standardized Loadings | | |
|---|---|---|---|
| **Item "I drink ..."** | **General $\Lambda_1$** | **Specific $\Lambda_{2-4}$** | **Error diag($\Theta_\varepsilon$)** |
| Social | | | |
| Because it makes gatherings fun | 0.92 | 0.09 | 0.15 |
| Because it improves parties | 0.92 | 0.29 | 0.07 |
| Because it helps enjoy party | 0.86 | −0.13 | 0.24 |
| To be sociable | 0.73 | −0.28 | 0.38 |
| To celebrate special occasions | 0.65 | 0.12 | 0.56 |
| Conformity | | | |
| To get in with a group | 0.19 | 0.89 | 0.17 |
| To be liked | 0.00 | 0.89 | 0.17 |
| So I will not be kidded | 0.12 | 0.81 | 0.34 |
| Because friends pressure me | 0.17 | 0.71 | 0.46 |
| So I will not feel left out | 0.16 | 0.65 | 0.55 |
| Enhancement | | | |
| Because it gives a pleasant feeling | 0.64 | 0.62 | 0.22 |
| Because I like the feeling | 0.64 | 0.54 | 0.29 |
| To get high | 0.51 | 0.49 | 0.5 |
| Because it is fun | 0.72 | 0.43 | 0.3 |
| Because it is exciting | 0.61 | 0.45 | 0.42 |

would be appropriate if, for example, the researcher believed that the general factor corresponded to a method factor corresponding to the method used to assess the items.

## Hierarchical Factor Models

Hierarchical factor models apply the basic pattern of a factor analysis to a hierarchy of latent variables. Essentially, this amounts to treating the three latent variables of motives as if they were items of a larger test. As such, all the rules for model identification of latent variables with manifest variable indicators apply to the hierarchical model as well, meaning that in this case, covariances between the three factors are completely explained by the three items. In terms of model specification, it is no longer possible to specify the three motive factors as having unit variance, as they are now endogenous. To identify the model, the variable with the highest loading was fixed to one for identification purposes (in this case, the first item in each category). The schematic is shown in Figure 12.4, and results of the analysis are shown in Table 12.4.

## Recap of Parsimonious Models

As stated in the introduction, our objective is to produce some parsimonious set of parameter values that correspond to more theoretically interpretable values. These patterns correspond to the general form of respecification of

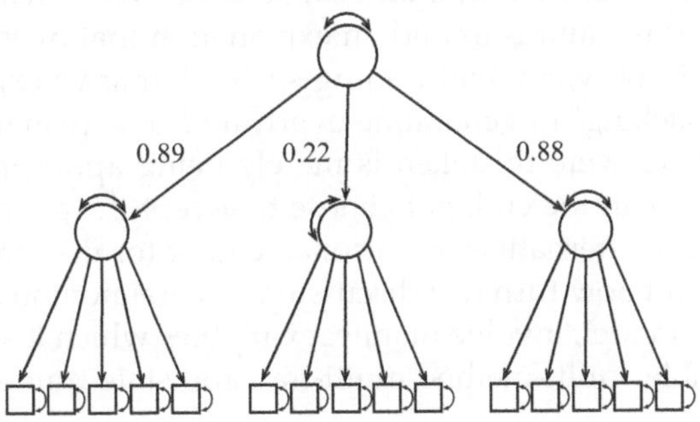


**FIGURE 12.4**
Schematic of hierarchical factor model.

**TABLE 12.4**

Factor Loadings and Errors for Hierarchical Factor

| Item "I drink ..." | Loading $\Lambda g$ | Error diag($\Theta_\varepsilon$) |
|---|---|---|
| General Factor | | |
| Social | 0.89 | 0.21 |
| Conformity | 0.22 | 0.95 |
| Enhancement | 0.88 | 0.23 |
| Social | | |
| Because it makes gatherings fun | 0.93 | 0.13 |
| Because it improves parties | 0.92 | 0.15 |
| Because it helps enjoy party | 0.84 | 0.3 |
| To be sociable | 0.69 | 0.52 |
| To celebrate special occasions | 0.66 | 0.56 |
| Conformity | | |
| To get in with a group | 0.91 | 0.17 |
| To be liked | 0.91 | 0.17 |
| So I will not be kidded | 0.81 | 0.34 |
| Because friends pressure me | 0.73 | 0.47 |
| So I will not feel left out | 0.67 | 0.55 |
| Enhancement | | |
| Because it gives a pleasant feeling | 0.86 | 0.25 |
| Because I like the feeling | 0.84 | 0.29 |
| To get high | 0.7 | 0.52 |
| Because it is fun | 0.84 | 0.29 |
| Because it is exciting | 0.77 | 0.41 |

a latent variable model taken up in Chapter 15 under rotations of exploratory factor analysis (to be considered later). Parsimonious models may be specified a priori if they correspond to a conceptual model. Exploratory techniques taken up later can be useful in finding parsimonious models, which, on hindsight, represent an alternate explanation that a reasonable skeptic could propose. If the aim is to find an explanation that overturns the initial model, we need not worry that finding such alternative explanations is an instance of "p-hacking" or generating overly ad-hoc explanations. Exploring ways in which we were mistaken is merely being appropriately cautious. Although we may, in the end, not choose to accept the alternative model, a discussion of such alternatives is reasonable grist for the limitations section of a study. We will now turn to a discussion of parsimonious models which, like the bifactor model, involve manifest variables which load on more than one latent variable, multi-method multi-trait and state-trait models.

## Multi-Trait Multi-Method Factor (MTMM) Models

MTMM models attempt to use the factor model to analyze data that systematically vary the traits under consideration and the methods used to assess those traits. These models are a direct response to Campbell and Fisk's (1959) observation that covariances between variables may exist because they share a common method of assessment, and that an exploration of convergent and divergent validity of constructs should therefore employ a variety of methods as well as traits. As such, these models represent an advance over the naming problem encountered for standard factor models in that the latent variables corresponding to proposed traits are no longer a conflation of the proposed trait and the method used to assess the trait. The naming problem, however, does not magically "go away" when this is done, as variability due to common attributes across traits will continue to be represented in the trait latent factors (or correlations across traits), and distinct variability associated with a given method of assessment may turn out, in a validity study, to be a more prescient predictor of a criterion of interest.

As an example of MTMM SEMs, Langer, Wood, Bergman, and Piacentini (2010), for example, consider a MTMM analysis of four child anxiety disorders in a sample of 174 children: Evaluations of the children were solicited from a diagnostician, the child's parents, as well as the child. The MTMM model for these data is presented in Figure 12.5. The factors on the left-hand side represent the four anxiety disorders, and the factors on the right represent the factors associated with the method of assessment. The figure assigns parameter names according to the LISREL All-Y notation to acquaint you with this convention and to make presentation of the matrices of the diagram clearer in Table 12.5.

Langer et al. (2010) report the overall model fit and the significant differences in fit between the model in Figure 12.5 and models that assume (1) no traits exist, (2) a single factor accounts for all traits, or (3) a single factor accounts for all method factors. It is somewhat puzzling, however, that the article reports no parameter values for their final model and offers no interpretation of the parameters.

A number of interesting characteristics emerge from the estimated parameters despite the relatively small sample under examination. Table 12.5 shows the parameter estimates for the final model of Langer et al., recalculated based on the summary statistics reported in the article. The first block of the table shows the factor loadings associated with each of the traits. The standardized loadings associated with Separation Anxiety appear larger than factor loadings for other traits, and within these, the child's ratings appeared to be larger than either the diagnostician or parent ratings. The only significant loading for Panic Disorder occurred for the diagnostician's ratings,

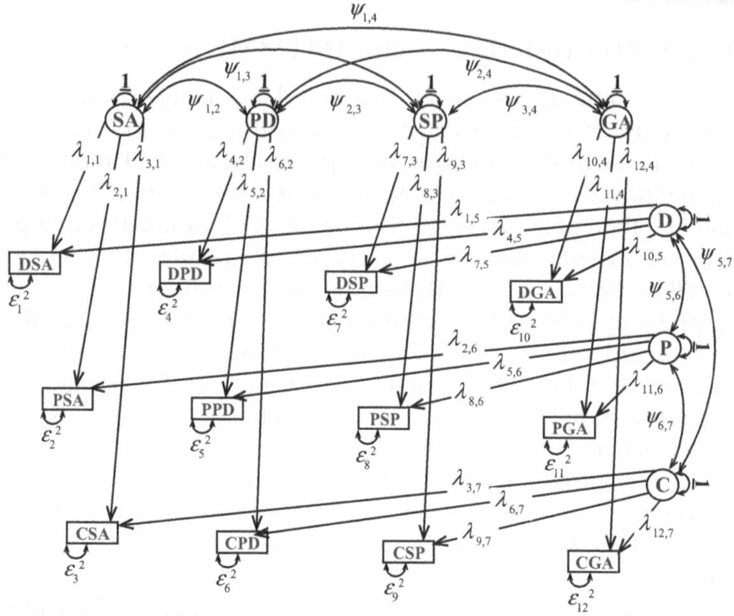


**FIGURE 12.5**
MTMM of Langer et al. (2010) of Separation Anxiety (SA), Panic Disorder (PD), Social Phobia (SP), and Generalized Anxiety disorder (GD) rated by Diagnostician ("D"), Parent ("P") and Child ("C").

suggesting that the construct may be poorly understood by the parents or children. Social Phobia loadings, although all statistically significant, were somewhat lower than those found for Separation Anxiety. Finally, the factor loading corresponding to the diagnostician's ratings of Generalized Anxiety was close to zero and not statistically significant. This is somewhat puzzling given that this diagnosis was the one most often assigned by the diagnostician in the study (28.7%). This may not necessarily mean that the parent and child ratings of Generalized Anxiety are necessarily valid, however. In line with comments earlier about the naming problem in latent variables, it may only signify that the parent and child ratings are similar to each other with respect to this disorder, reflecting a common lay understanding of the disorder rather than the clinical diagnosis.

Only one correlation between traits was statistically significant – the correlation between Separation Anxiety and Generalized Anxiety. Although the Langer et al. article correctly concludes that a model that assumes that all diagnoses are one common factor fits poorly, a model that constrains the correlation between the Separation Anxiety and Generalized Anxiety factors to one fits the data almost identically to their final model: $\chi^2(34) = 34.54, p = 0.44$

**TABLE 12.5**

MTMM Parameter Estimates for Langer et al. (2010)

**Factor Loadings of Traits ($\Lambda$)**

| **Method** | **SA ($\eta_1$)** | **PD ($\eta_2$)** | **SP ($\eta_3$)** | **GA ($\eta_4$)** |
|---|---|---|---|---|
| D | 0.47**($\lambda_{1,1}$) | 0.53**($\lambda_{4,2}$) | 0.43**($\lambda_{7,3}$) | −0.03($\lambda_{10,4}$) |
| P | 0.64**($\lambda_{2,1}$) | 0.20($\lambda_{5,2}$) | 0.49**($\lambda_{8,3}$) | 0.45**($\lambda_{11,4}$) |
| C | 0.79**($\lambda_{3,1}$) | 0.22($\lambda_{6,2}$) | 0.28**($\lambda_{9,3}$) | 0.54**($\lambda_{12,4}$) |

**Correlations Between Traits ($\Psi$)**

| | **SA ($\eta_1$)** | **PD ($\eta_2$)** | **SP ($\eta_3$)** | **GA ($\eta_4$)** |
|---|---|---|---|---|
| SA ($\eta_1$) | 1.00 | −0.05 | −0.21 | 0.91** |
| PD ($\eta_2$) | | 1.00 | −0.71 | 0.18 |
| SP ($\eta_3$) | | | 1.00 | −0.38 |
| GA ($\eta_4$) | | | | 1.00 |

**Error Variances ($diag(\Theta_y)$)**

| **Rater** | **SA** | **PD** | **SP** | **GA** |
|---|---|---|---|---|
| D | 0.73* ($\varepsilon_1$) | 0.56** ($\varepsilon_4$) | 0.60** ($\varepsilon_7$) | 0.85** ($\varepsilon_{10}$) |
| P | 0.19** ($\varepsilon_2$) | 0.49** ($\varepsilon_5$) | 0.49** ($\varepsilon_8$) | 0.41** ($\varepsilon_{11}$) |
| C | 0.14** ($\varepsilon_3$) | 0.31** ($\varepsilon_6$) | 0.47** ($\varepsilon_9$) | 0.56** ($\varepsilon_{12}$) |

**Method Factor Loadings ($\Lambda$)**

| **Disorder** | **D ($\eta_5$)** | **P ($\eta_6$)** | **C ($\eta_7$)** |
|---|---|---|---|
| SA | 0.21** | 0.64** | 0.48** |
| PD | 0.41** | 0.68** | 0.80** |
| SP | 0.46** | 0.52** | 0.68** |
| GA | 0.38** | 0.63** | 0.38** |

**Method Factor Correlations ($\Psi$)**

| | **D ($\eta_5$)** | **P ($\eta_6$)** | **C ($\eta_7$)** |
|---|---|---|---|
| D ($\eta_5$) | 1.00 | 0.58** | 0.33* |
| P ($\eta_6$) | | 1.00 | 0.22* |
| C ($\eta_7$) | | | 1.00 |

*Note*: SA = Separation Anxiety, PD = Panic Disorder, SP = Social Phobia, GA = Generalized Anxiety, D = Diagnostician, P = Parent, C = Child.

versus $\chi^2(33) = 33.74, p = 0.43$ for the original model. In line with the preceding comments, this may reflect a perception by both the parents and children that Separation Anxiety and Generalized Anxiety are the same constructs.

Method factor loadings for the diagnostician were uniformly lower than those for the parents and children, reflecting perhaps the greater conceptual distinctness of the disorders for the diagnostician. The children's method factors appeared to be characterized by a tendency to endorse both Panic Disorder and Social Phobia. Finally, the Diagnostician and Parent method factors appeared more correlated with each other than with the children's method factors, perhaps due to differences in the nature of the self versus other reports involved.

To summarize, the original conclusions of the MTMM analysis of the data are slightly different than the interpretations offered here based on the parameter estimates. Langer et al. (2010) concluded:

> results were supportive of both convergent and discriminant validity, even with the addition of GAD in the model (which had not shown strong evidence of validity by MTMM criteria in the simple correlation matrix).... the standard model of child anxiety disorders (as distinguishable but related entities) was a significantly better fit to the data than any alternative model.
>
> **(p. 557)**

By contrast, this re-analysis found no support for Generalized Anxiety disorder as a separate diagnosis based on the data, and that either the data under consideration did not contain children who met criteria for a Generalized Anxiety disorder or it could not be reliably diagnosed by the diagnostician if present. Panic Disorder, by contrast, could not be reliably assessed by parents or the child.

## State-Trait Models

State-trait models involve longitudinal data that are gathered in manifest or latent variable form across several measurement occasions (Eid & Diener, 1999; Eid et al., 2008; Steyer, Schmitt, & Eid, 1999). The basic idea of state-trait models is to extract one or more general factors with the addition of autoregressive paths between consecutive measurement occasions (termed lagged length one). The time-bound component may also involve larger windows of time by involving longer lags as well. Extensions of this basic idea have involved inclusion of multilevel models and questions of whether rater effects and methods are distinct or interchangeable (as discussed in Eid et al., 2008).

Such models have the advantage of considering that covariation between any two measurement occasions may be due either to a prospective effect or to a systematic inter-individual difference that spans all time measurements. We might think it is reasonable to explain in someone's report that they are stressed by saying that they seemed to still be stressed by something that happened yesterday, or alternatively, we might simply think that someone is usually more stressed than most people. This is not to say, however, that it is always the case that a single factor is the only possible factor structure for the model. Any of the variety of measurement models considered in Chapter 10 such as the tau equivalent, random intercept, or triangular models could be considered, as well as the parsimonious alternatives considered in this chapter.

As an illustration, consider the data from Sher and Rutledge's (2007) study of first-time college students who were later assessed twice a year for four years. One study variable, the General Severity Index of the Brief Symptom Inventory, is a measure of general psychological distress. The usual state-trait model for such a data set is shown in Figure 12.6. The Trait factor spans all measurement occasions, while the carryover effects from the previous measurement occasion are modeled via the paths labeled $\gamma$ in the model.

For the IMPACTS data, this model fits appreciably well. The $\chi^2_{13} = 39.10, p < 0.001$. RMSEA (90% CI) = 0.04 (0.03-0.06) Prob(RMSEA < 0.05) = 0.82, TLI = 0.99, CFI = 0.99, SRMR = 0.015, and BIC = 11622.713.

Based on this solution, the standardized parameter values shown at the right of Table 12.6 show that a relatively moderate percentage of variance is explained during Waves 4 and 5 (the second semester of the third year of college and first semester of the fourth year), with lower trait loadings at the earlier and later waves.

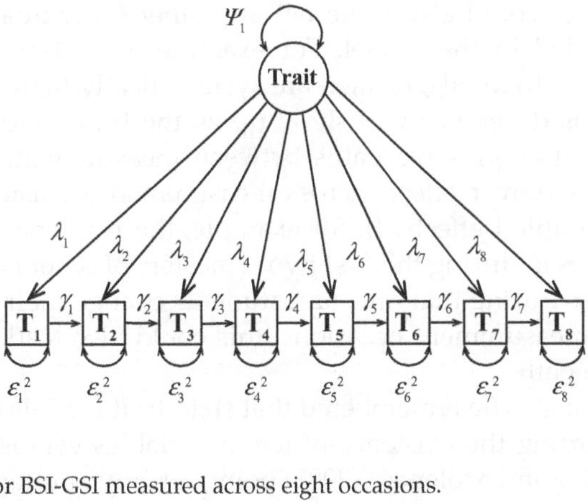


**FIGURE 12.6**
State-trait model for BSI-GSI measured across eight occasions.

**TABLE 12.6**

Parameters and Fit Statistics for Figure 12.6[a]

| | Unstandardized | | | Standardized | | |
|---|---|---|---|---|---|---|
| Var. | $\lambda$ | $\gamma$ | $\varepsilon^2$ | $\lambda$ | $\gamma$ | $\varepsilon^2$ |
| $T_1$ | 0.35 | | 0.24 | 0.59 | | 0.66 |
| $T_2$ | 0.28 | 0.22 | 0.16 | 0.52 | 0.24 | 0.52 |
| $T_3$ | 0.36 | 0.14 | 0.14 | 0.65 | 0.14 | 0.44 |
| $T_4$ | 0.40 | 0.05 | 0.12 | 0.74 | 0.05 | 0.41 |
| $T_5$ | 0.41 | −0.03 | 0.13 | 0.76 | −0.03 | 0.45 |
| $T_6$ | 0.31 | 0.15 | 0.13 | 0.60 | 0.16 | 0.49 |
| $T_7$ | 0.30 | 0.20 | 0.18 | 0.53 | 0.19 | 0.54 |
| $T_8$ | 0.30 | 0.19 | 0.15 | 0.55 | 0.19 | 0.50 |

[a] Factor variance, $\Psi_1$, fixed to 1.

Autoregressive paths document a significant carryover effect between contiguous measurement occasions except for the case of Waves 4 (first semester of the fourth year) and 5 (second semester of the fourth year). Carryover effect seems most pronounced for Wave 2.

## Cautions to Interpretation

Although state-trait models are often fit to data with the interpretation that the trait represents the trait or construct under consideration with the autoregressive paths representing the effects of the manifest variable, it bears remembering that, as with all structural models, alternative explanations exist for both of these parameters and the naming problem remains for these types of parameters as well.

Concerning the trait component of the model, it bears remembering that the general factor could also represent any other individual difference not explicitly included in the model. For example, if first-generation college students, older individuals, or men are systematically higher in psychological distress, this difference will show up as the trait factor in the model. Similarly, the auto-regressive paths between measurement occasions may not represent carryover effects across measurement occasions but may be due to time-delimited effects. If, for example, the residence hall that a college student lives in during the first two semesters of school has a significant effect on distress during that time and students change their living situation at subsequent measurement occasions, this could also lead to a significant auto-regressive path.

Finally, it must also be remembered that state-trait models make no definitive claim regarding the existence of latent variables versus autoregressive effects. As Rovine and Molenaar (2005) point out, any autoregressive process (such as that characterized by the causal arrows between measurement occasions) can be fit equivalently as a moving average process (i.e., a latent variable spanning those measurement occasions). This means that the trait factor of a state-trait model could be equivalently fit using a series of autoregressive paths that span multiple measurement occasions or, alternatively, that the carryover effects of the model could be modeled as additional "trait" factors. It is worth noting that such "sub-traits," if you will, will not consist of only two measurement occasions, but may span more measurement occasions due to the covariances explained by compound paths of the auto-regressive paths. In the end analysis, although conceptually elegant, the status of a state-trait model, as with any confirmatory technique, rests on the larger context of subsequent theory testing and the utility of predictions made under such a model.

### Right-Sizing State-Trait Models

As with any structural model, it is possible to consider structural alternatives to the state-trait model. Such alternatives follow the discussion of measurement models in Chapter 12. It is also possible to compare the relative fit statistics resulting from the assumption that an autoregressive structure between adjacent measurement occasions exists (in which progressively weaker covariances across measurement occasions that are not adjacent), versus simple pairwise correlated error between measurement occasions (which would, by the tracing rules, model only additional covariances between adjacent measurement occasions).

Many possible alternatives exist, of course, but it would seem reasonable to exclude some models from exploration. For example, it seems implausible to believe that extremely long lagged effects are present in the data, e.g., that a lagged effect exists that spans more than two years. An exploratory factor analysis as described in Chapter 15 reveals that at most three factors (without autoregressive paths) successfully account for the data, making it reasonable to limit our explorations of trait factor structure to at most three latents. Because the traditional state-trait model for these data revealed several statistically significant lagged effects between measurement occasions, it seems reasonable to continue to include these effects.

Reasonable measurement models for the trait factor, then, would appear to be the strictly equivalent, single factor tau equivalent measurement, congeneric measurement, the random intercept factor model, and the two-factor models in versions both with and without the lagged effects. It is also worth exploring whether correlated errors between adjacent measurement conditions rather than lagged effects fit the data better. Of these alternatives, a random intercept measurement model with no lagged effects appears to fit the data better than the original state-trait model. Both models demonstrate excellent fit based on RMSEA, TLI, and CFI models. The random intercept model with no lagged effects also fits better than the traditional model based on comparison of BIC values (BIC = 11,597.90, compared with 11,622.71 for the traditional state-trait model, a difference of 24.81). Models with correlated errors between adjacent measurement occasions also did not fit the data because they had higher BIC values and $\chi^2$ statistics. Under these considerations, the random intercept model would be slightly preferable to the single trait model usually specified in state-trait models (Table 12.7).

The random intercept state-trait model is shown in path diagram form in Figure 12.7.

The purpose of alternatives to an initially entertained state-trait model highlights the fact that the measurement models introduced in Chapter 10 can be introduced as components of any larger structural model and can be applied to either measurement occasions or different manifest variables. In this case, it would seem that the choices open to the researcher are between the traditional single factor model and one that assumes that a stable, tau equivalent factor exists, which explains somewhere between 42% and 56% of the variability in psychological distress scores and a more time-specific factor that is initially negative at earlier measurement occasions and positive at

**TABLE 12.7**

Parameter Estimates and Fit Statistics for Random Intercept Model

| Variable | Random Intercept Factor | Trait | Error Variance |
|---|---|---|---|
| Unstandardized Estimates | | | |
| T1 | 1 | −0.17 | 0.19 |
| T2 | 1 | −0.18 | 0.11 |
| T3 | 1 | −0.05 | 0.14 |
| T4 | 1 | −0.02(n.s.) | 0.13 |
| T5 | 1 | 0.08 | 0.13 |
| T6 | 1 | 0.11 | 0.11 |
| T7 | 1 | 0.15 | 0.14 |
| T8 | 1 | 0.09 | 0.14 |
| Standardized Estimates | | | |
| T1 | 0.65 | −0.27 | 0.51 |
| T2 | 0.72 | −0.32 | 0.38 |
| T3 | 0.72 | −0.10 | 0.46 |
| T4 | 0.73 | −0.05 | 0.46 |
| T5 | 0.73 | 0.15 | 0.44 |
| T6 | 0.75 | 0.20 | 0.39 |
| T7 | 0.70 | 0.26 | 0.45 |
| T8 | 0.72 | 0.16 | 0.46 |

*Note*: Trait Factor Variance fixed to 1. Random Intercept Variance = 0.15, $\Psi_{1,2} = 0$. Underlined values are not statistically significant.

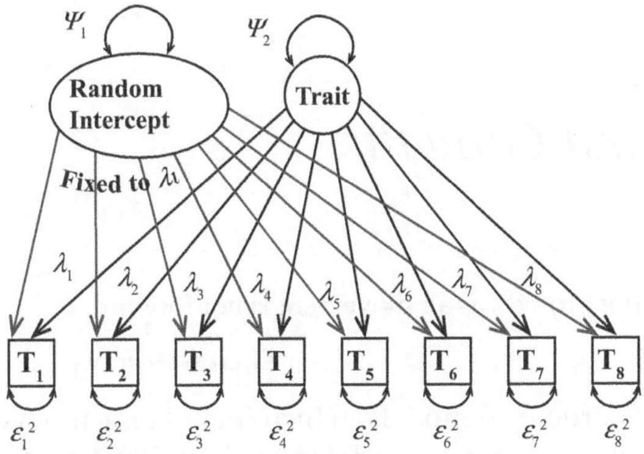


**FIGURE 12.7**
Random intercept state-trait model.

later ones. As such, it may be that some individuals are simply systematically higher in psychological distress or, alternatively, that psychological distress is a growth process. Specifically, the loadings here suggest that a latent basis growth model discussed in Chapter 13 may underlie the data. As mentioned earlier, other models could be explored, such as correlated simple structure factors or hierarchical models. More parsimonious models could involve tests of whether subsets of freely estimated factor loadings or autoregressive paths are equal across measurement occasions.

## Summary

Complicated latent variable models, in which manifest variables have multiple factor loadings may not be interpretable. Unrotated exploratory factor models, discussed in Chapter 15, represent such factor models. Specification of latent variable models that are parsimonious represent a variety of possible conceptual alternatives. Although alternatives are sometimes nested, other comparisons would require use of the BIC or other information-based fit measures. The Bayes factor, for example, may provide some insight as to whether the statistically significant differences between models document substantial differences, although adjudication between models is a trade-off between conceptual interpretability and model fit.

The general idea of specifying various parsimonious models as summaries of variation and covariation can also be extended to models seeking to explain mean level as well. These models, which include growth curve models, hierarchical linear models, and extensions of the general linear model, are taken up in the next chapter.

# 13

# *Change and Growth*

Change is not growth, as all movement is not forward.

**– Ellen Glasgow**

We now turn to structural models, which include mean level in addition to variation and covariation, i.e., models based on the average sum of squares and cross-products. These models are used to explain the general elevation of scores, change, and growth. Although these terms are often used interchangeably in informal usage within the context of this presentation, they describe different mathematical phenomena. Often longitudinal data are characterized by heteroscedasticity (i.e., differences in variability) across measurement occasions. Individual differences in trajectories across measurements characterize growth and decline processes, such as the number of individuals adopting a new technology (Banks, 2013), or increased use of screen time by adolescents (Rosenberg et al., 2018).

Within longitudinal SEMs with latent variables, three possibilities exist for variation across measurement occasions: (1) No appreciable differences exist across measurement occasions, in which case mean differences across occasions are associated with the manifest variable intercepts. (2) Differences in variation across measurement occasions are due to differences in estimated error variances associated with each measurement occasion, i.e., "error heteroscedasticity." And (3) when factor means are present in the model, factor loadings parsimoniously summarize the mean level of the measurement occasion (i.e., the compound path from the column of ones multiplied by the respective factor loadings of the model) as well as variation due to the latent variable (i.e., the compound path from the estimated factor variance multiplied by the factor loading). The word "change" is used to describe intercept-level effects, and the terms "growth" or "decline" are used to describe the third case of some linkage between estimated elevation and variability across measurement occasions.

Conceptually, growth figures in much research involving growth. Shadish, Cook, and Campbell (2002) referred to as the "rich get richer" phenomenon, in which individuals with initially higher levels of performance appear to increase even more across measurements. By contrast, the opposite pattern can be observed if large initial variation is due to deprivation in some individuals that is subsequently remediated, a phenomenon termed


DOI: 10.1201/9781003595267-13

"catch-up growth" (e.g., Womack, Beam, Giangrande, Turkheimer, & Emery, 2023). Similarly, in patterns of decline across measurement occasions, the rate or onset of such decline can be different across individuals, leading to increased variability in scores and lower average performance across measurement occasions (Corley et al., 2023).

Models of change and growth SEMs are illustrated using data from a four-wave longitudinal study of 110 African American and 163 Caucasian children who were administered the WISC vocabulary test during their first-, second-, fourth-, and sixth-grade years (Osborne & Lindsey, 1967; Osborne & Suddick, 1972). From these data, responses from 204 children with complete records are used. This data set has been used as an example in several papers dealing with longitudinal structural models (e.g., Ferrer, Hamagami, & McArdle, 2004; Ghisletta & McArdle, 2012; McArdle & Epstein, 1987; Wood, Steinley, & Jackson, 2015). A variety of models for testing hypotheses about growth and change exist and, in addition to comparing these as candidate SEMs, selection of SEMs involves assumptions about the nature of time-bound individual differences as well as measurement assumptions.

## Models of Change

Perhaps one of the most widely used change models is repeated measures analysis of variance (RM). The RM model is appropriate for testing hypotheses of change for relatively small sample studies because it is a parsimonious SEM, and therefore a more powerful statistical test than MANOVA (multivariate analysis of variance, described later). The measurement model for RM is the strictly parallel measurement model outlined in Chapter 10. In repeated measures models, the hypothesis of equal covariances and error terms is referred to as "compound symmetry" (CS). The RM model tests the hypothesis of no change versus change by comparing the relative fit of a model which constrains mean level to be the same across measurement occasions with a model in which mean level is allowed to vary.

### Repeated Measures Null Hypothesis

The null hypothesis in repeated measures for the children's vocabulary example is shown in Figure 13.1. As can be seen in sub-figures (a) and (b), the assumption of equal means can be modeled equivalently by either estimating a factor mean or by estimating individual intercepts set to equality. Picturing the null hypothesis in these two ways helps to make relationships between the null hypothesis and other structural models of time-bound mean effects.

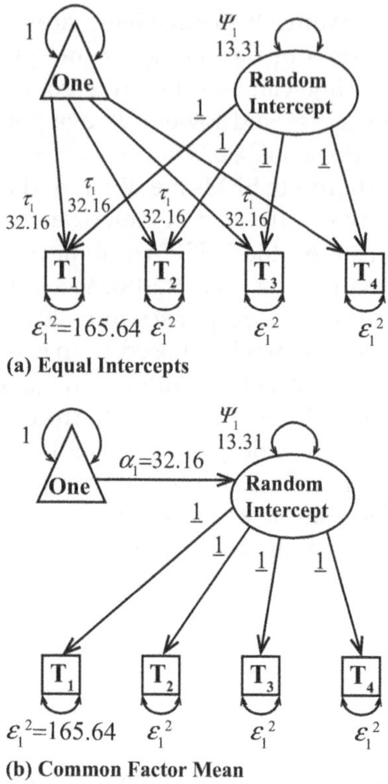


**FIGURE 13.1**
Repeated measures null hypothesis. (a) Equal intercepts. (b) Common factor mean.

## Repeated Measures Alternate Hypothesis

The change hypothesis that mean differences exist across the measurement occasions under the assumption of compound symmetry is shown in Figure 13.2. As can be seen, sub-model (a) differs from the corresponding sub-model in Figure 13.1 by allowing the intercepts for vary over time. It is also possible to estimate the mean differences relative to a particular measurement occasion. Sub-model (b) in Figure 13.2 demonstrates how a structural model can be expressed that estimates the difference among $T_2$ through $T_4$ relative to $T_1$.

The model fit test comparing the alternate to the null hypothesis is the difference $\chi^2$ between the null and alternate hypothesis, which in this case is $\chi^2_3 = 1{,}520.48$, which is highly significant. The difficulty, however, is that, from an SEM perspective, the repeated measures alternate hypothesis is also a poorly fitting model, with RMSEA of 0.33 (90% CI: 0.289–0.371) and TLI of 0.84. Given that the assumption of compound symmetry is unlikely, statistical adjustments to the $p$-values of the model are used to adjust for this violation such as the Geisser-Greenhouse and Hyunh-Feldt adjustments.

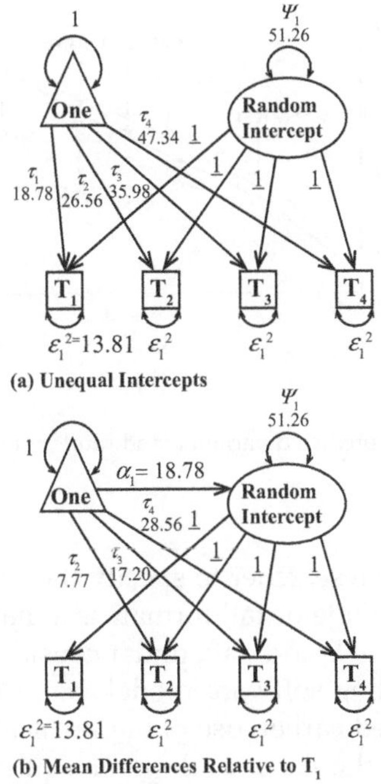


**FIGURE 13.2**
Repeated measures alternate hypothesis. (a) Unequal intercepts. (b) Mean differences relative to $T_1$.

The RM alternate model assumes that individual differences apart from measurement error are due only to differences in the manifest variable intercept across measurement occasions. Figure 13.3 shows the parallel lines associated with ±2.58 standard deviations above or below the average curve based on only the random intercept factor as well as to random intercept and error variances predicted under the model. The message of the figure is that the RM alternate model assumes that, apart from measurement error, differences between observations are the same across measurement occasions.

## "Wide" Versus "Tall" Data

The longitudinal data considered in the models here are classified as being in "wide" format, meaning that each row of the dataset corresponds to an individual with variables corresponding to measurement occasions. Alternatively, the data may be written in "tall" format in which each observation corresponds to an individual measurement occasion, with variables consisting of a number indexing the measurement occasion and a subject identifier. Statistical software for hierarchical linear models and mixed

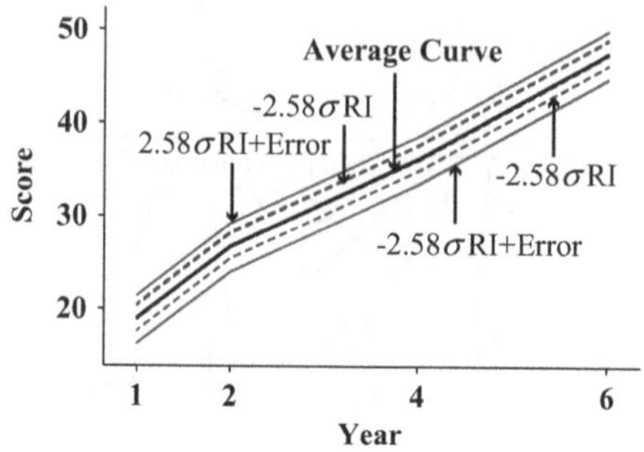


**FIGURE 13.3**
Predicted mean, ±2.58 SD for predicted variation and random intercept variation based on RM alternate model.

models often allows the researcher to specify their models in either of these formats, and the use of wide or tall formats is a matter of convenience relative to the particular models and data under consideration. In this sense, just as general linear modeling software model the just-identified SEMs of the simpler models discussed earlier, use of hierarchical linear and mixed modeling software produces the same results as an equivalent specification of the model using SEM software apart from the adjustments to degrees of freedom mentioned earlier. Grimm, Ram, and Estabrook (2017) present several examples of growth curve models expressed as such mixed and hierarchical linear models using Mplus and a variety of packages in R. Hierarchical linear model versions of such change scores relax the assumption of compound symmetry by imposing other error structures on the data (such as autoregressive relationships) or by allowing the researcher to include other random effects in the model.

## MANOVA

Rather than assume a different covariance structure for the data, it is also possible to compare models, testing whether mean differences exist while making no assumptions about variation and covariation across measurement occasions. In large sample studies comparing a model in which no mean effects exist across time with a model in which means are free to vary is a corresponding test of the main effect of time in the data. For a simple one-way design, the model with freely estimated means, variances, and covariances is the saturated model, making the $\chi^2$ difference test simply the $\chi^2$ for the null model. For the WISC data, the $\chi_3^2 = 625.00, p < 0.001$. Thus, the hypothesis of no mean change is also rejected using this agnostic model.

## Models of Growth

### Factor Mean (FM) Model

One alternative to the change model of RM is to consider the repeated measurements are ratio-level measurements conducted in the same scale and that evidence for different trajectories of growth exist in the data. The Factor Mean (FM) model, first proposed by McArdle and Epstein (1987), represents such measurement assumptions (introduced as the CURVE model). Figure 13.4 shows the FM model for the WISC data set. For purposes of parsimony, the error variances at the first and second measurement occasions were fixed to equality, given that they appeared similar across the measurement occasions, but were allowed to vary at the third and fourth measurement occasions.

Conceptually, the FM model is characterized by two properties. First, having "none" of the latent variable (i.e., having a factor loading of zero on the factor) is associated with a score of zero, in keeping with the assumption that the manifest variables document a ratio level of measurement. Second, mean levels at each measurement occasion are coupled with changes in variability in scores. Figure 13.5 demonstrates this general shape for example simulated curves based on the FM estimates for these data. The average curve is denoted by the indicated solid line. The solid lines denote the shapes of curves which are 2.58 and −2.58 standard deviations away from the mean. It should be noted, however, that in real data, such data points may not exist, as no individuals necessarily will fall above or below the average curve to such an extent.

The FM model makes no assumptions regarding the functional form of growth over time, in contrast to the linear or quadratic growth models to

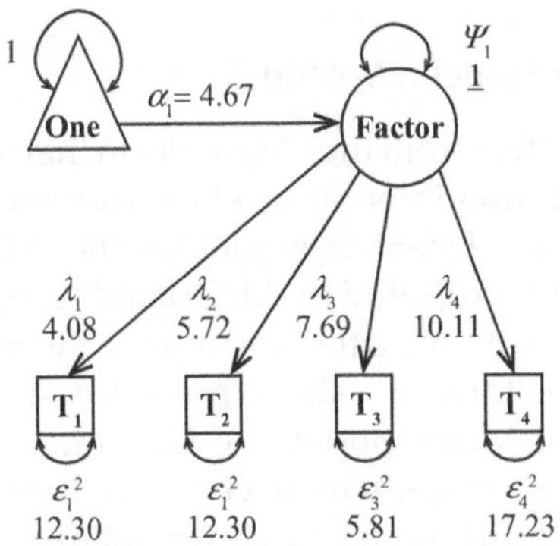


**FIGURE 13.4**
Factor Mean (FM) model.

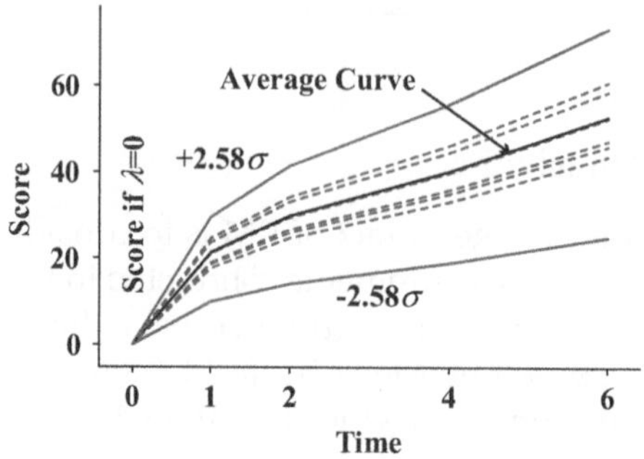


**FIGURE 13.5**
FM example curves.

be discussed shortly. It is entirely possible to fit FM models with linear or quadratic constraints, although in practice such models often do not fit well. In practice, however, the FM model may be appropriate in situations where a zero intercept as a function of time is reasonable – for example, if measuring the number of cells in a sample of organisms or branches on a sample of trees as a function of time, thinking that the process started at zero is reasonable. The FM model may be preliminary to a parametric form, such as linear, quadratic, or other nonlinear models (see for example Grimm et al., 2017; Wood et al., 2015). However, if random events occur across measurement occasions such as weather events, or for psychological data changes in school or residential environment, the nonparametric form of growth over time may be reasonable given that the form of growth will reflect these factors. For the WISC data, it appears that a markedly increased rate of growth occurs between the first and second measurement occasions with relatively linear growth after that.

### The Factor Mean-Shift Model (FM-Shift)

The FM-Shift model differs from the FM model in that it assumes that the manifest variables are common interval-level measurements. Figure 13.6 shows the path diagram for the FM-Shift model for the WISC data. The FM-Shift model was actually first proposed by McDonald (1967), who referred to it as the Linear Factor model because the loadings of the manifest variable on the latent variable have the form of a single predictor regression model.

Similar to the Factor Mean model, the FM-Shift model assumes that the manifest variables across measurement occasions represent quantitative measurements. In the FM-Shift model, however, manifest variables are thought to be assessed in a common interval scale. For example, if temperatures are recorded across measurement occasions, these assessments constitute ratio-level scales only if the units are expressed in a scale with an absolute zero,

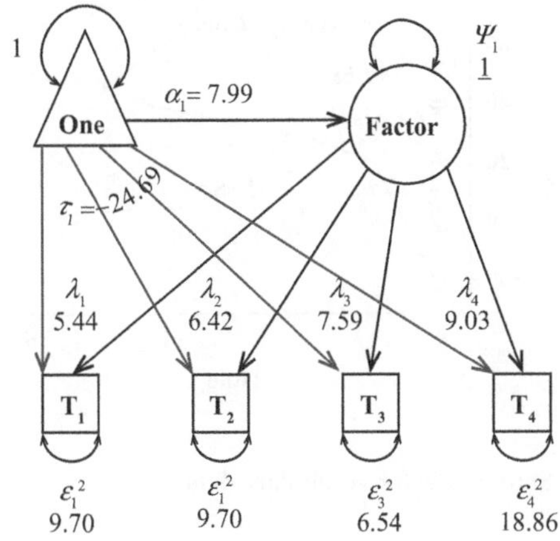


**FIGURE 13.6**
Factor Mean-Shift model (FM-Shift).

as it is with the Kelvin temperature scale. If, by contrast, measurements are conducted in Celsius or Fahrenheit scale, changes in temperature do not represent a change in the ratio of molecular motion, but only equally spaced units. A researcher, for example, who wishes to study how the temperature of gases affects their pressure needs to re-express any temperature readings into Kelvin in order to make measurements in which temperature is proportional to gas pressure. For many psychological scales, the FM-Shift model would seem a reasonable alternative to the FM model, due to the arbitrary nature of the item scales used to assess abilities and the fact that the specific items that compose scales are not designed to construct a continuous ratio level measurement.

The FM-Shift for the Vocabulary data is shown in Figure 13.6. Paths from the intercept to the manifest variable are constrained to equality, and the value −24.69 is the adjustment to the WISC vocabulary scores, which would bring them into a ratio scale. The $\alpha_1$ value of 7.99 represents the mean of the unit variance latent variable with loadings to each of the measurement occasions. The $\tau_1$ value of −24.69 represents the estimate of the value needed to adjust the vocabulary scores to bring them into ratio-level measurement.

Figure 13.7 shows exemplar curves under the FM-Shift model, beginning with the (extrapolated) initial measurement occasion at Time 0, which corresponds to the measurement occasion associated with a loading of zero. As can be seen, the pattern is similar to the example curves for the FM model shown in Figure 13.5, in that a general pattern of increasing variance accompanied by increases in overall level are observed. In addition, all curves mathematically converge to a common (extrapolated) point corresponding to a loading of zero. However, for the FM-Shift model, this zero point on the scale corresponds to a raw score value of −24.69 instead of zero as for the FM model.

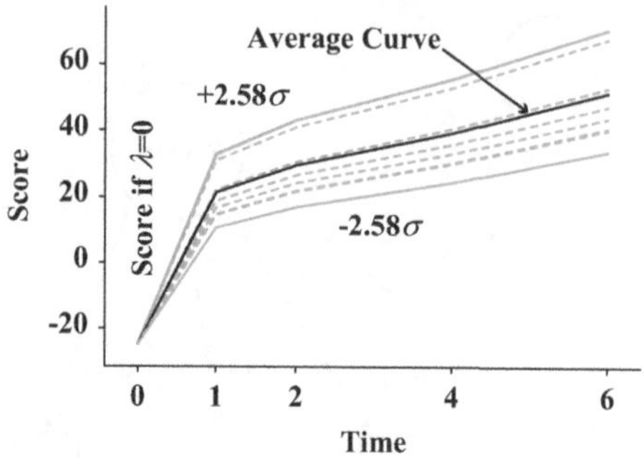


**FIGURE 13.7**
Example curves for FM-Shift model for vocabulary data.

## Orthogonal Free Curve Slope Intercept Model (FCSI)

Suppose that an FM-Shift model was fit to data, but that the shift value for the scale differed across individuals. The resulting model can be characterized as a Free Curve Slope Intercept model, also referred to as a Latent Basis growth curve model (e.g., Meredith & Tisak, 1990). I do not use the term "Latent Basis" for this model, as the FM and FM-Shift models may also be described as having a latent basis function in that no parametric form is assumed for growth across measurement occasions.

As with any factor model, it is possible to specify the FCSI model equivalently in orthogonal or oblique forms with identical model fit (Meredith & Tisak, 1990). In its orthogonal form shown in Figure 13.8, the mean of the Intercept factor corresponds to the average shift across all individuals

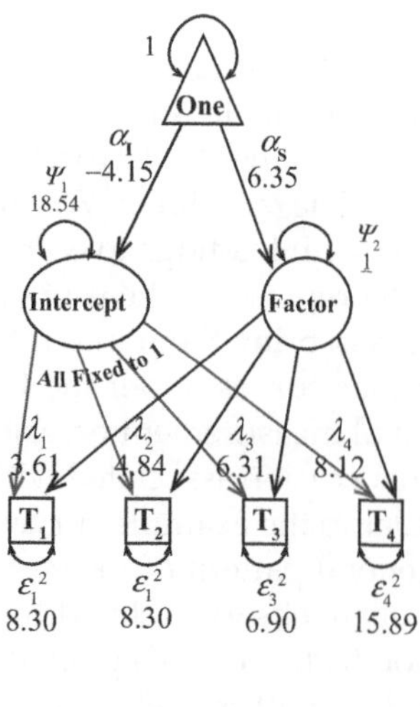


**FIGURE 13.8**
Orthogonal free curve slope intercept (FCSI) growth.

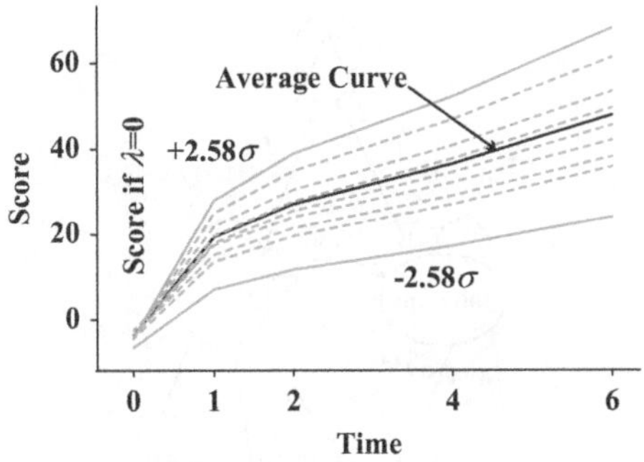


**FIGURE 13.9**
Examples of free curve slope intercept curves.

required to bring the measure into ratio scale. As was done for the other growth models, exemplar curves for the model are shown in Figure 13.9. As can be seen, the shape of these curves is similar to those shown for the FM and FM-Shift models in Figure 13.5 and 13.7, except that there is variability in the shift values associated with the measurement occasion at Time = 0, meaning that the value needed to bring scores into a ratio-level measurement is different for the different children or, alternatively, that some unmeasured variable(s) exist that systematically elevate or depress scores across measurement occasions.

## Oblique Free Curve Slope Intercept Model

It is also possible to specify the FCSI model to understand how growth/decline is correlated with the Intercept factor at a particular measurement occasion. For example, a researcher might be interested in knowing if overall level of score at the first measurement occasion is related to the steepness of the subsequent trajectory. Alternatively, a researcher might be interested in knowing if overall level of elevation at the last measurement occasion is associated with the steepness of the pattern of growth/decline. To specify this model, the slope factor loading associated with an individual measurement occasion is zeroed out, and a covariance is estimated between the intercept and slope factors as shown in Figure 13.10.

In this figure, the covariance between the intercept and slope factors, 3.61, corresponds to a correlation of 0.64, $r_{I,S} = \dfrac{3.61}{\sqrt{31.59(1)}}$. Note that the error variances and predicted means as well as the predicted means and covariances of this model are identical to those of the orthogonal FCSI model in Figure 13.8 and that, as shown later, the fit of the model is exactly the same as for the orthogonal solution shown in Figure 13.8.

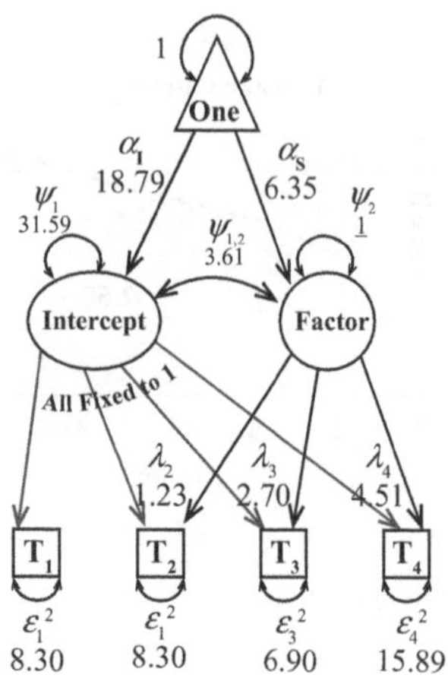


**FIGURE 13.10**
Oblique free curve slope intercept model.

## Orthogonal Versus Oblique Growth

Many growth curve studies using an SEM frequently use an oblique factor growth model. Often, in studies of learning or problem behavior, researchers want to interpret their solution as informing the value of early interventions in promoting or inhibiting individuals' subsequent trajectories. It is important that there is a clear justification for specifying the growth model relative to a particular measurement occasion. For example, if the researcher is interested in performance on a memory or reasoning task introduced on the first measurement occasion, it would seem reasonable to document performance relative to the first exposure to the stimulus materials. It is probably less justified to apply such a model to the vocabulary test used for the example here, as the performance of children at the first measurement occasion probably reflects growth, which is already ongoing by the time of the first assessment. Use of an oblique factor model relative to the first measurement occasion is useful as a conjecture as to what might happen if one were able to somehow intervene at the first measurement occasion in hopes of elevating the random intercept factor and, indirectly, affecting the subsequent trajectory. Intervening at the first measurement occasion may only result in a temporary occasion-specific elevation in scores, i.e., an increase in the manifest variable intercept associated with this measurement occasion, and this would have no effect whatever on subsequent growth or course of development. The oblique growth measurement model is an optimistic prediction of the value of intervention at the initial time period which highlights an avenue for future research.

Conceptually, specifying the last measurement occasion as the reference point may also have value. Such equally well-fitting models explain how individuals came to have their final scores as a function of their overall level and prior trajectories (Rovine & Molenaar, 1999; Wood et al., 2015). Such questions involve teleological explanations of growth, i.e., how a behavior "came to be." For example, in the case of a behavior such as problematic alcohol use, one might be interested in explaining how final level of problematic alcohol use is a function of the slope and intercept latent variables and whether the prior trajectory is related to the random intercept factor.

Finally, using the orthogonal growth model may have greater statistical power than oblique forms when these factors are exogenous variables in a larger model. Similar to the discussion of orthogonal versus correlated predictor variables in Chapter 4, use of orthogonal predictors eliminates covariances between parameter estimates associated with the Intercept and Slope exogenous latent variables. Side analyses of oblique growth models which zero out different measurement occasions can produce different patterns of statistical significance in such larger models, but orthogonal specification of growth provides the most statistically powerful tests of these two components.

## Linear Slope Intercept Model (SI)

In many analyses, the form of growth is represented as a straight line rather than the unconstrained factor models with estimated mean. Although few theoretical models explicitly predict that any change necessarily increases in only linear fashion, such models have three advantages. First, for smaller data sets, the reduction in the number of parameters may afford greater statistical power for tests of hypotheses. Second, the notion of growth as a constant rate of change over measurement occasions is approximated by a growth factor, which is why the growth factor is commonly referred to as a slope. Third, in some computer programs such as hierarchical linear modeling software it is possible to write growth models in a "tall" format for studies in which assessments are unevenly spaced over time across individuals. Provided that at least three measurement occasions are present for each individual, it is possible to estimate intercept and slope variance components for each individual. This assumes, of course, that the functional form of the slope is linear.

The most commonly form of the Linear Slope Intercept model is similar to the Oblique FCSI model in that it fixes a reference measurement occasion to zero. In Figure 13.11, for example, the first measurement occasion is taken as a reference point. Subsequent loadings on the slope factor for the other measurement occasions are fixed to the amount of time between the reference measurement occasion and the occasion. In Figure 13.11, the remaining factor

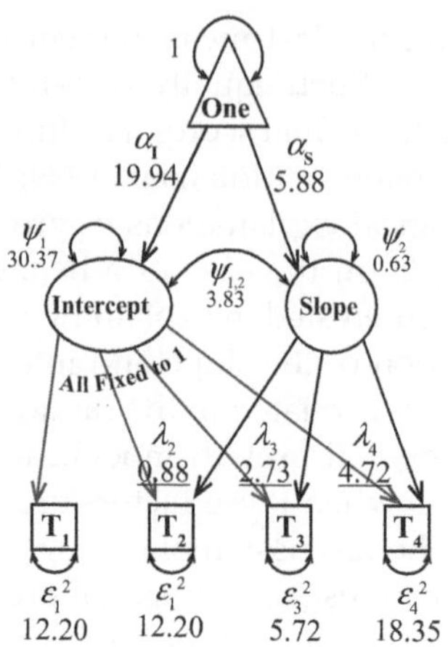


**FIGURE 13.11**
Linear slope and intercept growth.

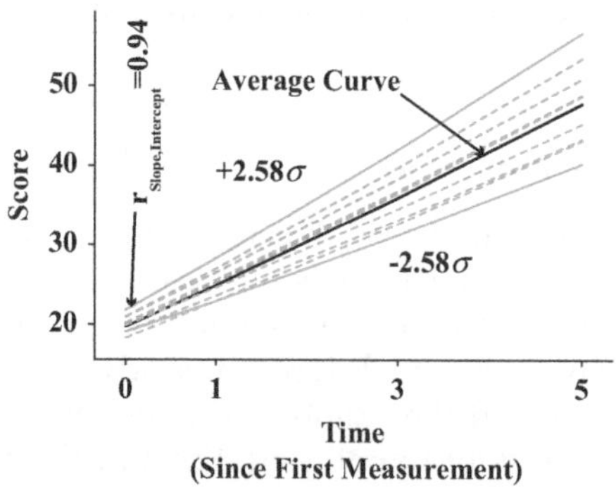


**FIGURE 13.12**
Example linear slope intercept growth.

loadings are fixed to represent that the other assessments occurred 0.88, 2.73, and 4.72 years after the initial assessment.

The linear functional form of growth assumed by the model is shown in Figure 13.12. Note that the horizontal axis shows elapsed time since the first assessment and not age. A substantial correlation between the intercept and slope factors at the first measurement occasion is observed.

## Hierarchical Linear Models (Scaled to Age)

When hierarchical linear modeling software is used, data are often stored in "tall" format, with each observation consisting of a subject identifier, dependent variable, and age of participant. The resulting model is equivalent to the Linear Slope Intercept model, except that the estimated variances and covariances associated with growth are scaled to the zero point of age, e.g., birth age. This is most often a substantial extrapolation from the observed data, and such data are perhaps better scaled to age relative to age relative to the earliest observed age. Hierarchical linear models are, however, no different form growth SEMs and can be represented as a path diagram. The diagram for this data set is shown in Figure 13.13. As before, the correlation between the random intercept and slope factors can be calculated as $0.21 = \frac{0.31}{\sqrt{3.44^{*}0.63}}$. Fit measures for this model are identical to those obtained for the Linear Slope Intercept growth model. Figure 13.14 shows the form of growth under this model, highlighting the fact that the correlation between the random intercept and slope factors is now taken relative to chronological age = 0.

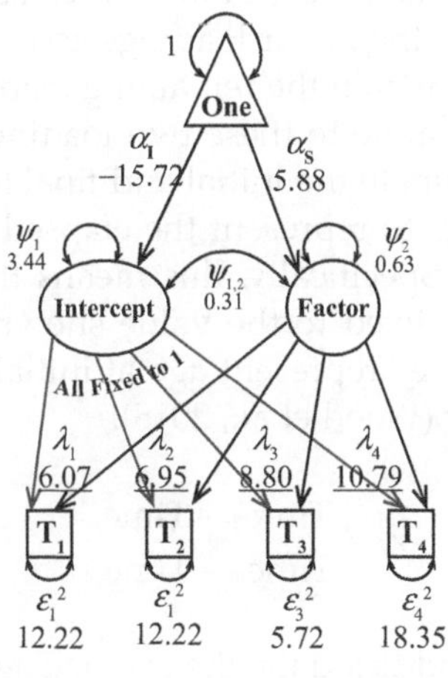


**FIGURE 13.13**
Hierarchical linear model (age as time).

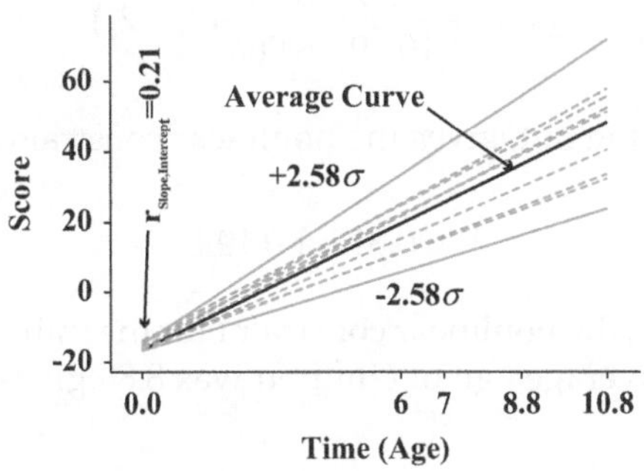


**FIGURE 13.14**
Example hierarchical linear model growth (age as time).

## Orthogonal Linear SI Model

Although most presentations of linear growth are parameterized relative to a particular measurement occasion such as birth (in the case of HLM models scaled in terms of age), or the initial or final measurement occasion, it is also possible to estimate a linear growth model scaled relative to that point where scores on the growth lines converge to their point of smallest variance and for which the correlation between the random intercept and slope factors is zero. Hancock and Choi (2006) discuss how this zero reference point of growth functions as an "aperture" i.e., that measurement occasion which documents a flip in the covariance between the slope and intercept factors. If reference measurement occasions are chosen at a value greater than the aperture, the covariance between intercept and slope will be opposite in sign to that estimated from a reference measurement occasion less than the aperture.

To estimate an Orthogonal Linear SI model scaled relative to the aperture, it is necessary to estimate the factor loadings associated with two measurement occasions and to constrain the remaining loadings to be linear interpolations/extrapolations relative to these two loadings. For convenience, it is often easiest to freely estimate the initial and final loadings and to constrain the intermediary loadings to represent the elapsed time between these two measurement occasions. Specifically, this means that for any intermediate loading $\lambda_i$, the loading is fixed to the value shown in Eq. (13.1), assuming that $\text{Time}_1, \text{Time}_i,$ and $\text{Time}_T$ represent age at initial, intermediate, and final assessments, respectively (Wood et al., 2015):

$$\lambda_i = \frac{\text{Time}_T - \text{Time}_i}{\text{Time}_T - \text{Time}_1} \tag{13.1}$$

For the Osborne and Suddick data, the average age at first assessment was 6.066 and at the last measurement occasion was 10.79. The constraint associated with the second measurement occasion when the average was 6.946 is then:

$$\lambda_2 = \lambda_1 + \frac{10.97 - 6.946}{10.79 - 6.066}(\lambda_t - \lambda_1)$$

which, on multiplying out, gives the nonlinear constraint:

$$\lambda_2 = 0.81\lambda_t + 0.19\lambda_1$$

In similar fashion, the nonlinear constraint for the third measurement occasion at which the average age of children was 8.80 gives the nonlinear constraint of

$$\lambda_3 = 0.42\lambda_t + 0.58\lambda_1$$

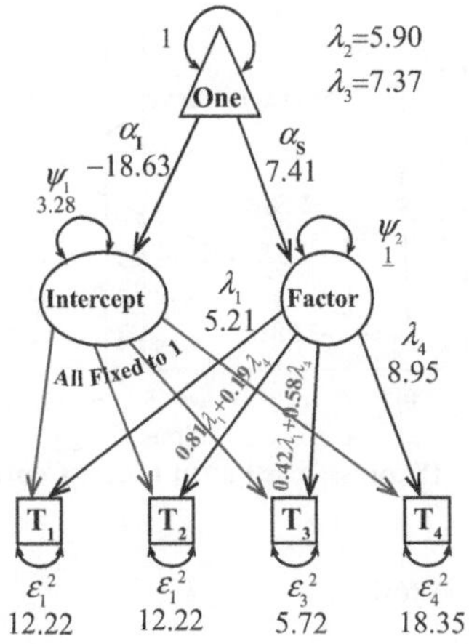


**FIGURE 13.15**
Orthogonal linear SI model.

These nonlinear constraints are shown in gray in Figure 13.15 with the estimated values shown in their usual positions. In order to estimate the Orthogonal Linear SI model, the statistical software package used either needs to be able to impose such constraints on the loadings (such as Mplus) or, alternatively, phantom variables will need to be included in the diagram to incorporate these constraints See Rindskopf (1984) for a description of how inclusion of a series of latent variables with zero estimated variance can be used to express multiplicative constraints on path parameters.

Although this model has not been presented in substantive research applications (other than Wood et al., 2015), it is worth noting that the factor loadings from this model represent a latent continuum of time at the level of the developing individual, sometimes termed "organismic time" (Riegel, 1977; Salthe, 1985; Slife, 1993). If this is the case, and if it were reasonable to assume that growth in vocabulary followed a linear latent trajectory, the estimated time loadings of the orthogonal linear SI model appear to be about one year earlier than the chronological age of the children, suggesting that the point of origin for such a scale occurs when the children are approximately one year old. Alternatively, one could, if desired, respecify the model by fixing one of the loadings to represent its chronological age, free the slope factor variance, constrain the intermediary measurement occasions as before, and interpret the remaining loadings as latent ages scaled relative to that particular time point. This is shown graphically in the example individual trajectories shown in Figure 13.16.

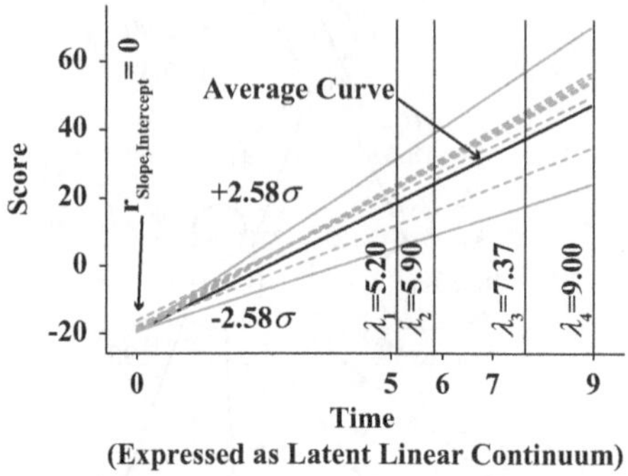


**FIGURE 13.16**
Orthogonal linear SI model trajectories.

## Quadratic Slope Intercept Growth

As a final growth model, researchers have often employed a quadratic growth model composed of random intercept, linear, and quadratic latent variables. In these models, an additional latent variable is included with factor loadings constrained to be the square of their respective linear slope values, with covariances between all latent variables freely estimated. The path diagram for the children's vocabulary example is shown in Figure 13.17.

As can be seen from the diagram, the estimated parameters are improper, with estimated negative variances for the slope and quadratic factors. In some situations, such improper solutions can be dealt with by rescaling the time variable to fall on the interval zero to one, with intermediate loadings

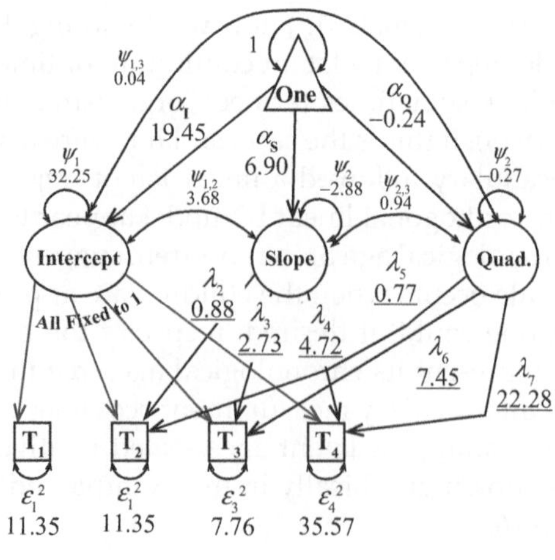


**FIGURE 13.17**
Quadratic linear slope intercept model.

**TABLE 13.1**

Fit Statistics for Growth and Change Models for Vocabulary Data

| | RM | FM | FM-Shift | FCSI | Oblique FCSI | SILinear | Orthogonal SI Linear | SI Quadratic |
|---|---|---|---|---|---|---|---|---|
| $\chi^2$ | 184.94 | 64.82 | 13.74 | **7.72** | **7.72** | 79.49 | 79.74 | 55.48 |
| (*df*) | 8 | 6 | 5 | **4** | **4** | 6 | 6 | 2 |
| prob($\chi^2$ \| Ho) | 0 | 0 | 0.02 | **0.1** | **0.1** | 0 | 0 | 0 |
| TLI | 0.84 | 0.93 | **0.99** | **0.99** | **0.99** | 0.91 | 0.91 | 0.8 |
| CFI | 0.79 | 0.93 | 0.99 | **1** | **1** | 0.91 | 0.91 | 0.94 |
| RMSEA | 0.33 | 0.22 | 0.09 | **0.07** | **0.07** | 0.25 | 0.25 | 0.362 |
| (90% CI) | (0.29-0.37) | (0.17-0.27) | (0.04-0.15) | **(0.0-0.14)** | **(0.0-0.14)** | (0.20-0.30) | (0.20-0.30) | (0.284-0.447) |
| PRMSEA $\leq$0.05 | 0 | 0 | 0.1 | **0.28** | **0.28** | 0 | 0 | 0 |
| AIC | 5033.7 | 4917.58 | 4868.50 | **4864.50** | 4864.48 | 4932.50 | 4932.24 | 4916.236 |
| BIC | 5053.61 | 4944.13 | 4898.37 | **4897.70** | 4897.66 | 4959.04 | 4958.79 | 4956.053 |

reflecting the proportion of elapsed time relative to the initial assessment. Unfortunately, in this situation estimated factor variances continue to be negative. In these situations, some researchers modify the model by respecifying the offending factor variances to be zero as well as all covariances involving it. Such models will fit the data slightly worse than the original improper solution. Given that, in this situation at least, it appears that the form of growth does not follow a quadratic form, we will compare the relative fit of the models introduced so far.

### Growth and Change Model Fit and Comparison

Having introduced these various growth curve alternatives, a comparison of their fit is of interest. Table 13.1 shows fit statistics associated with the models. Best fitting models according to each fit measure are in bold. As can be seen the FCSI model in either its orthogonal or equivalent oblique form appears to fit the data best. Given that the alternate hypothesis of the MANOVA model is a fully saturated model and the $\chi^2$ test for the FCSI model is not statistically significant, this can be taken as evidence that the MANOVA model does not fit better than the FCSI model.

## Summary

A variety of candidate growth models have been presented here, but they do not exhaust all possible models. For example, if factor loadings appear similar to the magnitude of error variances, the equal precision models discussed in Chapter 10 can lead to the generation of parsimonious alternatives. Change models which assume only tau-equivalent models or precision models are also possible. SEMs are also possible which assume multivariate growth processes (e.g., Bishop, Geiser, & Cole, 2015) or multiple growth curves due to the presence of known (Flora, 2008) or unknown (Liu & Perera, 2022) "knots" which indicate the onset of a new growth process.

This variety of possibilities may lead to the question of whether selection of an appropriate growth model is impractical. A conceptually appropriate model based on the rationale for the study is the safest starting point. However, exploration of other models as side analysis is important, especially if these correspond to the counterarguments of a reasonable skeptic. To keep the number of candidate models tractable, researchers may well wish to consider the general process of "right-sizing" their models to the data at hand (Wood et al., 2015). Under this approach, an assessment is first made of the dimensionality of the series to gain insight into the number of latent variables necessary. Exploration then turns to evaluation of patterned factor models such as the multifactor, random intercept, or other of the parsimonious confirmatory factor models outlined in Chapter 12. Following that, growth models involving mean effects are then added to the model. Occasion-specific "history effects" can then be explored to model measurement occasion effects associated with uniform gains or losses which are unrelated to the latent variable mean effects.

Although the parametric forms of linear and quadratic growth did not fare well in this example, they can be useful conceptual and parsimonious models in other settings. Other parametric functional forms of growth, however, may be more appropriate. See Panik (2014), Wood et al. (2015), and Grimm et al. (2017) for examples. Such models reintroduce an added degree of parsimony to the data but also form the basis for a more precise understanding of the mathematical nature of growth and change patterns in the data.

All models considered here, however, assume that the same SEM parameters underlie all individuals in the dataset. It may be, however, that different factor loadings, variances, and covariances as well as error variances may characterize subgroups of individuals in the data. Such multigroup structural models explore the external validity of a SEM to subpopulations under investigation. For this reason, we will now consider multigroup SEMs.

# 14

# *Multiple Groups*

The purpose of multigroup SEMs is to examine the external validity of a structural equation model to subgroups of individuals. Although the model comparisons involved in multigroup SEMs involve tests of whether parts of the structural model (such as loadings, intercepts, error variances, and factor variances and covariances) are different across groups, the goals of multi-group SEMs can extend beyond simple theory testing of whether an SEM generalizes across groups to questions of moderation (interaction) effects in an SEM as a function of group membership. At a general level, multigroup SEMs can be used to assess whether interventions or theories need to be tailored to specific groups.

Deciding how to specify multigroup SEMs involves the same types of questions of psychometric measurement models discussed in Chapter 10 and the exploration of possible parsimonious models described in Chapter 12. Mean-level differences at the latent or manifest variable level can also be tested. Specification of candidate multigroup models requires taking care to estimate multigroup models according to the identification constraints in Chapter 8.

The psychometric and conceptual questions in multigroup SEMs differ as a function of the variables and research questions. Questions of invariance in studies consisting of repeated measures, as discussed in Chapter 13 on change and growth, are different from those associated with multiple variables measured in different scales, as discussed in Chapter 8 on confirmatory factor models. Some studies are interested only in describing patterns of covariation – i.e., using the deviation score matrix – while other studies seek to describe mean differences across groups. As a result, in the research literature there are various ways to test multigroup differences. Default behavior of SEM software can also reflect these differences, and no assumption should be made that these defaults are the single best way to proceed in all situations. It seems most reasonable to consider multigroup models in increasing order of complexity.

1. If the model contains only manifest variables as in a regression or manifest variable path diagram, do structural parameters such as regression weights obtain across groups? Are there differences in mean level in either the predictor or criterion variables? Even if these


DOI: 10.1201/9781003595267-14

are present, a researcher is simply interested in knowing whether the same unstandardized regression weight(s) hold across groups and takes this as evidence of external validity even if error variances or predictor variances and covariances vary across groups.

2. If the model contains latent variables, is the measurement model comparable across groups? Often it is hoped that manifest variables will at least take the form of the same congeneric measurement across groups, meaning that the latent variable(s) represent the same construct. If this is not observed, this could indicate that the trait itself is different across groups or that some other third-variable explanations are present in one group but are absent or different in another. Finding different manifest variable error variances means that the accuracy of measurement varies across groups but does not constitute a conceptual threat to the external validity of the multigroup SEM.

   For example, if testing a multigroup latent variable of alcohol consumption, a researcher who finds that factor loadings are the same across men and women may have more confidence that the naming problem associated with a latent variable in one group is the same as (or at least no worse than) that associated with the other group. If, however, celebratory drinking is more common among men than women, this could cause the factor loadings between alcohol consumption variables that do and those that don't involve celebratory drinking to mean that the construct of alcohol use varies across group.

3. Can a multigroup model be specified to parsimoniously characterize group differences in mean level and variability in the latent variable? Multigroup SEMs in which latent variable means and, possibly, latent variable variances and covariances can be a useful way to describe group differences. Using the alcohol latent variable example, the multigroup model may summarize how men's alcohol use is higher and more variable than women's and that, perhaps, unstandardized regression weights involving the latent variable may still be found.

4. Can differences in growth, change, or developmental course be characterized via a multigroup model? In prospective designs, research questions may involve tests of whether the functional form of growth/decline is the same across groups and whether inter-individual differences in such growth/decline vary across groups. These multigroup SEMs may differ from the models described in previous chapters in that a growth model is thought to underlie the data rather than that used for a confirmatory factor model in which the manifest variables are assumed measured in different scales.

## Multiple Group Regression

### Invariance Models for Single Predictor Regression

It is easier to reason about multigroup SEMs by considering the simple case of one-predictor regression via scatterplots. For example, suppose the relationship between endorsement of the drinking motive "I drink because it makes gatherings fun" and past three-month alcohol consumption is of interest.

Figure 14.1 shows three possible invariance possibilities across a study of males and females. A model of complete invariance (i.e., a model in which the predictor means, variances, slopes, and intercepts) is not shown, given that the two groups would consist of overlapping pictures. Figure 14.1(a), the same slope, intercept, and conditional variability (i.e., error variance) of the criterion, obtains across the two groups, but the predictor means and variability of the predictor are different across the two groups. If the assumptions of regression are met (as introduced in Chapter 5), one may argue that the same functional relationship exists across groups. Further, the fact that the regression intercept is the same in the two groups suggests that observed

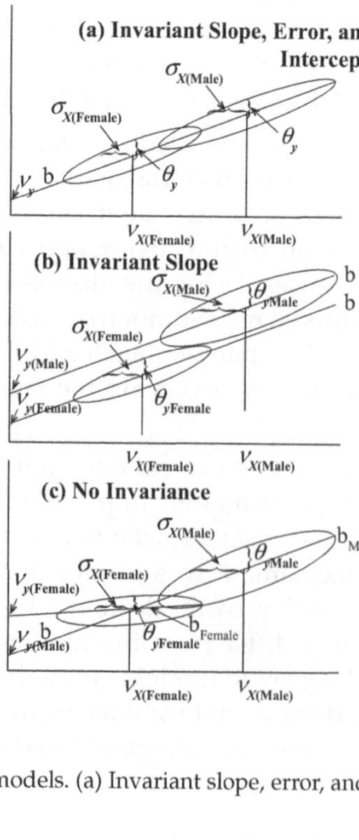


**FIGURE 14.1**
Single predictor invariance models. (a) Invariant slope, error, and intercept. (b) Invariant slope. (c) No invariance.

mean differences in the criterion are accounted for by differences in the predictor variable.

In Figure 14.1(b), a common regression slope still holds between the groups, but the intercepts now vary. As with Figure 14.1(a), a common functional relationship obtains between predictor and criterion across the two groups – i.e., if one were to increase the predictor by a unit, we would expect to see a gain of *b* points on the criterion. However, the fact that the regression intercepts differ across groups implies that this predictor is an insufficient explanation for why men drink more than women; that is, other variables may be present that also cause men to drink more than women. Additionally, it appears that there is greater variability in errors of prediction in men than in women, meaning that standardized regression weights from the two groups may mislead the researcher regarding the functional form in the data (in addition to differences induced by predictor variable variance differences across groups). In a general sense, however, finding that a common slope across groups exists suggests the same relationship between the predictor and criterion across the two groups.

Finally, consider the hypothesis of no invariant parameters as presented in Figure 14.1(c). The general conclusion from this model is that separate regression models would be required across the two groups. It could be, however, that a common functional form exists across groups, but that it is nonlinear in form. It may not be possible to distinguish between a model of between-group differences in regression from a general nonlinear functional form.

## Worked Example: Galton Height Data

We can compare invariance hypotheses for a single regression model using the Galton height data considered in Chapter 3. In that discussion we noticed that the correlation between Child Height and MidParent Height was 0.32. Examination of the scatterplot of the data by sex, however, reveals that a single predictor SEM regression model does not seem reasonable, as shown in Figure 14.2. The figure shows a rather dramatic difference in the regression intercepts with the regression intercept for boys, about 5.2 inches, which is similar to the 4.7-inch difference in height today. It's reasonable to speculate that a difference in slopes could possibly exist, given that in today's third-world countries, nutrition differences exist between men and women (e.g., Nubé & Van Den Boom, 2003), which could affect the relationship.

To test the invariance hypotheses, we can compare the following multi-group SEM regressions:

1. All parameters of the regressions by sex are different (a just-identified model with $\chi^2$ of zero).
2. The regression slopes predicting boys and girls are equal.
3. Model 2 with Midparent Variance also set equal.

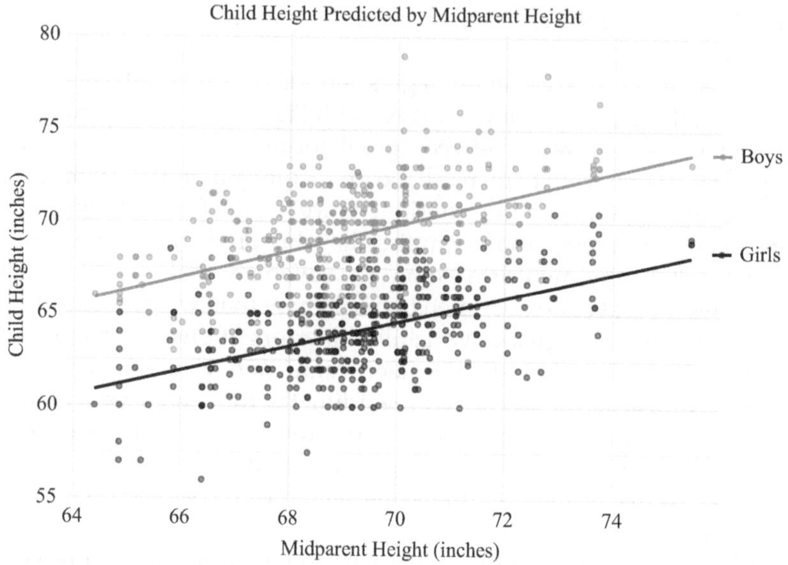


**FIGURE 14.2**
Child height predicted from Midparent Height for boys and girls.

4. Model 3 with Midparent Mean also set equal.
5. Model 4 with Error Variance also set equal
6. All regression parameters set equal.

This particular ordering of models is not unique, but based on Figure 14.2, it seemed reasonable to test the intercept equality last. Table 14.1 shows the $\chi^2$, $\chi^2$ difference test between successive models, AIC, BIC, TLI, CFI, and RMSEA for these models. As can be seen from the table, the model $\chi^2$ for Model 5 is not statistically significant, which would lead us to accept this as the most reasonable model by that criterion. Model 4, in which all parameters are invariant except for the regression intercepts and error variance of Child Height, appears to fit best based on the AIC, BIC, CFI, and RMSEA. The statistically significance $\chi^2$ difference between Models 4 and 5 would also suggest Model 4 is preferable. However, the advantage of estimating separate error variances over constraining them to equality appears slight, with the BIC difference between Models 4 and 5 being negligible according to Raftery's criteria for BIC differences. Taken together, it would appear reasonable to assume that only the regression intercepts differ by sex for these data.

**TABLE 14.1**

Fit Statistics for Sex Differences in Galton Regression Data

| Model | $\chi^2\,(df)$ | $\chi^2$ diff | Prob($\chi^2$ \| $H_0$) | AIC | BIC | TLI | CFI | RMSEA (95% CI) | $P_{(\text{Close Fit})}$ |
|---|---|---|---|---|---|---|---|---|---|
| 1 | 0.00 (0) | | | 7855.20 | 7903.60 | 1.00 | 1.00 | | |
| 2 | 0.45 (1) | 0.46 (1) | 0.499 | 7853.70 | 7897.20 | 1.00 | 1.00 | 0.00 (0.00–0.11) | 0.70 |
| 3 | 0.86 (2) | 0.33 (1) | 0.567 | 7852.10 | 7890.80 | 1.00 | 1.00 | 0.00 (0.00–0.07) | 0.86 |
| 4 | 1.96 (3) | 1.1 (1) | 0.294 | **7851.20** | **7885.10** | 1.00 | 1.00 | 0.00 (0.00–0.07) | 0.87 |
| 5 | **9.58 (4)** | 6.35 (1) | 0.012 | 7856.80 | 7885.80 | 0.99 | 0.98 | **0.05 (0.00–0.10)** | **0.36** |
| 6 | 844.83 (5) *** | 509.81 (1) *** | < 0.001 | 8690.00 | 8714.20 | –0.27 | 0.00 | 0.60 (0.57–0.63) | 0.00 |

Note: *** = $p < 0.001$

### Invariance Models for Multiple Regression Models

Consider now a two-predictor regression model in which the dependent variable is self-reported alcohol consumption in the past three months, and two motives for drinking are used as predictors: "because it gives me a pleasant feeling" and "because it makes gatherings fun." In the discussion of analysis of covariance, we mentioned that interactions (i.e., "moderator effects") can be tested under a traditional regression model by including two additional interaction vectors consisting of the product of sex and the two reasons, and testing whether this model additionally predicted alcohol use. Notice, though, that such a model would assume that the predictor variable variation and covariation as well as variability of errors of prediction are the same for men and women. While this may be acceptably parsimonious for small-sample studies, it is unlikely to be the case for larger-sample research.

The data for the analysis is given in Table 14.2. As can be seen from the table, men appear to consume more alcohol and to be more variable in their

**TABLE 14.2**

Descriptive Statistics for Male and Female Freshmen's Past Three Month Alcohol Consumption and Two Reasons for Drinking

| | Past 3-Month Alc | Makes Gatherings Fun | Pleasant Feeling |
|---|---|---|---|
| Males $(N = 1{,}724)$ | | | |
| Means | 4.894 | 3.125 | 2.983 |
| Standard Deviations | 3.425 | 0.854 | 0.908 |
| | Variance Covariance Matrix | | |
| Past 3-Month Alc | 11.730 | | |
| Makes Gatherings Fun | 0.788 | 0.729 | |
| Pleasant Feeling | 0.745 | 0.452 | 0.825 |
| | Correlations | | |
| Past 3-Month Alc | 1.000 | | |
| Makes Gatherings Fun | 0.350 | 1.000 | |
| Pleasant Feeling | 0.306 | 0.582 | 1.000 |
| | Females $(N = 1{,}991)$ | | |
| Means: | 3.402 | 3.052 | 2.784 |
| Standard Deviations | 2.531 | 0.875 | 0.948 |
| | Variance Covariance Matrix | | |
| Past 3-Month Alc | 6.408 | | |
| Makes Gatherings Fun | 0.770 | 0.765 | |
| Pleasant Feeling | 0.703 | 0.426 | 0.899 |
| | Correlations | | |
| Past 3-Month Alc | 1.000 | | |
| Makes Gatherings Fun | 0.404 | 1.000 | |
| Pleasant Feeling | 0.341 | 0.514 | 1.000 |

reported consumption than women. Men appear to endorse the reasons for drinking a bit more than women but appear to be slightly less variable in their endorsements of the reasons for drinking. The correlation between the two reasons appears to be slightly higher for men (0.582) than for women (0.514).

Analogous to the case of a single predictor regression, we might be interested in knowing if the same functional relationships exist between these two predictors and alcohol use across men and women. Conceptually, we might speak of the same predicted policy implications of interventions designed to decrease the drinking motivation "because it gives me a pleasant feeling" regardless of sex of participant. Variability in the predictors and the covariance between the variables may be different across groups. The standardized regression coefficients in the two groups could be substantially different even though the unstandardized regression weights are the same because, as mentioned in Chapter 3, the formula for a standardized regression weight is $\beta = \frac{b}{\sigma_X \sigma_Y}$. Analogous to the Galton height data, the criterion intercept may be different across men and women. Alternatively, the observed differences in mean level of alcohol consumption for men and women could be explained by differences in motivations to drink.

If we are interested in knowing whether the correlation between the predictor variables is different for men and women, we can calculate standardized versions of the predictor variables by means of latent variables with no error variances as discussed in Chapter 3 and shown in Figure 14.2. It may be the case, however, that the correlation between the two predictor variables is the same but that men, for example, may be more variable in their responses to these questions than women.

Multiple group models in which invariant regression weight, correlation, intercept, and predictor variable standard deviations were then examined. Invariance in predictor means was not explored, as the interest of the study was to determine the invariance of predicting alcohol consumption. Examination of the 32 models revealed that assumption of equal variation in the predictor variables resulted in poor model fit (all $\chi^2$ values larger than 20, all highly significant and no other fit measures choosing such models). Assuming intercept invariance also resulted in poorly fitting models ($\chi^2$ values ranging between 9 and 247, all highly significant and selected by no other fit measures). This then left investigation of models, in which the regression weights and correlation coefficient between the predictors was invariant (Table 14.3).

Based on the nonsignificant $\chi^2$ statistic and the lowest BIC value, it would seem reasonable to assume that the regression weights for both items are the same for men and women and that the correlation between the two predictors is the same, although substantial differences exist in variances, with men's reports being substantially more variable than women. Substantial

**TABLE 14.3**

Invariance of Prediction of 3-Month Alcohol Consumption: Two Reasons for Drinking

| Invariant Parameters | $\chi^2(df)$ | Prob($\chi^2 \mid H_0$) | AIC | BIC | TLI | CFI | RMSEA | $P_{(\text{Close Fit})}$ |
|---|---|---|---|---|---|---|---|---|
| None | 0.00 (0) | NA | 18270.07 | 18370.62 | 1.00 | 1.00 | — | NA |
| $b_1$ | 0.32 (1) | 0.570 | 18268.39 | 18363.36 | 1.00 | 1.00 | 0.00 (0.00–0.07) | 0.86 |
| $b_2$ | 0.10 (1) | 0.747 | 18268.17 | 18363.14 | 1.01 | 1.00 | 0.00 (0.00–0.06) | 0.92 |
| $b_1b_2$ | 0.94 (2) | 0.626 | **18267.00** | 18356.38 | 1.00 | 1.00 | 0.00 (0.00–0.05) | 0.95 |
| $r$ | 4.13 (1) | 0.042 | 18272.19 | 18367.16 | 0.98 | 1.00 | 0.06 (0.01–0.12) | 0.32 |
| $rb_1$ | 4.45 (2) | 0.108 | 18270.52 | 18359.9 | 0.99 | 1.00 | 0.04 (0.00–0.08 | 0.64 |
| $rb_2$ | 4.23 (2) | 0.121 | 18270.3 | 18359.68 | 0.99 | 1.00 | 0.03 (0.00–0.08) | 0.66 |
| $rb_1b_2$ | 5.06 (2) | **0.167** | 18269.13 | **18352.90** | 1.00 | 1.00 | 0.03 (0.00–0.07) | 0.81 |

differences in the regression intercept were also found, with men reporting more consumption than women. The second best-fitting model, in which only the regression weights were the same, was preferred by the AIC. The difference between the BIC value for that model and the model that constrained the correlations to be the same was 3.46, which would meet Raftery's general guidelines for positive support for the model assuming invariant correlation. The $\chi^2$ difference between the two models was 4.13 with 1 degree of freedom, with $p = 0.02$. Given that we are comparing two models with nonsignificant $\chi^2$ values, however, it does not seem reasonable to ascribe much meaning to the test in this situation.

The final invariance model constraining the unstandardized loadings and correlation between the reasons to be invariant is shown in Figure 14.3. The All-Y notation for estimated parameters is used.

Assumption checking of this model would be necessary. Separate LOESS regressions for the two groups to investigate whether the relationship between the predictors and criterion, or between the predictors, appeared to be linear, and scatterplots of the relationship between the two predictor variables would also provide further information about the reasonableness of assuming the correlation is the same for men and women.

## Multigroup Latent Variable Analysis

As mentioned at the beginning of the chapter, the appropriate multigroup invariance models are different depending on the data and research questions involved. As a very brief historical perspective, Jöreskog (1971) presented multigroup factor models that did not include mean effects. Sörbom (1974) considered multigroup factor models with mean effects but assumed that

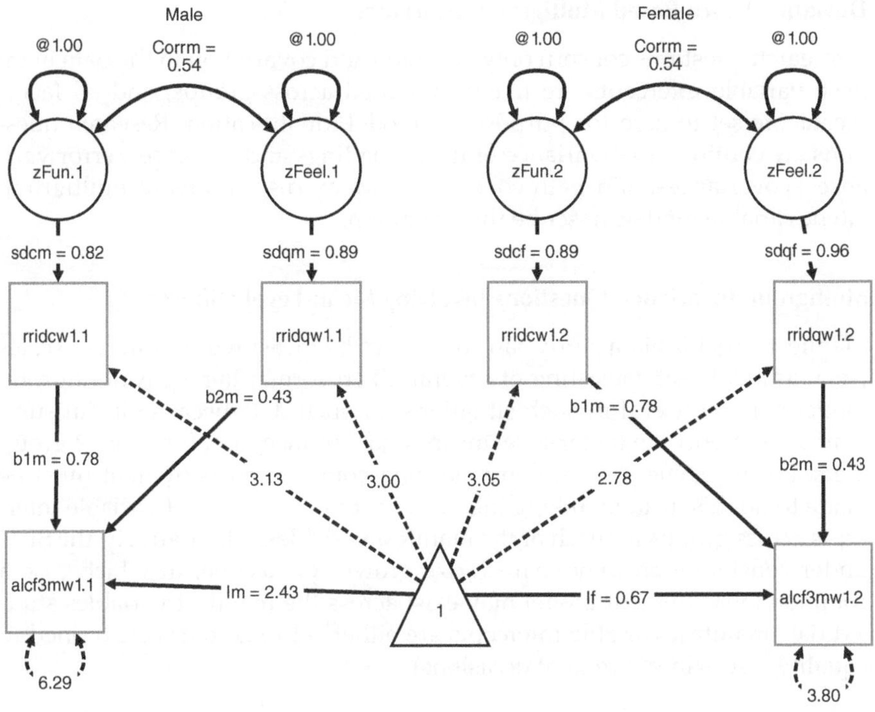


**FIGURE 14.3**
Multigroup two predictor invariance model.

the manifest variables were ratio-level measurement (i.e., manifest variable intercepts were constrained to zero). Millsap (2011) provides an extended discussion of multigroup measurement invariance in which intercepts are estimated at the manifest variable level.

Questions of multigroup differences in SEMs depend on the SEM under consideration. As discussed earlier, some SEMs are summaries of patterns of covariation (i.e., the deviation score matrix), while others explain elevation as well as variation (i.e., the adjoined score matrix). Chapters 10, 12, and 13 discussed different measurement models, parsimonious models, and models of growth and change. Similar choices regarding assumed measurement, parsimony, and assessment of change versus growth/decline also apply to multigroup analysis. As a result, no one set of steps can be specified for multigroup models in all situations. Vandenberg and Lance (2000) survey the various approaches to multigroup models and Van de Schoot, Lugtig, and Hox (2012) provide a convenient checklist and description of the invariance programming options across various software platforms.

### Deviation Score-Based Multigroup Invariance

If research questions concern only variation and covariation in the data manifest, variable intercepts are freely estimated across groups, and all factor means are set to zero for purposes of model identification. Research questions are confined to invariance of factor loadings and variances, error variances/covariances, and path coefficients. Many discussions of multigroup latent variable models describe this condition.

### Multigroup Invariance Questions Involving Mean-Level Differences

Manifest variables in a study can sometimes be measured in different scales (such as a data set consisting of several Likert-format items). A multigroup model may be useful in such situations for such data because it can summarize differences in factor-level means and variances as a function of group membership. Some assumption regarding common measurement must be made to do this, usually taking the form of common manifest variable intercepts across groups for each of the manifest variables. Alternatively, the SEM under consideration may characterize growth or decline, in which case a common measurement model may exist across the manifest variables studied (i.e., manifest variable intercepts are either set to zero or constrained to equality across measurement occasions).

### Tests Prior to Invariance Comparison

#### *Comparisons of Variance/Covariance Matrices*

Some have suggested, as a first step, a test of whether the variance/covariance matrices across groups are different (i.e., Alwin & Jackson, 1981; Bagozzi & Edwards, 1998; Byrne, Shavelson, & Muthén, 1989). The thinking is that if the covariance matrices are not significantly different, no further systematic investigation is necessary. This is not often done but is extremely easy to test by specifying all variances and covariances in the data and then comparing the relative fit of a model in which all respective variances and covariances are constrained to equality with a (just-identified) model in which they are allowed to vary. Analogously, if both mean level and variability are of interest, it would be relatively straightforward to consider mean manifest variable mean constraints across groups as well.

#### *Dimensionality*

In similar fashion, the techniques of exploratory factor analysis can be used to assess the dimensionality of the variables under consideration and, if the sample size is adequate, whether this is the same across groups. In longitudinal studies of college and noncollege populations, for example, it may be the case that a general factor of, say, alcohol use may exist while the college population demonstrates a developmentally limited factor of alcohol use corresponding to assessments during the college years. If this is the case, invariance analysis of factors or growth curves would be based on the general factor common to both groups.

### *Choice of Measurement Model*

It is also necessary to evaluate the appropriate measurement model for the data. Even for items with relatively similar content, candidate models would include single factor, random intercept, or oblique factor models. If the research questions also involve estimation of mean levels, a decision must be made as to whether the FM, FM-Shift, or some growth curve model is appropriate. As with the dimensionality discussion, if the sample size is adequate, this could be explored separately for each group but may need to be evaluated using the entire dataset. Examination of the resulting model can also inform the researcher as to whether error variances should be freely estimated or whether more parsimonious models of equality of error variance or a precision measurement model is appropriate.

### *Invariance Models*

Table 14.4 presents the terms associated with various invariance models in terms of structural equation model parameters following Vandenberg and Lance's (2000) invariance discussion. A checkmark (✓) is used to indicate parameters that are constrained to equality across groups, and **X** is used to indicate parameters that are freely estimated across groups.

*Configural Invariance.* Meredith (1964) applied Lawley's (1943) selection theorem for regressions to the question of how similar factor solutions would be if selection effects were present in either or both groups. The basic message is that estimated factor loadings could, indeed, be quite different, but it should be the case that factor loadings present in one group should be the same sign or at least zero in the other group. This is described as "configural invariance," and Horn and McArdle (1992) described how configural measurement is different from the metric invariance considered in more formal constraints of parameters.

*Weak Factorial Invariance (Pattern Invariance).* Conceptually, the default invariance model for a factor model assumes that the manifest variables represent (invariant) but interval-level measurements across the groups. If it is the case that only factor loadings are invariant across groups, and that manifest variable intercepts and the variances and covariances are different across groups, this is referred to as "weak" invariance. Factor means, variances, covariances, intercepts, and error

**TABLE 14.4**
Types of Measurement Invariance

| Invariance Label | $\Lambda(\lambda_i)$ Loadings | $T(\tau)$ Intercepts | $\theta_\varepsilon$ Error Var/Cov. | $A(\alpha_i)$ Factor Means | diag $(\Psi)$ Factor Variances | Offdiag $(\Psi)$ Factor Covariances |
|---|---|---|---|---|---|---|
| Configural | Patterns of zero & nonzero loadings same | | | | | |
| Metric (Weak Factorial Invariance) | ✓ | X | X | X | X | X |
| Scalar Invariance with Free Latent Means Variances[a] | ✓ | ✓ | X | Fixed to 0 for reference group, free in others | Fixed to 1 for reference group, free in others | X |
| Strict | ✓ | ✓ | X | ✓ | ✓ | ✓ |
| Total | ✓ | ✓ | ✓ | ✓ | ✓ | ✓ |

*Source*: Adapted from Vandenberg and Lance (2000).
[a] Subgroup invariance questions can consider factor means and variances separately.

variances that are part of a mathematically identified solution can vary across the groups, however. The idea comes from Thurstone (1947).

*Scalar Invariance with Free Latent Means/Variances ("Strong Invariance").* If the manifest variable intercepts are constrained to equality across groups, it is then possible to impose identification constraints on one group (e.g., factor means of zero and unit factor variance) and freely estimate the means and variances in the remaining group(s). Submodels in this category can impose equality of factor variation/covariation while estimating different factor means as well as equality of factor means with variant factor/covariation. These models were first introduced by Meredith (1993). This level of invariance is generally the most sought after as it permits statements of group differences using the structural equation model.

*Strong Invariance.* All parameters are equal across groups except for error variances. This model is useful as a null model of no group differences in comparison with the Scalar Invariance models, or, alternatively, as establishing a level of measurement invariance for the construct(s) under consideration.

*Strict Invariance.* If the SEM's latent variable means, variances, and covariances are the same across groups, the model is said to demonstrate strict invariance. Such a level of invariance would be useful in establishing a measure as being a parallel form. As described at the end of the chapter, examination of free latent means/variances can also be used to summarize group differences under the assumption of parallel forms, although this is not frequently encountered in practice.

*Additional Hypotheses: Partial Invariance Models.* Invariance may not be an "all or none" phenomenon in the data. As such, local violations of invariance may be present, but a minimum number of equality constraints in the model may be necessary to specify an invariant model. For example, in a study of reasons for drinking, it could be that some reasons, such as to celebrate special occasions, may be a better indicator of the factor for men than for women.

*Longitudinal Invariance.* In some situations, as when the manifest variables represent repeated measurement of growth over time, multigroup models can be specified that assume either a ratio or common interval-level measurement across the manifest variables within a group. The second example presented considers such a situation where it seems reasonable to make these assumptions, given that the same measurement instrument is repeated across occasions.

As an example of a simple multigroup model, let's take a set of items designed to measure social reasons for drinking in a sample of men and women. Specifically, consider the items taken from the Social motives for drinking considered in Table 12.1. Summary statistics for the two groups are shown in Table 14.5. As mentioned earlier, the first step is to explore an appropriate measurement model for the data. Given the small number of items, it seems reasonable to explore the random intercept factor model and the single factor model. When analyzed as a single group model, the fit of the random intercept model was appreciably better than for the single factor model (RI, $\chi_4^2 = 234.38$; Single Factor, $\chi_5^2 = 247.93$; $\chi_{\text{diff}(1)}^2 = 9.55, p = 0.001$). This is similar to the results

**TABLE 14.5**

Social Reasons for Drinking "I Drink Because …"

| | Makes Gatherings Fun | Improves Parties | Helps Me Enjoy Party | Helps Me Be Sociable | Celebrate Special Occasions |
|---|---|---|---|---|---|
| Men $(N = 881)$ | | | | | |
| Means | 3.022 | 3.014 | 3.125 | 3.082 | 3.143 |
| Standard Deviations | 0.857 | 0.836 | 0.854 | 0.869 | 0.831 |
| Variance/Covariance Matrix | | | | | |
| Makes Gatherings Fun | 0.735 | | | | |
| Improves Parties | 0.493 | 0.698 | | | |
| Helps Me Enjoy Party | 0.543 | 0.493 | 0.729 | | |
| Helps Me Be Sociable | 0.554 | 0.469 | 0.645 | 0.755 | |
| Celebrate Special Occasions | 0.392 | 0.359 | 0.444 | 0.500 | 0.691 |
| Correlations | | | | | |
| Makes Gatherings Fun | 1.000 | | | | |
| Improves Parties | 0.689 | 1.000 | | | |
| Helps Me Enjoy Party | 0.741 | 0.693 | 1.000 | | |
| Helps Me Be Sociable | 0.743 | 0.648 | 0.869 | 1.000 | |
| Celebrate Special Occasions | 0.552 | 0.518 | 0.625 | 0.692 | 1.000 |
| Women $(N = 1296)$ | | | | | |
| Means | 2.904 | 2.968 | 3.052 | 2.970 | 3.113 |
| Standard Deviations | 0.907 | 0.854 | 0.875 | 0.916 | 0.820 |
| Variance/Covariance Matrix | | | | | |
| Makes Gatherings Fun | 0.823 | | | | |
| Improves Parties | 0.568 | 0.729 | | | |
| Helps Me Enjoy Party | 0.616 | 0.527 | 0.765 | | |
| Helps Me Be Sociable | 0.634 | 0.540 | 0.720 | 0.838 | |
| Celebrate Special Occasions | 0.369 | 0.351 | 0.440 | 0.470 | 0.673 |
| Correlations | | | | | |
| Makes Gatherings Fun | 1.000 | | | | |
| Improves Parties | 0.733 | 1.000 | | | |
| Helps Me Enjoy Party | 0.776 | 0.705 | 1.000 | | |
| Helps Me Be Sociable | 0.763 | 0.690 | 0.900 | 1.000 | |
| Celebrate Special Occasions | 0.496 | 0.501 | 0.614 | 0.625 | 1.000 |

from a comparison of the two measurement models under the totally variant two-group model as well (RI, $\chi^2_8 = 215.16$; Single Factor, $\chi^2_{10} = 271.00$; $\chi^2_{\text{diff}(2)} = 55.82, p < 0.001$).

Table 14.6 shows fit statistics associated with the configural model, the three weak factorial invariance models, and the strong invariance and strict invariance models. In the first weak factorial invariance model, only loadings are constrained to be the same across the two groups. All other parameters are free to vary and are identified as one would for a single group analysis. The purpose of the second weak factorial invariance model is to summarize group differences using identification constraints for a single group model in one group, which serves as a reference group and estimating factor means, variances, and covariances as freely estimated parameters in the other group. This requires constraining manifest variable intercepts to equality across the groups. Using women as a reference group in this model, the random intercept and factor covariance is constrained to zero (as it would be in a single group analysis) but is freely estimating the covariance for men. If, for example, men who are more likely to endorse the items as a reason for drinking but were also more likely to uniformly endorse all items regardless of question, this would cause a nonzero covariance between the factors for men. In the third weak factorial invariance model, factor variances and covariances are constrained across groups. In the strong invariance model, factor means are constrained to equality, and in the strict invariance model, all parameters are constrained to equality.

These models illustrate the issue of conflicting recommendations from fit statistics. Based on the $\chi^2$ difference test, one would conclude that only configural invariance applies. Based on the BIC and lowest RMSEA, however, the totally invariant model is favored. Based on the TLI, all models except the configural model fit well, while all models fit well according to the CFI. Finally, based on Cheung and Rensvold's (2002) recommendations regarding multigroup invariance, a case could be made that the weak invariance model with estimated difference in RMSEA greater than 0.015 could indicate a potential invariance violation. One defensible interpretation of the data is that they are generally in support of an invariant random intercept model but that differences, if they exist, are overall relatively modest and that the weak invariance model could be used to characterize differences between variability and elevation across men and women at the level of the latent variables involved.

Figure 14.4 shows the weak invariance model, which estimates the latent variable means, latent variable variances, and covariance for men treating women as a reference group. As can be seen, the estimate of the mean of the random effect factor and the covariance between the random intercept and factor are not significantly different across men and women. It appears, though, that the mean of the factor is higher for men than for women, and that the variability in the factor is substantially less than for women. Using

**TABLE 14.6**

Fit Indices for Multi-Group Invariance Hypotheses[a]

| Model | $\lambda$ | $\tau$ | $\psi$ | $\alpha$ | $\theta$ | $\chi^2(df)$ | AIC | BIC | TLI | CFI | RMSEA |
|---|---|---|---|---|---|---|---|---|---|---|---|
| Configural | | | | | | **215.15 (8)** | **19201.12** | 19382.77 | 0.94 | 0.98 | 0.15 (0.14–0.17) |
| Weak Invariance | x | | | | | 262.99 (11) | 19242.96 | 19407.57 | 0.95 | 0.97 | 0.15 (0.13–0.16) |
| Scalar Invariance, Male $\Psi$, $\alpha$ Estimated | x | x | | | | 270.73 (14) | 19245.70 | 19376.26 | 0.96 | 0.97 | **0.12 (0.11–0.13)** |
| Scalar Invariance, Male $\alpha$, Estimated | x | x | x | | | 272.89 (17) | 19242.86 | 19379.09 | 0.96 | 0.97 | 0.12 (0.11–0.13) |
| Strong Invariance | x | x | x | x | | 284.58 (19) | 19248.54 | 19367.75 | 0.97 | 0.97 | 0.11 (0.10–0.13) |
| Total Invariance | x | x | x | x | x | 299.59 (24) | 19253.56 | **19344.38** | **0.97** | 0.97 | **0.10 (0.09–0.11)** |
| Partial Invariant | x | x | | | | 298.55 (23) | 19254.52 | 19351.02 | 0.97 | 0.97 | 0.11 (0.09–0.12) |

[a] $P_{CloseFit} = 0$ for all models, $x$ denotes parameters constrained to equality across groups.

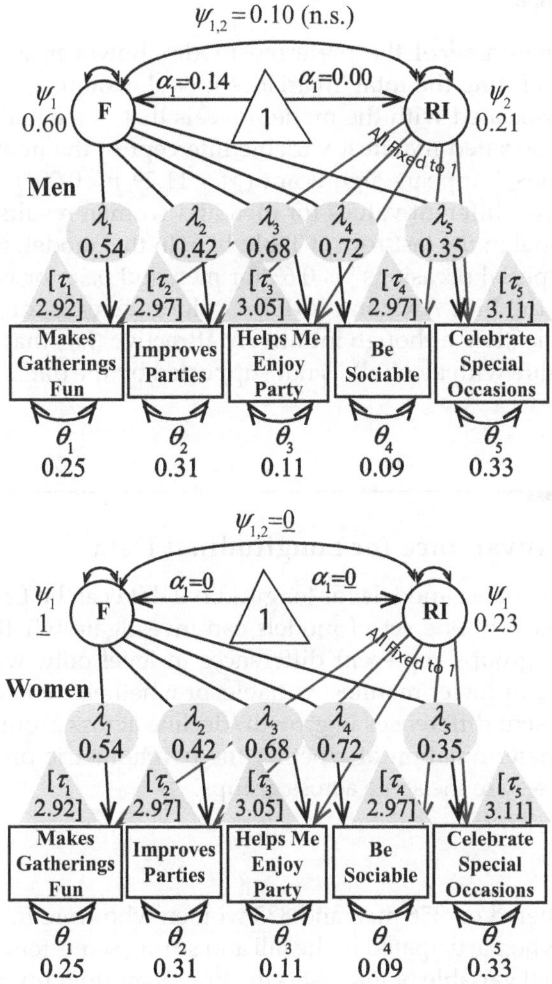


**FIGURE 14.4**
Weak multi-group invariance model.

the standard error of the factor variance for men gives the following test: $(\frac{1-0.60}{0.16} = 2.53, p < 0.001)$.

It would be possible to specify a model in which the factor covariance and the random intercept mean were both set to zero, but specifying doing so is a post-hoc comparison, given that the parameter estimates have been examined.

### *Partial Invariance*

Additional fine-tuning of the preferred model, however, is possible, if in the service of refuting the total invariance model. Examination of modification indices associated with the model reveals that one equality constraint, namely that associated manifest variable intercept of the item "to celebrate special occasions," appears significant ($\chi_1^2 = 11.29, p < 0.001$). Allowing this intercept to have different values for men and women results in the partial invariance model in the last row of Table 14.6. In this model, the loading for "to celebrate special occasions" is 0.63 for men and 0.36 for women. Given, however, that the BIC, AIC, and RMSEA values increase for this modification, the modification, although intuitively attractive (i.e., that men are more likely to celebrate with alcohol), is not supported by the data.

## Multigroup Invariance for Longitudinal Data

Multigroup invariance models for longitudinal data can be fit the same way as described earlier. One set of models can investigate whether the differences between groups represent differences in level only, with one group having a higher or lower manifest variable, or whether the latent variable(s) involved represent differences in growth/decline across groups. If the latter is true, some determination of whether the psychometric properties of the growth processes are the same across groups.

### Example Data

Data were gathered on 356 men and 835 women who were first-time college students and who participated in the fall and spring semesters for four years. The longitudinal variable of interest is the Brief Symptom Inventory General Severity Index (BSI-GSI), a continuous measure of overall psychological distress calculated as a sum of 18 items. Means, standard deviations, variance/covariance matrices, and correlations are shown in Table 14.7.

The first step in considering an appropriate measurement model is to determine the dimensionality and patterning of the solution for the data set as a whole. Using parallel factor analysis (described in Chapter 15), two factors appeared to underlie the data. Attempts to rotate the two yielded no oblique simple structure solution, but an oblique geomin rotation revealed the presence of one factor with largely similar loadings and a second factor with a mix of smaller positive and negative loadings. Conceptually, an additional factor consisting of waves one through four could be possible could exist given that students are younger than 21 during those years and students at

**TABLE 14.7**

Summary Statistics for Eight Waves of Longitudinal Data on Psychological Distress

| | BSIW1 | BSIW2 | BSIW3 | BSIW4 | BSIW5 | BSIW6 | BSIW7 | BSIW8 |
|---|---|---|---|---|---|---|---|---|
| Men | | | | | | | | |
| Means | 0.520 | 0.421 | 0.453 | 0.421 | 0.429 | 0.430 | 0.458 | 0.391 |
| Std.Dev. | 0.620 | 0.487 | 0.569 | 0.511 | 0.569 | 0.531 | 0.614 | 0.534 |
| Variance/Covariances | | | | | | | | |
| BSIW1 | 0.384 | | | | | | | |
| BSIW2 | 0.145 | 0.237 | | | | | | |
| BSIW3 | 0.136 | 0.137 | 0.324 | | | | | |
| BSIW4 | 0.117 | 0.090 | 0.150 | 0.261 | | | | |
| BSIW5 | 0.137 | 0.109 | 0.164 | 0.135 | 0.324 | | | |
| BSIW6 | 0.081 | 0.075 | 0.130 | 0.113 | 0.147 | 0.282 | | |
| BSIW7 | 0.082 | 0.080 | 0.124 | 0.127 | 0.145 | 0.154 | 0.377 | |
| BSIW8 | 0.104 | 0.095 | 0.143 | 0.127 | 0.161 | 0.125 | 0.169 | 0.285 |
| Correlations | | | | | | | | |
| BSIW1 | 1.000 | | | | | | | |
| BSIW2 | 0.495 | 1.000 | | | | | | |
| BSIW3 | 0.397 | 0.555 | 1.000 | | | | | |
| BSIW4 | 0.419 | 0.416 | 0.557 | 1.000 | | | | |
| BSIW5 | 0.415 | 0.463 | 0.560 | 0.515 | 1.000 | | | |
| BSIW6 | 0.285 | 0.359 | 0.481 | 0.476 | 0.558 | 1.000 | | |
| BSIW7 | 0.239 | 0.314 | 0.419 | 0.426 | 0.496 | 0.511 | 1.000 | |
| BSIW8 | 0.317 | 0.425 | 0.509 | 0.474 | 0.521 | 0.501 | 0.551 | 1.000 |
| Women | | | | | | | | |
| Means | 0.558 | 0.539 | 0.514 | 0.511 | 0.475 | 0.475 | 0.460 | 0.463 |
| Std. Dev. | 0.593 | 0.565 | 0.553 | 0.559 | 0.533 | 0.515 | 0.554 | 0.543 |
| Variance/Covariances | | | | | | | | |
| BSIW1 | 0.352 | | | | | | | |
| BSIW2 | 0.190 | 0.319 | | | | | | |
| BSIW3 | 0.164 | 0.181 | 0.306 | | | | | |
| BSIW4 | 0.166 | 0.183 | 0.189 | 0.313 | | | | |
| BSIW5 | 0.131 | 0.142 | 0.152 | 0.166 | 0.284 | | | |
| BSIW6 | 0.132 | 0.147 | 0.149 | 0.153 | 0.166 | 0.265 | | |
| BSIW7 | 0.144 | 0.141 | 0.146 | 0.151 | 0.167 | 0.161 | 0.307 | |
| BSIW8 | 0.142 | 0.134 | 0.145 | 0.161 | 0.136 | 0.149 | 0.167 | 0.295 |
| Correlations | | | | | | | | |
| BSIW1 | 1.000 | | | | | | | |
| BSIW2 | 0.571 | 1.000 | | | | | | |
| BSIW3 | 0.511 | 0.575 | 1.000 | | | | | |
| BSIW4 | 0.494 | 0.573 | 0.606 | 1.000 | | | | |
| BSIW5 | 0.419 | 0.479 | 0.544 | 0.577 | 1.000 | | | |

*(Continued)*

**TABLE14.7 (Continued)**

| | BSIW1 | BSIW2 | BSIW3 | BSIW4 | BSIW5 | BSIW6 | BSIW7 | BSIW8 |
|---|---|---|---|---|---|---|---|---|
| BSIW6 | 0.435 | 0.517 | 0.548 | 0.550 | 0.611 | 1.000 | | |
| BSIW7 | 0.448 | 0.478 | 0.512 | 0.520 | 0.586 | 0.591 | 1.000 | |
| BSIW8 | 0.443 | 0.448 | 0.506 | 0.551 | 0.487 | 0.561 | 0.581 | 1.000 |

this campus have fewer options to live off campus at that time. Because a two-factor solution was suggested by the parallel factor analysis, four single group factor models were explored as appropriate measurement models: (1) a single factor model; (2) a two-factor model that included an additional factor consisting of the first four measurement occasions; (3) a random intercept model; and (4) a two-factor model such as just described but including a random intercept model. Of these models, the random intercept model appeared to be the best fitting, with the highest TLI and CFI values, the lowest RMSEA, and the lowest BIC value. Table 14.8 shows the fit indices for these competing models. Based on these models, it was decided to adopt the random intercept model based on the BIC value and the RMSEA fit. The two-factor model, although also having a low RMSEA value, had a higher BIC value; and the two-factor random intercept model, although having a significant $\chi^2$ different from the random intercept model $\left(\chi^2_{\text{diff}(4)} = 13.70, p = 0.0008\right)$, contained small loadings on the second factor of which only one loading was statistically significant.

Turning to a multigroup analysis, the following invariance hypotheses were examined for the random intercept model:

1. Configural invariance (separate random intercept models for each group).
2. Weak invariance (only factor loadings constrained to equality across groups, factor means set to zero).

**TABLE 14.8**

Fit Statistics for Single Group Models

| Model | $\chi^2(df)$[a] | AIC | BIC | TLI | CFI | RMSEA (95% CI) | $P_{\text{CloseFit}}$ |
|---|---|---|---|---|---|---|---|
| One Factor | 211.86 (20) | 11664.57 | 11745.89 | 0.94 | 0.96 | 0.09 (0.08–0.10) | 0.000 |
| Two Factor | 54.12 (16) | 11514.83 | 11616.48 | 0.98 | 0.99 | 0.04 (0.03–0.06) | 0.726 |
| Random Intercept | 56.79 (19) | 11511.50 | **11597.90** | 0.99 | 0.99 | 0.04 (0.03–0.05) | **0.882** |
| Two Factor Random Intercept | 43.09 (15) | **11505.80** | 11612.53 | 0.99 | 0.99 | 0.04 (0.03–0.05) | 0.881 |

[a] All p-values < 0.001.

3. Scalar invariance: intercepts (intercepts equal across groups, male general factor mean fixed to zero, and male general factor variance fixed to 1 corresponding female factor means and variances freely estimated).
4. Scalar invariance: intercepts and factor variances (intercepts equal across groups, male general factor mean fixed to zero, corresponding female factor means free, factor variances constrained to equality).
5. Strong scalar invariance (all parameters set to equality across groups except for manifest variable error variances).
6. Strict invariance (all parameters set to equality across groups).

However, because the data are longitudinal it is also possible to investigate whether multi-group models fit the data in which manifest intercepts are fixed to zero and factor means are estimated for both groups. This leads to the following latent basis growth curve models in which manifest variable intercepts are constrained to zero:

7. Configural latent basis (both groups fit as qualitatively different growth processes with different loadings and random intercept variances).
8. Weak latent basis (both slope factor loadings constrained to equality across groups).
9. Scalar latent basis invariance (factor variances set to equality, factor means freely estimated across groups).
10. Strong latent basis invariance (all parameters set to equality except for error variances).
11. Strict latent basis invariance (all parameters set to equality across groups).

Fit statistics for these models are shown in Table 14.9. As can be seen, none of the random intercept models fits the data as well as the weak or strict latent basis models. This suggests that history effects unique to each measurement occasion which would cause manifest variable intercepts to vary do not appear present in the data. Of the latent basis models, the BIC appears to favor the strict invariance model of no differences in growth/decline across the measurement occasions. The $\chi^2$ difference test between the strict and weak invariance models is statistically significant ($\chi^2_{\text{diff}(13)} = 61.44, p < 0.001$) and the BIC difference between the two models (30.63) indicates a better fit for the strict model according to Raftery's criteria for substantial difference between the models. Alternatively, the weakly invariant model is preferred based on the AIC. The remaining fit statistics all appear largely identical.

**TABLE 14.9**

Fit Statistics for Longitudinal BSI Data

| Invariance Model | $\chi^2\,(df)$ | AIC | BIC | TLI | CFI | RMSEA (95% CI) | $P_{\text{CloseFit}}$ |
|---|---|---|---|---|---|---|---|
| Random Intercept Models | | | | | | | |
| Configural | 241.83 (38) | 11642.57 | 11896.69 | 0.93 | 0.95 | 0.09 (0.08–0.11) | 0.00 |
| Weak | 252.52 (45) | 11639.26 | 11857.81 | 0.94 | 0.95 | 0.09 (0.08–0.10) | 0.00 |
| Scalar: Intercepts | 172.87 (50) | 11549.61 | 11742.75 | 0.97 | 0.97 | 0.06 (0.05–0.07) | 0.01 |
| Scalar Intercepts and Factor Variances | 125.26 (53) | 11495.99 | 11673.88 | 0.98 | 0.98 | 0.05 (0.04–0.06) | 0.61 |
| Strong | 172.86 (48) | 11553.59 | 11756.90 | 0.97 | 0.97 | 0.07 (0.06–0.08) | 0.01 |
| Strict | 152.06 (56) | 11516.80 | 11679.44 | 0.98 | 0.98 | 0.05 (0.04–0.06) | 0.26 |
| Latent Basis Models | | | | | | | |
| Configural | 162.95 (50) | 11539.68 | 11732.82 | 0.97 | 0.97 | 0.06 (0.05–0.07) | 0.03 |
| Weak | 126.12 (56) | **11490.85** | 11653.50 | 0.98 | 0.98 | **0.05 (0.04–0.06)** | **0.73** |
| Scalar | 135.87 (59) | 11494.60 | 11641.99 | 0.98 | 0.98 | 0.05 (0.04–0.06) | 0.68 |
| Strong | 145.74 (61) | 11500.47 | 11637.70 | 0.98 | 0.98 | 0.05 (0.04–0.06) | 0.59 |
| Strict | 187.56 (69) | 11526.30 | **11622.87** | 0.98 | 0.97 | 0.05 (0.04–0.06) | 0.24 |

Even though the strict invariance latent basis model fits the data best, the weak invariance model may be a useful way to characterize differences between men and women in the course of psychological distress across occasions. Based on comparison of the individual parameters associated with factor means and variances, significant differences between men and women in the means, variances, and covariance of the slope and intercept factors appear to exist and persist regardless of whether the weak invariance model is fit or if a modified strict invariance model is fit (which includes equal error variances across groups), with factor means and variances free to vary. The path diagram for the weakly invariant latent basis model, displayed in Figure 14.5, shows that men appear to demonstrate a relatively uniform lower pattern of psychological distress during this time, with relatively little change over the measurement occasions as evidenced by the lower factor means on both factors; while women demonstrate a higher level of psychological distress that declines across measurement occasions, as shown by the higher factor mean associated with the slope factor.

These differences can be represented graphically by generating predicted values for each occasion for each individual in the data set by saving factor scores from the solution and multiplying the factor scores by their respective loadings for each measurement occasion. A line graph of all individuals would likely be uninformative given the clutter imposed by the number of observations, however. As a remedy, observations can be binned into values

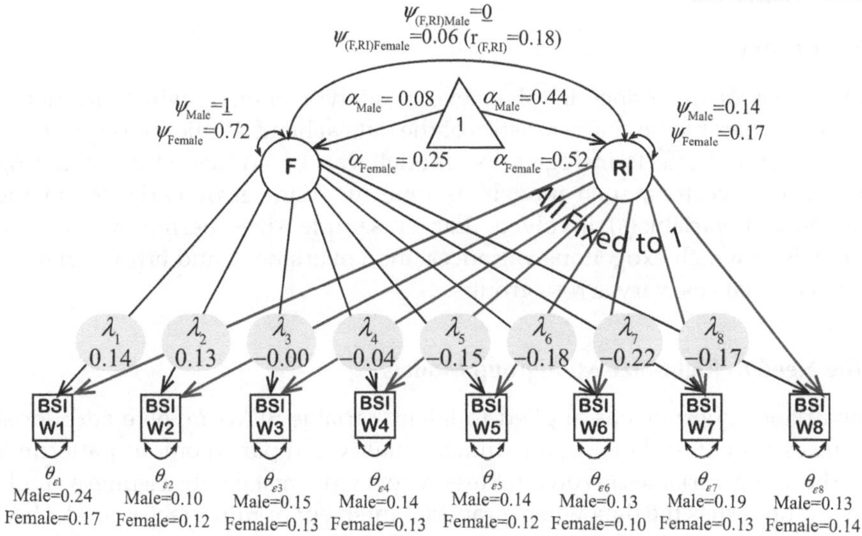


**FIGURE 14.5**
Weak invariance latent basis model for BSI data.

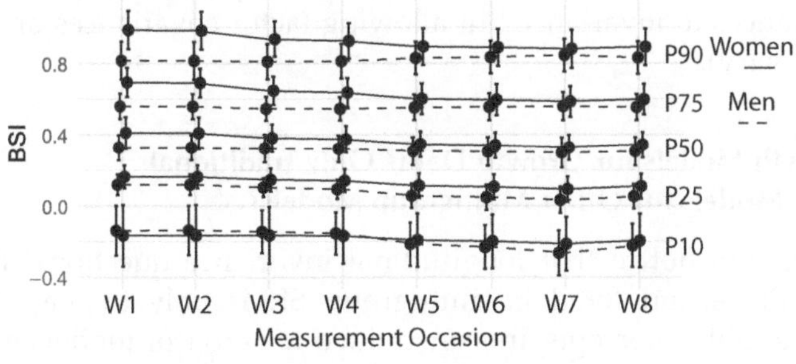


**FIGURE 14.6**
Binned percentiles of predicted BSI scores based on weak invariance model.

representing the 10th, 25th, 50th, 75th, and 90th percentile groups of the predicted scores averaged over all measurement occasions separately for the two sexes. The resulting Figure 14.6 shows a generally higher level of psychological distress and decreasing pattern of scores across measurement occasions for women as well as the slightly higher variability in scores for women who are low in psychological distress.

## Summary

Questions of invariance involve several analytic choices, which are determined in part by the characteristics of the data at hand. While some questions of moderated effects in regression models can be addressed by including interaction vectors, such models assume invariant error variances in the dependent variable(s) involved. Larger sample sizes permit multigroup models in which exogenous variances and covariances, and error variances and covariances, vary across groups.

### The Need to Right-Size Multigroup Models

Invariance questions as applied to latent variable SEMs involve additional choices related to choice of the dimensionality and parsimonious patterning of the latent variables involved. Researchers who mistakenly assume a single factor underlies the data when a random intercept or multi-factor model better characterizes the data will likely conclude that only configural invariance holds. Exploration of the dimensionality of a set of variables is the topic of the next chapter. To the extent that the data allow, researchers should examine the dimensionality of the SEM and the appropriate parsimonious model for the data. After these initial steps, invariance comparisons can proceed as described. More fine-grained analyses are also possible (e.g., constraining factor variances to invariance but allowing factor covariances or regression weights to vary).

### Only Growth Models for Growth Data? Only Traditional Invariance Models for Other Multigroup Models?

The reader will notice that longitudinal invariance questions differ from those usually encountered in multigroup SEMs only to the degree that manifest variable intercepts are constrained to zero (for models involving a random intercept factor or ratio-level measurement) or equality (for factor mean-shift models). Traditional scalar invariance models in which respective manifest variable intercepts vary across manifest variables but are constrained to equality across groups are identical to growth curve models of data, which show substantial history effects across measurement occasions. Each measurement occasion, which is part of a larger pattern of growth or decline, is also subject to an occasion-specific change in elevation. For this reason, individuals who assess multigroup invariance of growth processes should consider estimating the corresponding multigroup scalar invariance model as well.

The converse is also true: Manifest variables in a traditional SEM may reasonably fit under the assumption that the manifest variable means are equal conditional on the latent variables of the model. Such an occasion could arise if the manifest variables constitute essentially parallel measurement instruments of the latent variables of the model. The latent variables of the model could be thought of as representing a general difficulty factor represented by continuously measured variables.

# 15

# *Exploratory Factor Analysis*

> Factor analysis is useful, especially in those domains where basic and essential concepts are essentially lacking and where crucial experiments are difficult to conceive ... In a domain where fundamental and fruitful concepts are already well formulated and tested, it would be absurd to use the factorial methods except for didactic purposes to illustrate factorial logic.
>
> **– L. L. Thurstone (1947, p. 56)**

A complete presentation of exploratory factor analysis (EFA) could constitute a separate class in its own right. Exploratory factor models are, however, similar to confirmatory models, and the differences between the techniques are in service of slightly different research goals. Exploratory factor analysis is covered across two chapters. This chapter introduces EFA terminology and focuses on the question of how many factors in an EFA reasonably summarize patterns of variation and covariation. This will consist of strategies for discovering patterns in the data, diagnostics, and statistical tests for the identification of the number of factors in the data. The next chapter discusses methods for transforming the initial factor loading matrix from an EFA to more parsimonious forms. In other words, EFA is like CFA in that it seeks to identify the number of latent variables responsible for variation and covariation in the data. In CFA models, model comparisons were done by explicit reference to known numbers of latent variables and parsimonious patterns in which some manifest variables loaded on some factors and not others. EFA, by contrast, begins with a matrix of numbers in which all variables load to some extent on all variables. In a second step, this matrix is transformed to represent some type of parsimonious pattern.

## EFA: Perceptions and Practice

Before continuing, a few words are in order given the initial mystique surrounding the technique, how it was performed by its early pioneers, and some of the common problems with using EFA given the ease with which several statistical software packages perform the analysis and the defaults often assumed by the software. First, it is sometimes observed that EFA is an "objective" technique, producing a model that does not require the several a


DOI: 10.1201/9781003595267-15

priori specifications and assumed theory required when performing confirmatory factor analysis. This is not so. First, by the time that data are collected, several subjective decisions have already been made on the research project, including construct selection, item wording, and selection of a subject population, just to name a few. As Popper (2002) observed: "All observation is theory laden."

Second, the process of EFA is an inherently inductive one, even though the technical terms used in the technique are often quite "mathematical" sounding. The rationale of confirmatory factor or structural models, by contrast, is explicitly deductive in nature: The researcher specifies structural models thought to represent relationships in the data and, based on this, evaluates which model best summarizes the data in terms of parsimony and model fit. In EFA, by contrast, the data are examined and EFA is used to summarize the relationships that seem evident from these summaries. The tools of EFA simply express these patterns in a general mathematical form.

The perception that EFA is deductive is particularly pernicious given that the analysis defaults of many software packages may make it seem that "some expert somewhere" has wisely chosen the decision rules regarding the number of latent variables and the rules for factor loading patterns that are optimal for any application. This is also not so. The point of this chapter is to discuss the necessary decisions an analyst should actively consider, and these will frequently involve overriding the default settings of any EFA computer program.

In an earlier time, individuals who pioneered EFA made judgments about analysis choices by directly looking at the correlation matrix or often re-evaluated their initial candidate factor model. For example, two variables may often be dropped if they are the only two representatives of a factor (called a "doublet factor"; for an example, see Mulaik, 2009b). Examination of patterns of correlations may lead a researcher to consider models containing method factors, hierarchical factor structures, or, in the case of longitudinal data, some ideas concerning error covariance structure over time. Similarly, scatterplots of factor loadings can be examined to determine if correlated or orthogonal factors are present. This inspection and these decisions were not usually reported, and the final research articles often gave the impression that the visually reasonable model was, in fact, conceived prior to collecting and looking at the data. While not a wholesale post-hoc generation of a theory after looking at the data (known as HARKing; Kerr, 1998), it does mean that these decisions, unless explicitly pre-registered and reported, will make the analysis look more convincing than it is.

For example, Guilford (1956) outlined one of the uses of exploratory factor analysis, seeking to show that intellectual abilities could be organized along three dimensions represented as a cube called the Structure of the Intellect model. As a researcher in ill-structured problem-solving at the time, I was intrigued to know how any real-world problem-solving measures fit into such a model. I found one real-world measure explored by Guilford in which army officers were asked to think about how to cross a bridge under a rather complex set of contingencies. On reading Guilford's original report to the army, however, I was dismayed to learn that the measure had been dropped, not on psychometric grounds as being unreliable, but because an initial factor analysis revealed it was "factorially complex" and therefore dropped from further consideration.

In an earlier time, these inductive decisions were made in statistical consultation with a recognized expert familiar with such "interocular" analysis and decisions. EFA in the present day, however, is more accessible, and graphical "heat maps" of associations between variables can be used to suggest a relevant factor structure and identify groups of variables or individual variables that either correlate with "too many" other variables or are unrelated. A wide variety of factor rotations are now more readily available than the few options previously available, and a better appreciation of possible nonlinear associations in the data and the possible artifacts induced by influential or outlying observations are explored in a more thorough understanding of the range of alternative EFA models. Finally, there are hybrid approaches to EFA, in which the researcher explores whether the correspondence of an EFA solution to a given theoretical structure is also possible. These techniques are often referred to as "target rotations" or "exploratory structural equation models" (ESEM).

These developments may leave the potential user of EFA somewhat uneasy. "How much can we trust a reported exploratory factor analysis," it may be wondered, "if, at every turn of the analysis we 'peeked' at the data?" Despite its inductive nature and the several (often underreported) decisions that accompany the technique, EFA has a scientific utility. First, from the critical rationalist perspective of Karl Popper mentioned in the first chapter, EFA can be used to generate alternative structural models for the data other than those initially considered by the researcher. As such, if a proposed EFA fits the data better than the original model proposed by the researcher, EFA results constitute an alternative, albeit tentative refutation of existing theory. Second, EFA and, more elegantly, ESEM approaches can be used to identify systematic refinements of a proposed structural model that offer an alternative to single parameter additions to a model suggested by individual modification indices.

In sum, Whewell (1837, 1840) and, later, Cattell (1988) noted that science may be described as an inductive-hypotheticodeductive spiral. Under this idea, observations about some topic generate some hypotheses on the part of the researcher. These hypotheses are then used to guide data collection to test these conjectures. The resulting conclusions based on these data then result in further inductive conjectures. Although EFA and ESEM fall heavily on the inductive arm of this spiral, they have their uses as potential unanticipated refutations of a proposed theory or as a way to discover new potential theories that can then be tested in subsequent research. The process of discovery of such models must be fully reported and checked for the reasonableness of their assumptions, and the existence of other plausible alternative EFA models are part of this research process.

## EFA Process and Terminology

EFA also differs from CFA in two ways: It analyzes correlations between variables rather than covariances or the adjoined average sums of squares, and it also assumes that error variation has been removed from the data prior to analysis. This is accomplished by replacing the ones on the diagonal elements of a correlation matrix with a value representing the proportion of variance in the variable due to the latent variables in the data, a value termed the "communality" and represented as $h^2$. This proportion is often improved upon by an iterative process. The complement of the communality, the proportion of variance in a manifest variable not accounted for by the latent variables, is equal to $1-h^2$ and referred to as the "uniqueness" (represented as $u^2$). Exploratory factor analysis usually proceeds in the following steps:

1. Some values are taken as initial estimates of communality and then used to decide on the number of factors. (Often the squared multiple correlation (SMC) of each variable using all of the remaining variables is taken as an estimate.)
2. Some decision is then taken on the number of factors present in the data. As with the issue of model fit in CFA, multiple decision rules exist for this decision.
3. An iterative procedure is then often used to improve the communality estimates.
4. Estimated factor loadings are re-expressed so that they resemble one of the types of parsimonious factor models presented in Chapter 12. These re-expressions are accomplished using the mathematical idea of "rotation," the mathematical equivalent of changing the perspective a photographer might use in taking a picture of something.

   The default behavior of most statistical software is to rotate factor loadings to represent "simple structure" in which each manifest variable loads on only one factor. Some software defaults to rotating to an orthogonal simple structure, while others default to rotation to a correlated (i.e., "oblique") simple factor structure. A variety of such rotations exist and, even within these two options, the behavior of some rotations differs from others. Some rotations result in a relatively equal number of items loading on each factor, while others result in a single factor composed of many variables and a number of additional factors composed of relatively few manifest variables.

Even this does not exhaust the possibilities, however. Recall that passing reference to rotation was made in Chapter 12 to the BI-CFQuartimax rotation procedure to rotate the factor solution to resemble a bifactor model consisting of one general factor loading on all variables with additional discrete subsets of items. Hierarchical factor models can be approximated using a technique known as the Schmid-Leiman transformation. Finally, as mentioned earlier, exploratory SEM (ESEM) can be used to attempt an approximation to some other initially preferred parsimonious model.

5. Appropriate assumption checking is made on the four steps presented. This can include cross-validation, in which data are split into separate groups to assess whether a consistent picture emerges regarding the decision about the number of factors and patterning of proposed loadings. Examination of whether influential observations exert an undue influence on the final solution or whether the linear relationship assumed between the loading between latent and manifest variables is reasonable as well.

## Example EFA

To begin, it might help to consider an example of a "final" printout of an exploratory factor analysis of the items involved in the Reasons for Drinking Scale mentioned in Chapter 12 (Table 15.1).

**TABLE 15.1**

Four Factor Solution Reasons for Drinking Scale Geomin Rotation

| Item | Factor | | | |
|---|---|---|---|---|
| "I drink..." | 1 | 2 | 3 | 4 |
| C because it makes social gatherings more fun | **0.96** | −0.01 | −0.02 | −0.01 |
| D because it improves parties | **0.90** | 0.01 | 0.02 | 0.00 |
| A because it helps me enjoy party | **0.76** | 0.02 | 0.08 | 0.02 |
| B to be sociable | **0.69** | 0.01 | −0.03 | 0.12 |
| E to celebrate | **0.55** | 0.07 | 0.11 | −0.05 |

(*Continued*)

**TABLE 15.1 (Continued)**

| Item | Factor | | | |
|---|---|---|---|---|
| G because it helps me when I feel depressed or nervous | −0.03 | **0.92** | −0.02 | 0.00 |
| J because it helps me forget about my problems | −0.01 | **0.89** | −0.01 | 0.02 |
| F because it helps me forget worries | 0.03 | **0.88** | −0.02 | −0.02 |
| H to cheer myself up | 0.01 | **0.83** | 0.07 | −0.02 |
| I because it helps me to feel more self-confident | 0.12 | **0.47** | 0.16 | 0.11 |
| Q because it gives me a pleasant feeling | −0.02 | −0.02 | **0.91** | −0.01 |
| K because I like the feeling | 0.05 | 0.05 | **0.79** | −0.06 |
| P to get high | −0.06 | 0.08 | **0.71** | 0.07 |
| O because it's exciting | 0.10 | 0.02 | **0.67** | 0.08 |
| R because it's fun | 0.26 | −0.06 | **0.67** | −0.02 |
| V to be liked | −0.02 | 0.03 | 0.03 | **0.91** |
| U to get in with a group I like | 0.00 | 0.00 | 0.01 | **0.90** |
| T so others won't kid me | 0.00 | −0.01 | −0.04 | **0.82** |
| W so I won't feel left out | 0.05 | 0.00 | −0.02 | **0.79** |
| S because my friends pressure me | −0.01 | 0.01 | 0.04 | **0.72** |
| | Factor Intercorrelations | | | |
| | 1 | 2 | 3 | 4 |
| 1 | 1 | | | |
| 2 | 0.41 | 1 | | |
| 3 | 0.73 | 0.48 | 1 | |
| 4 | 0.18 | 0.33 | 0.17 | 1 |

Examination of this table in which the variables have been organized somewhat reveals similarities and differences from a CFA. As with CFA, loadings map the latent variables onto manifest variables and, like CFA, correlations appear to connect the latent variables of the model. Note, however, that unlike CFA, each manifest variable has loadings on all factors, although it appears that each variable appears to load most heavily on only one latent variable.

## Questions of Connection Between CFA and EFA

This simple example, though, provokes a few questions: How is it possible that we did not need to impose some constraints on the latent variables similar to those described for confirmatory factor analysis to impose an identified solution, as discussed in Chapter 8? How, also, was it possible to estimate several loadings in a factor model when it was noted earlier that freely estimating loadings for all factors was not mathematically identified because a structural equation modeling program would not "know" whether to assign a particular loading to one of the multiple factors? (This was the rationale for considering a triangular decomposition, discussed in Chapter 10, in which some loadings were set to zero in order to secure a mathematically identified solution.) The short answer to these questions is that factor loadings in EFA are sequentially extracted one by one, assuming a factor variance of one. After extraction, these loadings are transformed into the simple structure shown in Table 15.1.

Why were correlations used in the analysis? Couldn't covariances or adjoined matrices be analyzed as well? Although not frequently encountered in psychology, EFA techniques can be just as easily applied to variance/covariance matrices as well (e.g., Jolliffe, 1986, 2002). The key point here, though, is that the variables should be measured in the same units so that differences in variance/covariance are in a comparable metric. This has been done in areas such as brain mapping (Colibazzi et al., 2008; Toga & Mazziotta, 2002). Although it is also possible to factor the average sums of squares and cross-products in order to use exploratory factor models of growth and decline over time (e.g., Tuckerized growth curves; Tucker, 1958), given that variables under analysis consist of variables in different scales, almost all discussions of EFA employ analysis based on correlations between variables to remove scaling differences across variables.

## Graphic Exploration of Variables

One way to detect the presence of latent variables is to examine the general patterns of associations in the data. Graphic exploration can also be used to determine whether the data meet the assumptions of linearity assumed by the estimated factor loadings of a factor model. As mentioned earlier, analysts used to start only with an examination of the correlation matrix of the data and some inspection of the univariate statistics of the manifest variables, which often overlooked the possibility of nonlinear relationships between variables.

Examination of many items in a correlation matrix can be a somewhat daunting task. One way to process many correlations is to consider a heat map of the correlations. Heat maps can be produced under any of the major statistic packages such as R, SAS, or SPSS. The graph shown in Figure 15.1 was generated in R but slightly post-processed using Adobe Illustrator.

As can be seen from the large squares on the diagonal in the figure, it seems that items in each content area seem to correlate highly. Additionally, it seems that the items from items A–E correlate with items K–R, possibly indicating

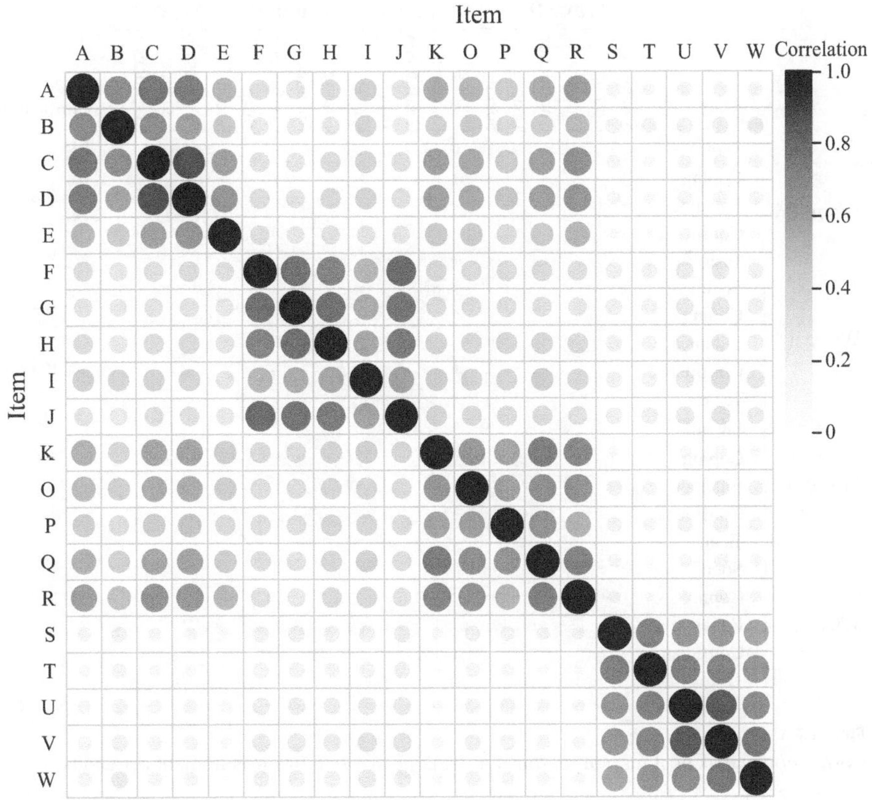


**FIGURE 15.1**
Correlation heat map of reasons for drinking in questionnaire order.

that latent variables for these content areas may correlate significantly. Items S–W, which deal with drinking so as not to be left out, do not seem to be correlated with the other items. The slightly smaller correlation for Item I in the plot shows that the analyst should expect a slightly lower factor loading for this item in a factor analysis, for example.

Heat maps can also be useful in identifying significant pairwise associations in the data. For example, Figure 15.2 shows the result of producing a heat map of the reasons for drinking subscales across waves. In addition to the substantial correlations between the Social and Enhance subscales within wave, substantial correlations across adjacent waves are seen within for each subscale with weaker correlations in subscales, a pattern sometimes referred to as a "simplex." Within each wave of assessment, the correlation between the Enhance and Social scales could be used to provide the justification for either combining these subscales within wave or, at the least, considering correlating these two scales within wave in the context of some

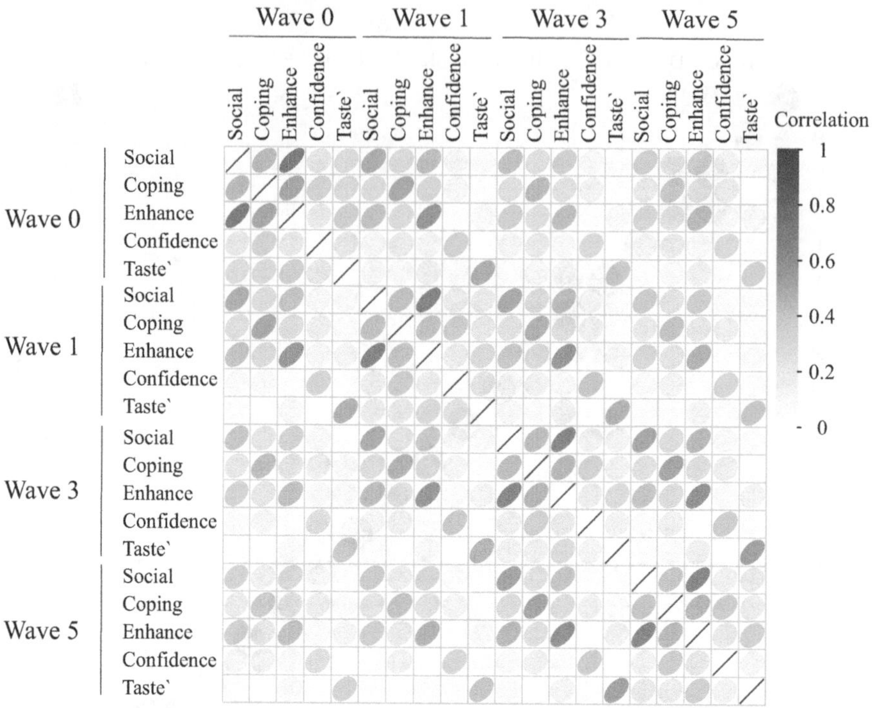


**FIGURE 15.2**
Correlation matrix heat map of reasons for drinking across four measurement occasions.

larger prospective model. If one were to entertain some kind of general factor across subscales and waves for these data, it would seem reasonable to think that only the Social, Coping, and Enhance subscales would form such a general factor, as the Confidence and Taste subscales seem not to share correlations with these scales across waves.

In cases where a large number of items are considered, or where the items have no particular order in terms of constructs, it is helpful to reorder the heatmap. Various clustering algorithms are possible, such as ordering the matrix by the first eigenvector or some combination of the first two eigenvectors (for a discussion of this latter option, see Friendly, 2002; for a discussion of various clustering algorithms in general, see Hurley, 2004). One of the more popular methods for reorganizing the correlation matrix is to reorganize the correlations using a simple hierarchical cluster analysis of the correlations. The resulting plot is shown in Figure 15.3.

Some information from the initial heat map in Figure 15.1 is still evident, such as the fact that the conformity items S, T, W, U, and V are relatively uncorrelated with the other items. It is easier to see, however, that the items having to do with experience enhancement items R "Because it's fun," K,

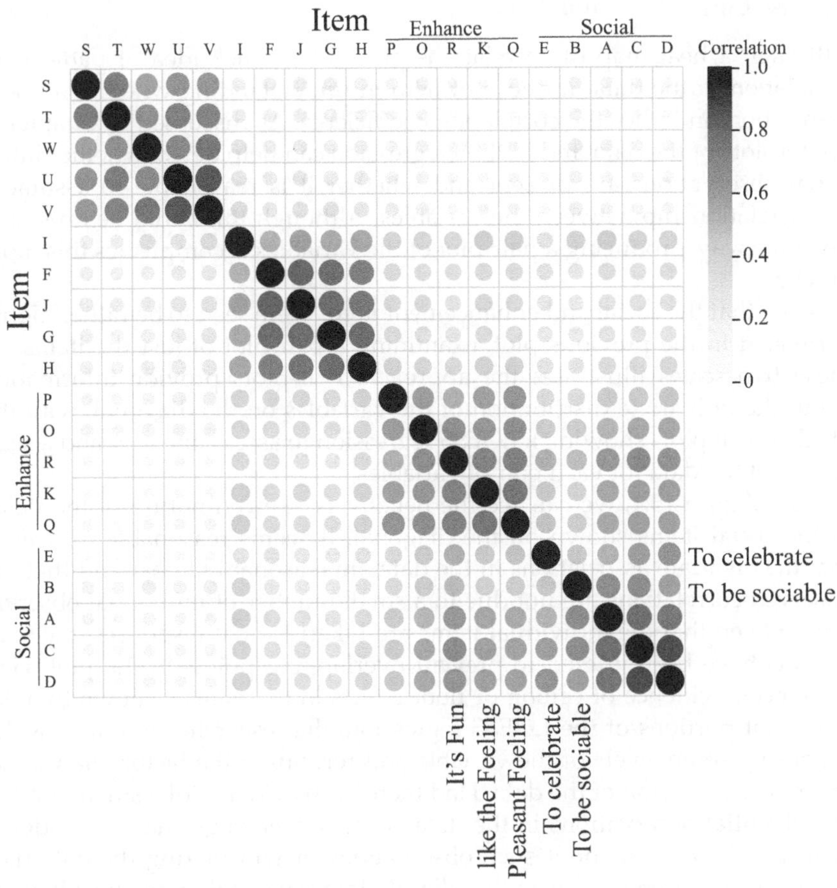


**FIGURE 15.3**
Reordered correlation matrix of reasons for drinking.

"I like the feeling," and Q, "it gives me a pleasant feeling," show substantial correlations with the Social items. In addition, two items in the social group – E, "To celebrate," and B, "To be sociable" – seem to have lower correlations than the remaining items in that group. In some situations, a researcher may consider simply dropping items that show such cross-correlations across areas or may consider aggregating items such as E and B by taking a mean with the goal of generating a pool of items that show a more homogeneous pattern of correlations. These options won't be pursued going forward in this chapter, but they are examples of the types of changes one might make the exploratory phase of evaluating the item pool.

### Pairwise Correlations and Plots

Although a heat map of correlations gives us a quick idea of patterns of correlation in the data, it does not help in understanding whether a linear association underlies the correlation coefficients. As discussed in Chapter 5, scatterplots of the data and LOESS regression can help detect whether influential observations are present and whether it is reasonable to assume a linear relationship between the variables. Although this could be done in a bivariate way, producing a pair-plot of the variables accomplishes this more quickly.

Given that the social and enhancement items were seen to be rather highly correlated in the previous plot, examining a pair-plot of just the items for these two scales illustrates the approach. Pair-plots provide information about the univariate distribution of the variables on the diagonal, with the off-diagonal plots showing a LOESS regression to assess linearity and a scatterplot of the data for each pair of variables.

From Figure 15.4, we can see the negatively skewed nature of the items in the social items shown on the diagonal. It seems reasonable to believe that the bivariate associations in the data are reasonably linear and that the observed correlations are not due to a small number of influential observations. Given that the individual items are Likert-format responses, the scatterplots have been jittered to prevent overlapping points in the plot. Had there been evidence of ceiling or floor effects in the data as shown by relatively flat portions of the LOESS regression, the researcher might consider collapsing some levels of the variable and rerunning the factor analysis on this rescaled version of the data. Had there been evidence of a small number of influential observations in the data, a researcher might have considered factoring the data without such observations or winsorizing the data (i.e., replacing an extreme value with a slightly larger value than the next highest observed score) to minimize the effects of such observations on the estimated bivariate correlations.

Because the shape of LOESS regressions is different depending on which variable is considered the criterion and which the predictor, they may appear rather different depending on which regression model is considered. Considering the items in reverse order accomplishes this alternative set of regressions. In this case, as shown in Figure 15.5, the LOESS regressions appear to be largely linear as well. Given the sample size and the fact that the individual items are Likert-format questions, the researcher may wish to consider the possibility of modeling each of the items as ordered categories, as discussed later in Chapter 17, but that will not be explored in this chapter.

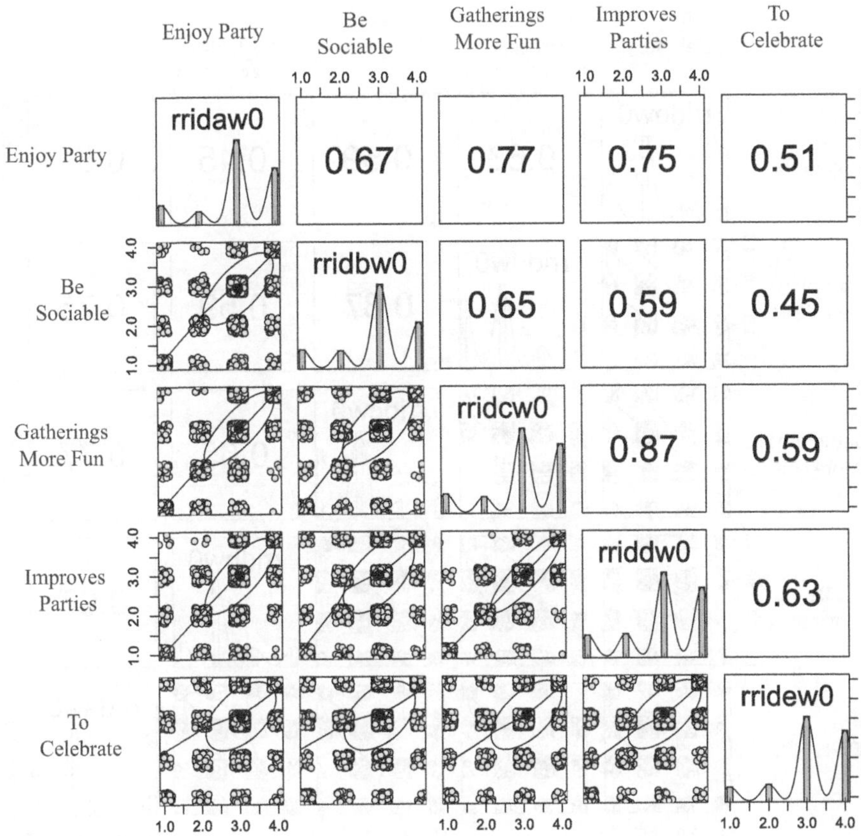


**FIGURE 15.4**
Pair-plot of social reasons in RFD measure.

## Principal Components Factor Analysis as Matrices

The output of an EFA is presented as a matrix of loadings because, unlike CFA, every manifest variable has a loading on every factor. For purposes of presentation, it is easiest to consider the decomposition of a correlation matrix, a model termed "principal components." To make the example manageable, consider a factor analysis of only eight variables, consisting of the two variables that showed the highest loadings on each of the four factors in Figure 15.1, named "rridcw0," "rriddw0," "rridgw0," "rridjw0," "rridkw0," "rridqw0," "rriduw0," and "rridvw0." Table 15.2 shows the correlations for these variables. The strategy for identifying the model will use the normalized basis approach mentioned in Chapter 8, where the sum of the squared

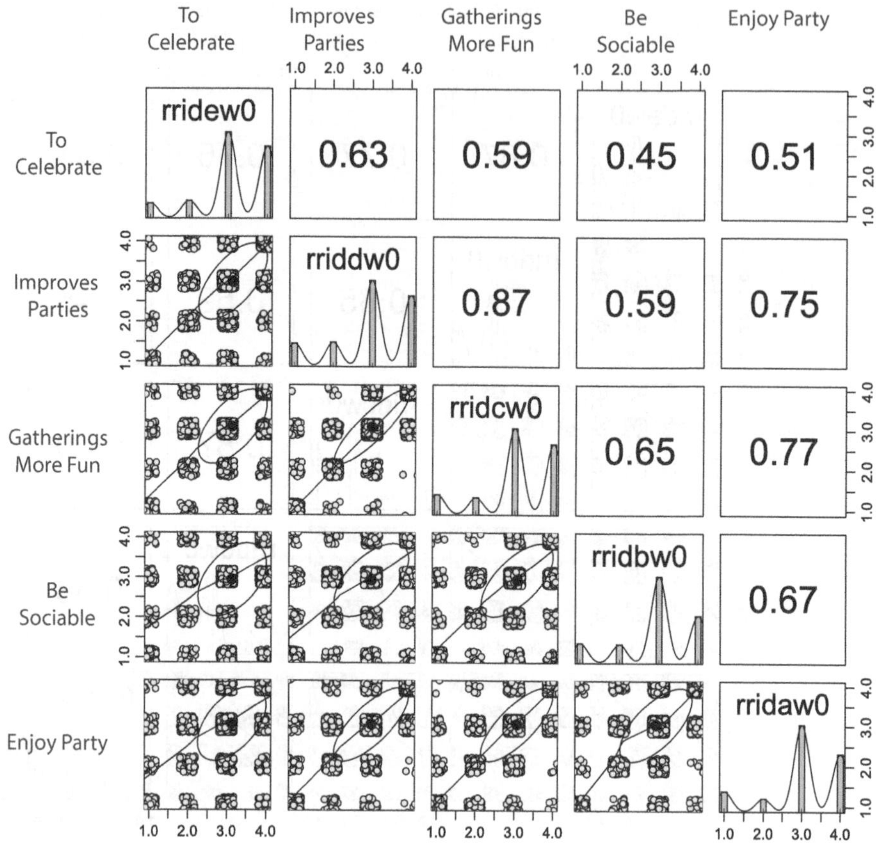


**FIGURE 15.5**
Pair-plot of social reasons in reverse order.

loadings within a factor is constrained to be one. The principal components approach is different from EFA in that it is mathematically possible to extract as many latent variables as there are manifest variables. As such, principal components is really only a re-expression of the original data, just with a different set of variables from the original manifest variables.

## Eigenvalue Decomposition of Principal Components

As mentioned earlier in "Identification via Normalized Basis." on page 129 of Chapter 8, it is possible for us to think of the loadings of a latent variable as being mathematically identified by constraining the sum of the squares to unity. If, for example, we conducted an exploratory factor analysis of this correlation matrix using this strategy, the set of loadings, termed eigenvectors, would look like those shown in Table 15.3. These can be automatically

**TABLE 15.2**

Correlation Matrix of Reduced Item Subset "I Drink Because …"

| Items | rridcw0 | rriddw0 | rridgw0 | rridjw0 | rridkw0 | rridqw0 | rriduw0 | rridvw0 |
|---|---|---|---|---|---|---|---|---|
| | It makes social gatherings more fun | It improves parties | It helps me when I feel depressed or nervous | It helps me forget about my problems | I like the feeling | It gives me a pleasant feeling | To get in with a group I like | To be liked |
| rridcw0 | 1.00 | | | | | | | |
| rriddw0 | 0.87 | 1.00 | | | | | | |
| rridgw0 | 0.30 | 0.30 | 1.00 | | | | | |
| rridjw0 | 0.33 | 0.33 | 0.78 | 1.00 | | | | |
| rridkw0 | 0.59 | 0.59 | 0.35 | 0.38 | 1.00 | | | |
| rridqw0 | 0.59 | 0.59 | 0.33 | 0.34 | 0.75 | 1.00 | | |
| rriduw0 | 0.15 | 0.16 | 0.27 | 0.29 | 0.11 | 0.13 | 1.00 | |
| rridvw0 | 0.15 | 0.16 | 0.29 | 0.31 | 0.12 | 0.15 | 0.85 | 1.00 |

**TABLE 15.3**

Eigenvectors of Correlation Matrix in Figure 15.2

| Variable | F1 | F2 | F3 | F4 | F5 | F6 | F7 | F8 |
|---|---|---|---|---|---|---|---|---|
| rridcw0 | 0.41 | 0.26 | −0.24 | 0.46 | 0.00 | −0.01 | 0.12 | 0.69 |
| rriddw0 | 0.41 | 0.25 | −0.24 | 0.46 | 0.01 | 0.03 | −0.10 | −0.70 |
| rridgw0 | 0.34 | −0.20 | 0.59 | 0.10 | 0.21 | 0.67 | −0.04 | 0.03 |
| rridjw0 | 0.35 | −0.20 | 0.56 | 0.10 | −0.17 | −0.70 | 0.02 | −0.02 |
| rridkw0 | 0.40 | 0.25 | −0.04 | −0.51 | −0.69 | 0.20 | 0.01 | 0.00 |
| rridqw0 | 0.40 | 0.24 | −0.10 | −0.54 | 0.67 | −0.18 | −0.04 | 0.01 |
| rriduw0 | 0.23 | −0.58 | −0.34 | −0.04 | −0.05 | −0.01 | −0.69 | 0.11 |
| rridvw0 | 0.23 | −0.58 | −0.31 | −0.07 | 0.01 | 0.04 | 0.70 | −0.11 |

generated by many statistical software programs and do not require the use of SEM software.

Notice a few things about these values, which will be explained in the next section. Although the sum of squares of the columns of the table add up to 1, the first factor appears to have several loadings that are lower in magnitude, while the remaining factors all seem to have two larger loadings and smaller loadings on the remaining items. The first factor also appears to consist of all positive loadings, while the remaining ones have a mix of positive and negative values. The factor variances, also known as "eigenvalues" for the data, are shown in Table 15.4.

Recall that the eigenvalues are a rough indicator of "how many variables' worth" a particular factor is. The sum of all the eigenvalues of a complete factorization of a matrix is equal to the sum of the diagonal elements of the covariance matrix. In the case of principal components analysis, the sum of the eigenvalues will equal the number of variables in the data.

## Principal Components Analysis

Recall that another way to secure a mathematically identified solution for a factor model is to freely estimate the loadings but to set the factor variance to unity, as in Chapter 8. These two solutions are mathematically equivalent representations of the same data. Specifically, the factor loadings from a principal components model can be found from the eigenvectors of an

**TABLE 15.4**

Eigenvalues of Reduced Data Set Principal Components

| Factor | F1 | F2 | F3 | F4 | F5 | F6 | F7 | F8 |
|---|---|---|---|---|---|---|---|---|
| Eigenvalue | 3.72 | 1.80 | 1.11 | 0.62 | 0.25 | 0.22 | 0.15 | 0.13 |

eigenvalue decomposition by dividing each eigenvector by the square root of its corresponding eigenvalue. For example, for the components loading of rridcw0 on the first factor, this is $0.79 = 0.41\sqrt{3.72}$. Notice that now the loadings from a communality analysis have the property that loadings on earlier factors appear larger than subsequent ones. This property means that some judgment about the importance of a factor in explaining the correlations can be made based on the loadings of that factor.

Principal components solutions is a data summary technique and not a factor model per se because it affords no parsimony beyond the original correlation matrix. Each manifest variable's variation and covariation is expressed in terms of a combination of latent variables. The method of principal factors discussed later, by contrast, seeks parsimony and replaces the ones on the diagonal of the correlation matrix with some estimate of the proportion of variability of the manifest variable, the communality ($h^2$). If, as mentioned previously, we choose to factor a covariance matrix, some estimate of the true score variance is used instead. Notice that the sum of squares of the loadings taken across all variables is still 8, the number of variables in the model, but that the sum of the squares of the loadings within a factor is the same as the eigenvalue from the original eigenvalue decomposition in Table 15.5.

## Estimation of Exploratory Component Loadings: A Path Diagram Approach

Although the eigenvalue and principal component decompositions show that, like confirmatory factor models, latent variables and factor loadings are involved in the estimation, they don't answer the question of how these numbers are estimated. One way to understand how these numbers were arrived at is to think of the process as estimating a single set of factor loadings across all manifest variables, fixing these loadings to their estimated values, and then estimating a second set of freely estimated factor loadings. This process is then continued as many times as there are manifest variables in the

**TABLE 15.5**

Principal Components Loadings

| Variable | F1 | F2 | F3 | F4 | F5 | F6 | F7 | F8 |
|---|---|---|---|---|---|---|---|---|
| rridcw0 | 0.79 | −0.35 | 0.25 | −0.36 | 0.00 | −0.01 | −0.05 | 0.25 |
| rriddw0 | 0.79 | −0.34 | 0.25 | −0.36 | 0.01 | 0.01 | 0.04 | −0.25 |
| rridgw0 | 0.66 | 0.26 | −0.62 | −0.08 | 0.11 | 0.31 | 0.01 | 0.01 |
| rridjw0 | 0.68 | 0.26 | −0.59 | −0.08 | −0.09 | −0.32 | −0.01 | −0.01 |
| rridkw0 | 0.77 | −0.34 | 0.04 | 0.40 | −0.35 | 0.09 | 0.00 | 0.00 |
| rridqw0 | 0.77 | −0.32 | 0.10 | 0.42 | 0.34 | −0.09 | 0.01 | 0.00 |
| rriduw0 | 0.44 | 0.78 | 0.36 | 0.03 | −0.02 | 0.00 | 0.27 | 0.04 |
| rridvw0 | 0.45 | 0.78 | 0.33 | 0.05 | 0.01 | 0.02 | −0.27 | −0.04 |

**TABLE 15.6**
Initial Observed Correlation Matrix

| Item | C | D | G | J | K | A | U | V |
|---|---|---|---|---|---|---|---|---|
| C | 1.00 | | | | | | | |
| D | 0.87 | 1.00 | | | | | | |
| G | 0.30 | 0.30 | 1.00 | | | | | |
| J | 0.33 | 0.33 | 0.78 | 1.00 | | | | |
| K | 0.59 | 0.59 | 0.35 | 0.38 | 1.00 | | | |
| Q | 0.59 | 0.59 | 0.33 | 0.34 | 0.75 | 1.00 | | |
| U | 0.15 | 0.16 | 0.27 | 0.29 | 0.11 | 0.13 | 1.00 | |
| V | 0.15 | 0.16 | 0.29 | 0.31 | 0.12 | 0.15 | 0.85 | 1.00 |

data. As an example, consider Table 15.6 in which the correlation matrix in Table 15.2 is shown again with shorter variable names.

The loadings for the first component are based on a one-factor model using least squares loss, as discussed in Chapter 8. The resulting path diagram is shown in Figure 15.6. As can be seen from the figure, this factor model is somewhat different from the single factor models considered in Chapter 8 in that no error variances are estimated. If we were to stop the process at this point and claim that only one factor existed in the data, we might more plausibly substitute these squared loadings on the diagonal of the correlation matrix of Table 15.6, re-estimate the factor loadings resulting from that

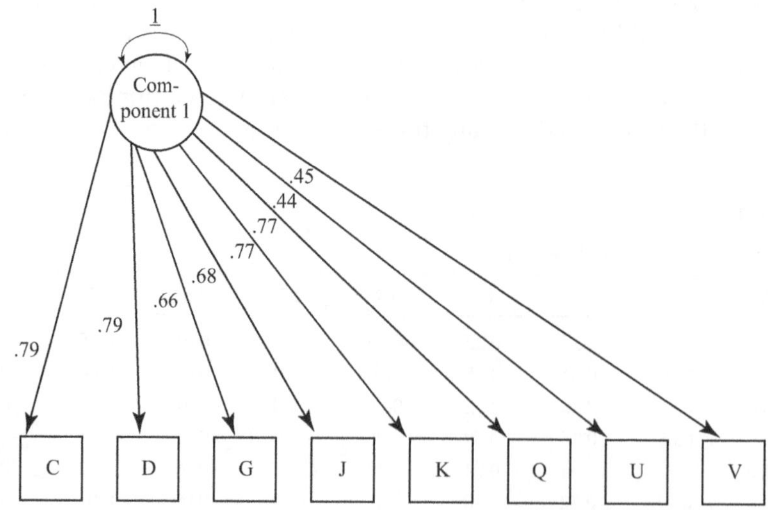


**FIGURE 15.6**
Path Model for First Principal Component

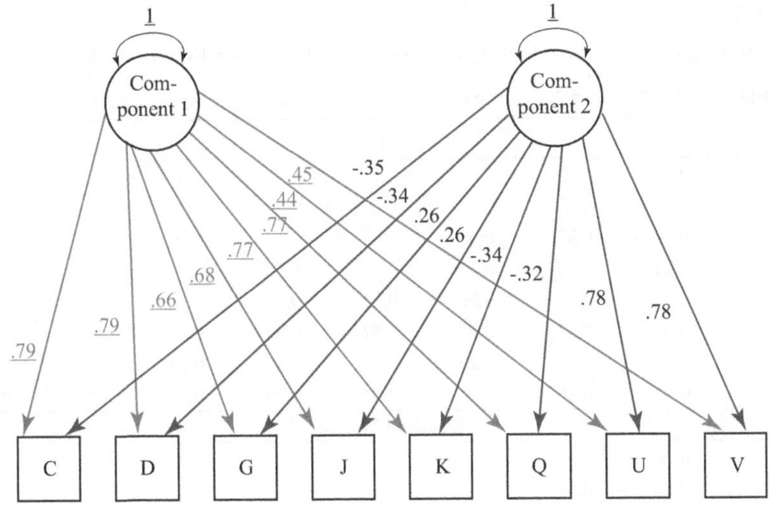


**FIGURE 15.7**
Extraction of second principal component.

matrix, and continue the process until no substantial change in the communality estimates is observed. This action is exactly the process described later in this chapter when describing iterative improvement of communalities.

To extract a second component, we then fix the loadings determined earlier to their estimated values and estimate a second set of loadings, as shown in Figure 15.7. As mentioned earlier, these loadings are generally smaller in absolute value than loadings on the first factor. This is due to the fact that the first factor explained as much of the pattern of covariation between variables as was possible, and the estimated loadings of the second factor explain only the residual variation remaining in the correlation matrix. To see how this happens, consider the residual matrix that results when the predicted matrix, based on the first factor shown in the top of Table 15.7, is subtracted from the original correlation matrix. As can be seen, each element of the predicted correlation matrix is equal to that predicted by the tracing rules (e.g., the variance of Item C is $0.62 = 0.79^2$, and the covariance of Item C with Item *V* is $0.36 = 0.79 * 0.45$. The corresponding residual matrix is calculated by subtracting these predicted values from the observed correlation matrix. As a result, the estimated loadings of this second component, being based on the positive and negative values, must necessarily be composed of some positive and some negative loadings given that errors of prediction are composed of some positive and some negative covariances. Given that it is based on what is "left over" after the first factor is extracted, we would also expect that these loadings be lower in absolute magnitude relative to the first component. The sum of the diagonal elements of the predicted matrix is equal to the

**TABLE 15.7**

Predicted Matrix Based on First Component and Resulting Residual Matrix

| Variable | C | d | g | j | k | q | u | v |
|---|---|---|---|---|---|---|---|---|
| | | | | Predicted Matrix | | | | |
| c | 0.62 | | | | | | | |
| d | 0.63 | 0.63 | | | | | | |
| g | 0.52 | 0.52 | 0.43 | | | | | |
| j | 0.54 | 0.54 | 0.45 | 0.46 | | | | |
| k | 0.61 | 0.61 | 0.50 | 0.52 | 0.59 | | | |
| q | 0.60 | 0.61 | 0.50 | 0.52 | 0.59 | 0.59 | | |
| u | 0.34 | 0.35 | 0.29 | 0.3 | 0.34 | 0.33 | 0.19 | |
| v | 0.36 | 0.36 | 0.3 | 0.31 | 0.35 | 0.35 | 0.2 | 0.2 |
| | | | | Residual Matrix | | | | |
| c | 0.38 | | | | | | | |
| d | 0.24 | 0.37 | | | | | | |
| g | −0.22 | −0.22 | 0.57 | | | | | |
| j | −0.21 | −0.21 | 0.33 | 0.54 | | | | |
| k | −0.02 | −0.02 | −0.15 | −0.14 | 0.41 | | | |
| q | −0.02 | −0.02 | −0.17 | −0.18 | 0.16 | 0.41 | | |
| u | −0.19 | −0.19 | −0.02 | −0.01 | −0.23 | −0.2 | 0.81 | |
| v | −0.20 | −0.20 | 0.00 | 0.00 | −0.23 | −0.19 | 0.65 | 0.8 |

eigenvalue for the first component, and the term "trace" is used to describe this summation over the diagonal of a matrix.

The process of estimating loadings and then fixing these values is then continued as many times as there are manifest variables in the data. Figure 15.8 shows the diagram for the extraction of the fourth component. As mentioned earlier, these correspond to the matrix values shown in Table 15.5, with the absolute magnitude of loadings associated with each successive component becoming increasingly smaller as the components are extracted.

## Principal Factors Using Communality Estimates

We are now ready to introduce the principal factor method, which is also sometimes referred to as the principal axis method. As with principal components, the method, although straightforward, makes the somewhat unrealistic assumption that all variation in the manifest variables can be accounted for extraction of as many components are there are variables. If only we knew exactly the proportion of variability in each manifest variable due to

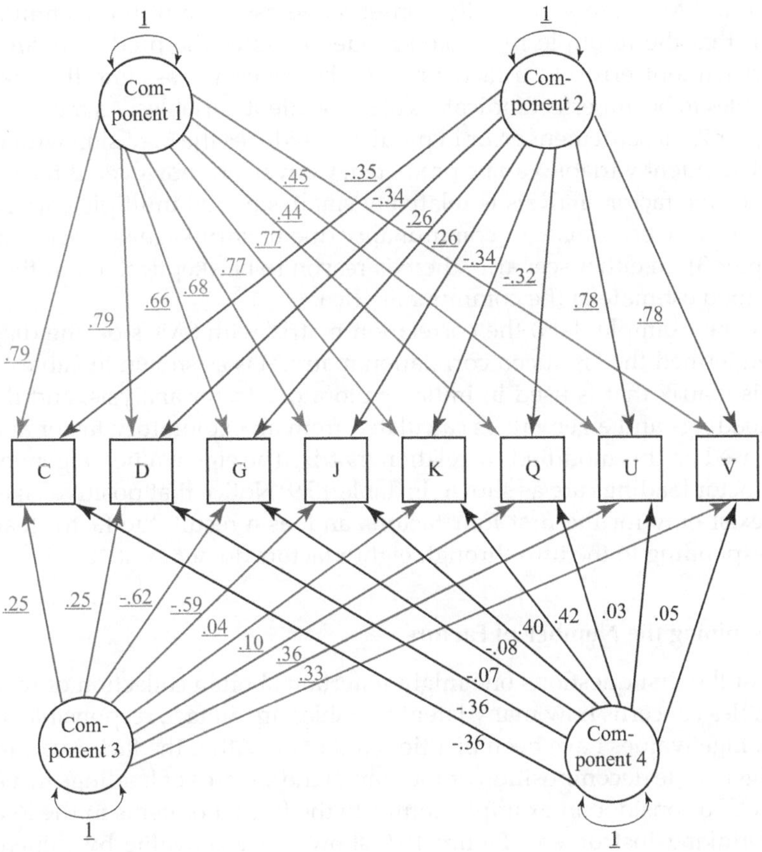


**FIGURE 15.8**
Extraction of first four principal components.

the latent variables (the communality), we could then extract factor loadings until the resulting residual matrix contained only zeros. The fundamental problem, then, is that we could determine the number of latent variables in the model if we knew the true communalities, or alternatively, that we could determine the communalities if we knew the number of factors. However, as we know neither of these, we have to begin by approximating the communalities as a way of making a decision about the number of factors.

Historically, many initial approximations of the communalities have been used, such as replacing the diagonal with the highest correlation of that variable with the remaining variables, or using estimated intraclass correlations. By far the most popular estimate of the communality estimate is the squared multiple correlation (SMC), predicting each manifest variable from the remaining variables in the data. SMC communality estimates are the default because computer software can easily calculate them.[1]

Use of SMCs are a generally conservative estimate of the communality, given that the multiple regression model assumes the predictors are measured without error. In a factor model, however, we assume the manifest variables to be imperfect indicators of some latent variables. As mentioned in Chapter 7, measurement error in predictors reduces the precision with which the dependent variable can be predicted. Conversely, however, if the sample size for the factor analysis is relatively small, squared multiple correlations may be an overestimate of communality due to shrinkage (as mentioned in Chapter 3). In either scenario, there is reason to be skeptical about the SMC as a good estimate of the communality ($h^2$).

For the example data, the correlation matrix with SMCs on the diagonal (often termed the "reduced correlation matrix") is as shown in Table 15.8. It is this matrix that is used in initial exploratory factor analysis, and the initial loadings and eigenvalues calculated from an exploratory factor analysis are based on this modified correlation matrix. The eigenvalues, eigenvectors, and factor loadings are as shown in Table 15.9. Notice that positive eigenvalues exist only for the first four factors, and, as a result, the factor loadings corresponding to the fifth through eighth factors do not exist.

## Determining the Number of Factors

One of the first questions one might entertain about a collection of manifest variables concerns how many latent variables are, in fact, responsible for the data. Eigenvalues have been mentioned that are either the factor variance of an eigenvalue decomposition or the sum of the squares of loadings in a factor model. To consider an example, return to the full set of items in the Reasons for Drinking instrument. Figure 15.9 shows an eigenvalue by dimensions plot, which is often used to decide the number of factors present in the data. The eigenvalues associated with the components analysis is shown as a black line in the figure.

### *Kaiser Criterion*

The default behavior of many statistical software packages is to retain all factors with an eigenvalue greater than 1, often referred to as the "Kaiser criterion" even though Kaiser and Rice (1974) were clear on the point that this criterion was not meant to identify the number of factors in the data, but it corresponds to Guttman's weaker lower bound on the number of factors. It should be noted that the Kaiser criterion is appropriate only for use in components analysis, and not in the EFA model, in which some estimate of the communality is made. Use of the Kaiser rule when communality estimates such as squared multiple correlations are used is inappropriate (Horn, 1965).

Graphically, the Kaiser criterion is shown as a reference line of 1, "Kaiser eigenvalue = 1." The number of factors to retain is four factors; based on this

**TABLE 15.8**

Reduced Correlation Matrix for EFA "I drink because …"

| Items | rridcw0 | rriddw0 | rridgw0 | rridjw0 | rridkw0 | rridqw0 | rriduw0 | rridvw0 |
|---|---|---|---|---|---|---|---|---|
| | It makes social gatherings more fun | It improves parties | It helps me when I feel depressed or nervous | It helps me forget about my problems | I like the feeling | It gives me a pleasant feeling | To get in with a group I like | To be liked |
| rridcw0 | 0.7645 | | | | | | | |
| rriddw0 | 0.8679 | 0.7665 | | | | | | |
| rridgw0 | 0.2986 | 0.3042 | 0.6155 | | | | | |
| rridjw0 | 0.3261 | 0.3266 | 0.7795 | 0.6292 | | | | |
| rridkw0 | 0.5889 | 0.5916 | 0.3503 | 0.3791 | 0.6074 | | | |
| rridqw0 | 0.5860 | 0.5888 | 0.3330 | 0.3442 | 0.7464 | 0.5976 | | |
| rriduw0 | 0.1510 | 0.1602 | 0.2661 | 0.2890 | 0.1065 | 0.1306 | 0.7239 | |
| rridvw0 | 0.1538 | 0.1576 | 0.293 | 0.3089 | 0.1216 | 0.1522 | 0.8498 | 0.728 |

**TABLE 15.9**

Eigenvalues/Vectors and Factor Loadings SMC Communality Estimates

| Eigen values | Items | Eigenvectors | | | | | | | | Factor Loadings ($\lambda\sqrt{Eigenvalue}$) | | | |
|---|---|---|---|---|---|---|---|---|---|---|---|---|---|
| | | F1 | F2 | F3 | F4 | F5 | F6 | F7 | F8 | Fl | F2 | F3 | F4 |
| 3.39 | c | 0.42 | −0.27 | 0.26 | 0.43 | 0.70 | 0.06 | −0.03 | 0.03 | 0.78 | −0.33 | 0.23 | 0.23 |
| 1.51 | d | 0.42 | −0.26 | 0.26 | 0.42 | −0.70 | −0.09 | 0.03 | −0.02 | 0.78 | −0.48 | 0.23 | 0.23 |
| 0.77 | g | 0.33 | 0.17 | −0.59 | 0.13 | −0.03 | −0.11 | −0.32 | 0.61 | 0.61 | 0.32 | −0.52 | 0.07 |
| 0.28 | j | 0.35 | 0.18 | −0.57 | 0.13 | 0.03 | 0.12 | 0.29 | −0.64 | 0.64 | 0.33 | −0.50 | 0.07 |
| −0.10 | k | 0.39 | −0.23 | 0.01 | −0.54 | 0.00 | 0.18 | 0.60 | 0.33 | 0.72 | −0.43 | 0.01 | −0.29 |
| −0.12 | q | 0.39 | −0.21 | 0.06 | −0.55 | 0.00 | −0.15 | −0.61 | −0.32 | 0.71 | −0.40 | 0.05 | −0.29 |
| −0.14 | u | 0.23 | 0.59 | 0.32 | −0.04 | −0.08 | 0.67 | −0.19 | 0.04 | 0.42 | 1.09 | 0.28 | −0.02 |
| −0.16 | v | 0.24 | 0.59 | 0.29 | −0.07 | 0.08 | −0.67 | 0.21 | −0.01 | 0.44 | 1.09 | 0.25 | −0.04 |

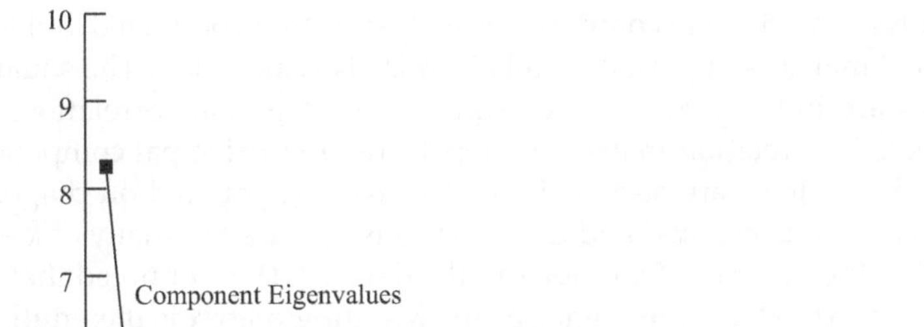


**FIGURE 15.9**
Eigenvalues by dimensions plot for example data.

criterion, the eigenvalue for five factors is smaller than 1. Although the criterion perhaps is appealing because of its simplicity, there are many reasons not to use it in practice, and Kaiser did not consider this criterion a valid way to identify the number of factors. Factors can be identified using this criterion using random simulated uncorrelated data (Loehlin, 2004). Others have noted that it hardly makes sense to judge a factor with an eigenvalue of 1.01 as a "major factor" and a factor with an eigenvalue of 0.99 as a "minor factor" (Fabrigar, Wegener, MacCallum, & Strahan, 1999). While there is some reason to think that an eigenvalue greater than 1 corresponds to the latent variable explaining at least one manifest "variable's worth" of variance, there are many examples of the rule not working well in practice (e.g., Cattell & Vogelmann, 1977; Tucker, Koopman, & Linn, 1969; Zwick & Velicer, 1982, 1986).

### *Velicer's MAP Test*

Velicer (1976) proposed a method of determining the number of factors in a principal components analysis commonly referred to as the MAP test. Under this analysis, a principal components analysis is first conducted, but in which a series of matrices of partial correlations is also calculated. The squared correlations are initially the average squared off-diagonal correlations for the unpartialed correlation matrix. In Step 1, the first principal component and its associated items are partialed out. The average squared off-diagonal correlation is then computed and this process is repeated as many as $k - 1$ times where $k$ is the number of variables in the data set. (Velicer noted that it is not possible to extract $k$ components, as the resulting matrix is then null.)

All average squared correlations for each step are then examined and the step associated with the lowest average squared partial correlation is chosen as the number of components or factors to retain. The MAP test has performed well in many simulation studies (e.g. Ruscio & Roche, 2012) and is widely available. For the Reasons for Drinking data set, the MAP test returns a value of four factors, shown in Table 15.10.

### *Cattell's Scree Test*

A third approach to determining the number of factors is based on the observation that the pattern of eigenvalues often shows a decided discontinuity across the number of factors. The term "scree" is taken from geography and is used to describe the rocky outfall of stones seen prior to arriving at an actual mountain. In the scree approach, the researcher looks for an elbow in the data and retains those factors that are one fewer than this elbow. In Figure 15.9, for example, one might visually entertain a three- or four-factor solution as reasonable. Gorsuch and Nelson (1981) proposed a formal test of the number of factors based on whether the regression line of eigenvalues changes significantly; however, often this decision is made visually. For these data, the nCng package in R returns a value of three factors.

### *Parallel Components Analysis*

Horn (1965) noted that, at the population level, eigenvalues of a correlation matrix composed of zero correlations yields eigenvalues all equal to 1. He noted however, that some eigenvalues will be larger than 1 and others less than 1 in sample data. This has led some to speculate that this fact, as well as the non-normality of real-world data, would lead to distortions of the eigenvalues by dimensions plot. Horn (1965) proposed that one could construct a

**TABLE 15.10**

Map Test Values for Example Data

| Factors | 1 | 2 | 3 | 4 | 5 | 6 | 7 | 8 |
|---|---|---|---|---|---|---|---|---|
| MAP | 0.08 | 0.06 | 0.03 | **0.02** | 0.03 | 0.03 | 0.04 | 0.05 |

pseudo data set by randomly permuting the rows of data, effectively scrambling the data by matching values of each variable for each observation with some other values found in the data. The plot of this permuted data set could then be compared to the eigenvalue plot of the observed data and those factors retained where the eigenvalue of the observed data is greater than that of the randomly permuted data set. The argument is that if non-normality of the data does affect the eigenvalues, the randomly permuted data would be similarly affected and the dimensionality recovered by the observation of eigenvalues greater than the randomly permuted data. Performing this analysis only once, though, still yields the problem of determining if the decision regarding dimensionality is the result of chance variation.

Given the advent of high-speed computing, two alternatives have been used. First, it is possible to conduct this random permutation several times (say 1,000) and plot the eigenvalues from the 95th percentile, thus giving some idea of the sampling error involved in the shape of the eigenvalue by dimensions curve by examining the mean, median, and 95th percentile points for such a permuted dataset. The second, computationally simpler approach, which does not involve the raw data, is to simulate a sample drawn from multivariate normal distribution with sample size and number of variables equal to the original data.

Figure 15.9 shows the eigenvalues of parallel analyses of the permuted and simulated data for the three-, four-, and five-factor solutions. Three factors are indicated, as the permuted and simulated eigenvalues for four factors are larger than that observed for the data, while the eigenvalues for the third factor based on the data are larger than that for the simulated or permuted data. Although the randomly generated and permuted approaches yielded almost practically identical results in this large sample, eigenvalues based on the permuted data composed of only 200 observations yielded larger values (1.46, 1.41, and 1.33 for the three-, four-, and five-factor eigenvalues, respectively) lending support to the idea that non-normality inflates eigenvalue estimation at least in this dataset.

### How Many Factors to Retain?

This discussion and example of the scree plots, nCNG, MAP, and parallel factor and component tests identify three and possibly four factors in the data. The reordered correlation matrix shown in Figure 15.3 looks as if four groups of items are present in the data, and this number also guided selection of the items in the measure. Generally, the various methods of identifying the number of factors responsible for the data are best viewed as general guidelines rather than "clear bright lines" of guidance.

Lim and Jahng (2019) surveyed many of the parallel factor model variants, many of which have to do with improving communality estimates, and limit their discussion to the normal theory data simulation approach. After presenting the rationales and criticisms of these alternatives, they present simulation data supporting the idea that the number of factors identified by parallel factor analysis should serve only as a general starting point for evaluating the dimensionality of the data. When factors in the data are highly correlated or when factor loadings underlying the data are weak, parallel factor analysis is more likely to falsely underestimate the number of factors in the data. They advise that factor solutions within a ±1 range should be explored, and that the final decision on factor structure should rest on the interpretability of the resulting solution. One must keep in mind that the factors or components portrayed here inform the researcher about the presence of successive orthogonal components. If the actual model underlying the data consists of correlated factors, or if models such as the random intercept model underlie the data, the presence of these additional dimensions are less likely to be detected. It is reasonable to assume that similar advice applies to the other tests presented in this chapter as well.

## Iterative Improvement of Communality

Once a decision has been made about the dimensionality of the data, it is then possible to improve the communality estimates of a candidate factor model. As mentioned before, an updated communality estimate for the manifest variables can be calculated as the estimated variance of each manifest variable. For the orthogonal unrotated factor models presented so far, this communality is simply the sum of the squared loadings for that variable. Factoring this matrix with the new set of communalities will then result in an updated set of factor loadings that can be inserted on the diagonal of the correlation matrix and refactored. This process of repeatedly factoring the data and replacing the diagonal with updated communality estimates is then repeated until no significant improvement in the communality values is observed. Although in theory the same factor loadings and communalities should be observed regardless of whether, say, a principal components or a principal factor model using squared multiple correlations is initially used, differences in the final estimated factor loadings can occur, especially in smaller data sets or factor analyses with relatively large numbers of variables relative to their sample size, making side analyses involving different initial communality estimates useful.

In some situations, this iterative approach may result in communalities which exceed 1, a situation referred to as a Heywood case. Although Heywood cases seem to affect variants of the exploratory factor model such as alpha factor analysis, it can also occur in standard exploratory factor models as well. This can be indicative of over factoring of the data, sampling variation due to a small sample size, lack of statistical power due to a large number of items under consideration relative to sample size, or the presence of influential observations in the data. Remedies for the model can result in imposing a constraint that communalities may not exceed one, examination of the data for influential outliers or other types of assumption checking discussed in Chapter 17.

## Note

1 Specifically, the first step is to calculate the inverse of the correlation matrix and save the diagonal values (i.e., the "trace") of this matrix $\left(\text{diag}\left(R^{-1}\right)\right)$. Squared multiple correlations result when the reciprocal of $\text{diag}\left(R^{-1}\right)$ is subtracted from 1, i.e., $1 - \text{trace}\left(\text{diag}\left(R^{-1}\right)\right)$.

# 16

# *Factor Rotation*

Once a decision has been made concerning the number of factors underlying the data, factor rotation techniques are used to transform the loadings into some form of parsimonious model such as those discussed in Chapter 12. This transformation is accomplished by matrix multiplication, and the adequacy of this transformation can be examined using scatterplots of factor loadings. As described in Chapter 12, these parsimonious models can involve uncorrelated (i.e., "orthogonal") or correlated ("oblique") factors as well as patterns of general factors with additional subsets of items. Just as there is nothing occult about the process of extraction of exploratory factor loadings, mechanical rotations are simply a way to mathematically arrive at a parsimonious model without having to specify how variables are organized. "Hand rotation," in which the researcher specifies how to reorganize the factor loadings from EFA to a different form, is a simple introduction to the idea and demonstrates how vector plots can provide a visual assessment of the adequacy of a rotation. Strategies for mechanical rotations in which an algorithm is used to transform the loadings to a different set of factors are then presented. These strategies are designed to represent Thurstone's idea of simple structure.

## Orthogonal "Hand Rotation" Using Direction Cosines

The heat map of the items in the Reasons for Drinking scale in the previous chapter showed that the items dealing with social motivations for drinking appeared relatively uncorrelated with items dealing with conformity reasons. For this reason, rotating the original EFA factors involving these two subsets of items to an orthogonal rotation is a simple example.

Consider the results of a maximum likelihood iterated communality two-factor solution for these items, shown in Table 16.1. Notice that each of the items appears to have large numbers on both factors. It's important to mention that this solution may be conceptually meaningful. It might be reasonable to think that each item represents a combination of a general factor differentiating individuals with many reasons for drinking from those with fewer, and a second factor that differentiates individuals who have positive affiliation versus need-based reasons for drinking. The pattern of positive


DOI: 10.1201/9781003595267-16

**TABLE 16.1**

Unrotated Two-Factor Solution for Social and Conformity Drinking Reasons

| Item "I Drink Because ..." | Factor I | Factor II |
|---|---|---|
| C because it makes social gatherings more fun | 0.80 | −0.49 |
| D because it improves parties | 0.80 | −0.48 |
| A because it helps me enjoy party | 0.74 | −0.35 |
| B to be sociable | 0.69 | −0.31 |
| E to celebrate | 0.56 | −0.38 |
| V to be liked | 0.58 | 0.72 |
| U to get in with a group I like | 0.57 | 0.72 |
| T so others won't kid me | 0.49 | 0.70 |
| W so I won't feel left out | 0.55 | 0.6 |
| S because my friends | 0.49 | 0.6 |

and negative loadings on the second factor, as well as the fact that the second factor's loadings are lower in magnitude, was due to the process by which the factors were extracted. An examination of the scatterplot of the factor loadings shown in Figure 16.1 suggests that two clusters of items are present in the data.

This plot of the unrotated factor loadings is known as a vector plot, as introduced in Chapter 6. It is a scatterplot of the factor loadings. The horizontal and vertical axes represent the factors, and the length of the lines conveys the square root of the communality ($h^2$) of each variable. If you have had a

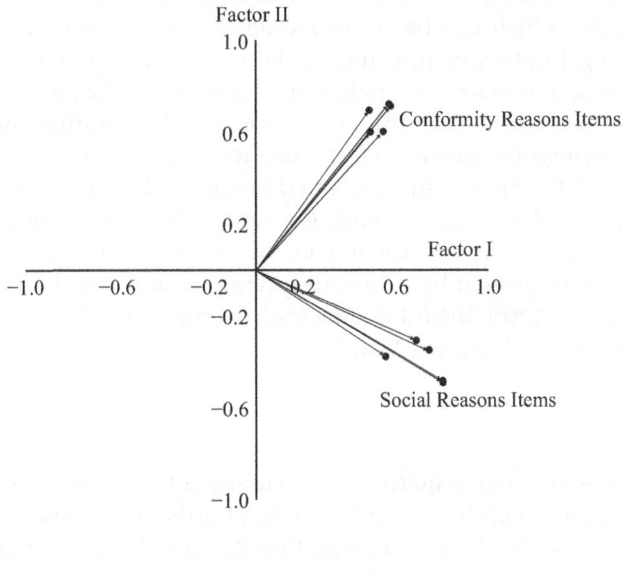


**FIGURE 16.1**
Vector plot of unrotated factor loadings for social and conformity drinking reasons.

course in linear algebra, you may recall that the length of this line is called the "norm" of the vector. The upper right-hand points represent the items relating to conformity reasons for drinking, and items in the lower right-hand part of the plot represent the social reasons. Graphically, it might make more sense to think of two latent variables that run through the clouds of variables in the vector space.

## Direction Cosines

Graphically, what is desired is to rotate the unrotated factor loadings to a new pair of (orthogonal in this case) reference axes so that each variable vector has a sizable loading on only one factor and a near zero loading on the remaining factor. For example, in these data, the Varimax rotation to be described later in this chapter suggests that a rotation of 56° would produce a simple structure solution. The angle of rotation showing the movement of each unrotated factor to its new rotated location is shown by the curved arrows in the Figure 16.2. There is nothing magical about the degree of rotation suggested by the Varimax rotation. Other rotations will represent the data equally well. For example, if one of the items in the instrument represented some kind of "gold standard," one could rotate the solution so that the item had a loading on only one factor and no loading on the remaining factor. This is the general idea behind target rotations, which will be described later.

## Transformation Matrices for Orthogonal Rotation

When doing a rotation of even just two factors, it is necessary to keep track of four angles, which can be cumbersome. For that reason, it's helpful to use matrix algebra to accomplish the rotation, and it is here where the idea of a transformation matrix, $T$, makes our work easier. Just as factor pattern matrices are organized with the latent variables in columns and variables in rows to express the arrows which run from latent to manifest variables, the columns of $T$ represent the unrotated factors and the rows of $T$ represent the destination rotated factors. Each element of $T$ is the cosine of the angle between the unrotated and rotated factors in counterclockwise direction. Rotation is accomplished by post-multiplying the unrotated factor loadings by $T$. In equation form, the rotated factor loadings $\Lambda_r$ can be calculated from the unrotated factor loadings $\Lambda$ and $T$ as:

$$\Lambda_r = \Lambda T \tag{16.1}$$

$T$ is a square, symmetric matrix with as many columns and rows as factors. For those familiar with linear algebra, $T$ is an orthonormal basis for the vector space of factor loadings, meaning that the new factors remain orthogonal ("Ortho") and the length of the vectors (i.e., the "norm," which is the distance from the origin to the end of the vector) remains the same. Oblique rotations, as we will see, are normal bases only, meaning that they retain the norm of the vectors but are not necessarily orthogonal.

$T$ consists of the cosines of the angles shown in Figure 16.2. These are:

$$T = \begin{bmatrix} \cos(56^\circ) & \cos(326^\circ) \\ \cos(146^\circ) & \cos(56^\circ) \end{bmatrix} = \begin{bmatrix} 0.559 & 0.829 \\ -0.829 & 0.559 \end{bmatrix}$$

Using the numbers from the example here, this is:

$$\Lambda_r = \Lambda T = \begin{bmatrix} 0.80 & -0.49 \\ 0.80 & -0.48 \\ 0.74 & -0.35 \\ 0.69 & -0.31 \\ 0.56 & -0.38 \\ 0.58 & 0.72 \\ 0.57 & 0.72 \\ 0.49 & 0.70 \\ 0.55 & 0.60 \\ 0.49 & 0.60 \end{bmatrix} \begin{bmatrix} 0.829 & 0.559 \\ -0.559 & 0.829 \end{bmatrix} = \begin{bmatrix} 0.94 & 0.04 \\ 0.93 & 0.05 \\ 0.81 & 0.12 \\ 0.75 & 0.13 \\ 0.68 & 0.00 \\ 0.08 & 0.92 \\ 0.07 & 0.92 \\ 0.01 & 0.85 \\ 0.12 & 0.80 \\ 0.07 & 0.77 \end{bmatrix} \quad (16.2)$$

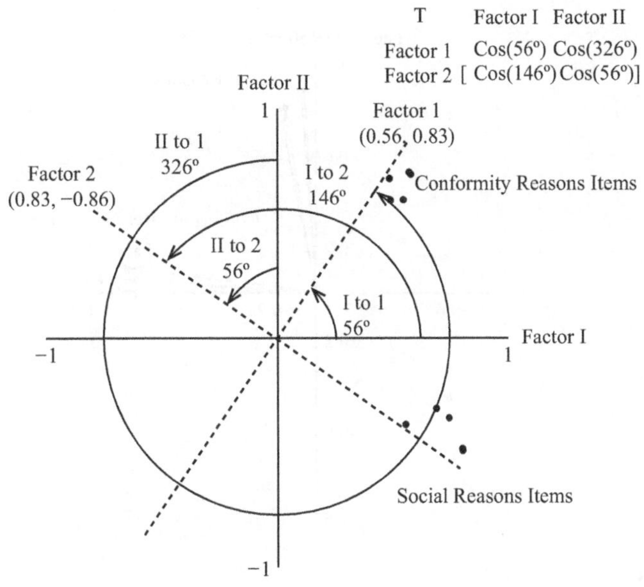


**FIGURE 16.2**
Direction cosines of an orthogonal varimax rotation.

$T$ has some useful properties. First, the rows of $T$ represent the endpoints of the new basis in terms of the original unrotated factor space. For example, the end of the unit vector for the first factor goes through the point (0.56, 0.83) in the metric of the unrotated factor space and the corresponding point for the second rotated factor goes through the point (0.83, –0.86) as shown in the figure. If calculating a rotation by hand, one check that can be done is to multiply $T$ by its transpox matrix, e.g.:

$$TT^T = \begin{bmatrix} 0.829 & 0.559 \\ -0.559 & 0.829 \end{bmatrix} \begin{bmatrix} 0.829 & -0.559 \\ 0.559 & 0.829 \end{bmatrix} = \begin{bmatrix} 1 & 0 \\ 0 & 1 \end{bmatrix} \tag{16.3}$$

The rotated factor loadings using the new rotated dimensions is shown in Figure 16.3. Under this rotation, the first five variables appear to load on a social motivation for drinking factor, while the last five items appear to load on a conformity motivation for drinking. When looking at this graph, an even cleaner picture would emerge if we could somehow have the two factors go more directly through the vector space. Doing so means that the requirement that the two factors be orthogonal would have to be relaxed, i.e., that the angle between the two factors would be less than 90°. In general, normal rotations that do not require that the resulting factors be orthogonal are referred to as "oblique rotations," a topic to which we turn next.

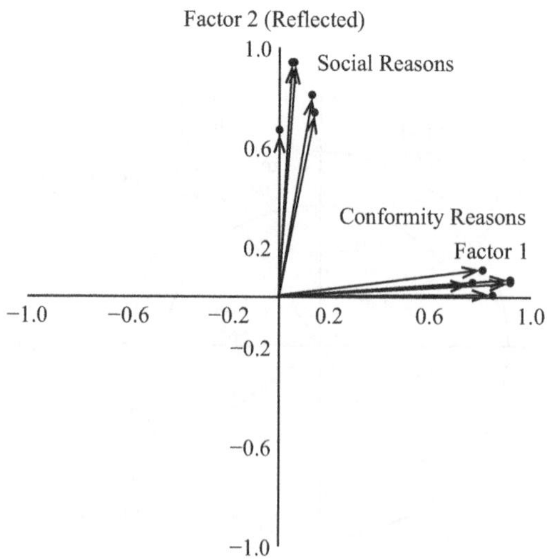


**FIGURE 16.3**
Rotated Varimax rotation.

## Oblique Rotation Using Direction Cosines

Recall from Chapter 15 and the heat map shown in Figure 15.1 that a rather significant correlation appeared to exist between items dealing with social motivations for drinking and enhancement motivations. An EFA example of these two sets of items is aninteresting example of how oblique rotation works. The items from these two sets and their unrotated factor loadings are shown in Table 16.2, and the vector plot of the data is shown in Figure 16.4.

**TABLE 16.2**

Unrotated Two-Factor Solution for Social and Enhancement Drinking Reasons

| Item "I Drink ..." | Factor I | Factor II |
|---|---|---|
| C because it makes social gatherings more fun | 0.92 | −0.2 |
| D because it improves parties | 0.92 | −0.19 |
| A because it helps me enjoy party | 0.81 | −0.11 |
| B to be sociable | 0.74 | −0.13 |
| E to celebrate | 0.67 | −0.08 |
| Q because it gives me a pleasant feeling | 0.70 | 0.52 |
| K because I like the feeling | 0.70 | 0.33 |
| P to get high | 0.57 | 0.43 |
| O because it's exciting | 0.61 | 0.39 |
| R because it's fun | 0.74 | 0.3 |

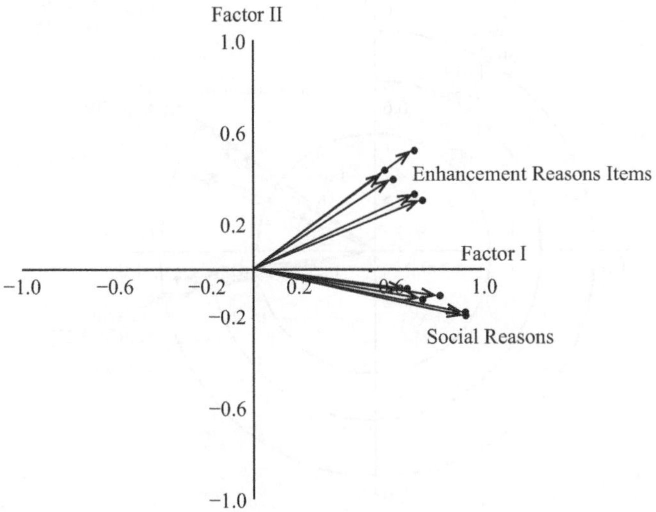


**FIGURE 16.4**

Unrotated vector plot of unrotated social and enhancement drinking reasons.

The vector plot for items from these two subscales looks rather different from the vector plot of Figure 16.1. As before, the unrotated Factor 1 of Table 16.4 could represent a latent variable in its own right, with some individuals simply having positive reasons to drink and, within that, some of these reasons are related to positive social experiences; and at the other end of that continuum, some individuals have positive individual experiences. Each item of the survey, though, provides information about both factors. Rotating to an oblique solution can provide a simple structure for these two sets of items in which the variables should have loadings for one only one latent variable. The angle between the rotated factors will no longer be a right angle, however. The cosine of the angle between the rotated factors will convey the degree of correlation between the two factors. The same strategy of using direction cosines organized into *T* will be used as before, with the rows of *T* representing the cosine from the unrotated to the rotated factor space as shown in Figure 16.5.

The mathematics of rotation is exactly the same as before, except that the new reference axes, although of unit length, are not orthogonal to each other. If we wish to find the factor loadings associated with the oblique Promax rotation (discussed in more detail below), we multiply the unrotated factor loadings by the *T* matrix provided by the program. This *T* matrix, however, will not correspond to the direction cosines shown in Figure 16.5 because the oblique rotation is not an orthonormal basis. It is used to produce factor loadings in the oblique solution. It can be thought of instead as a set of

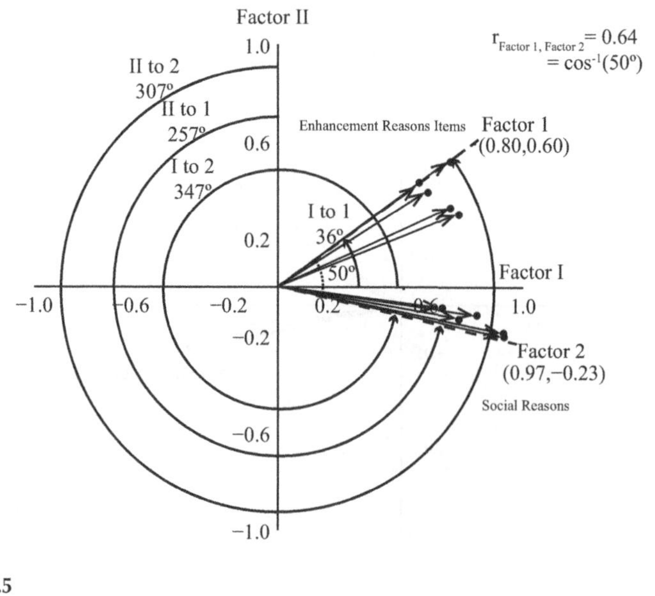


**FIGURE 16.5**
Direction cosines for oblique rotation.

standardized regression weights in which the predictor variables (i.e., the rotated factors) are correlated. For these data, this is:

$$\Lambda T = \begin{bmatrix} 0.92 & -0.20 \\ 0.92 & -0.19 \\ 0.81 & -0.11 \\ 0.74 & -0.13 \\ 0.67 & -0.08 \\ 0.70 & 0.52 \\ 0.70 & 0.33 \\ 0.57 & 0.43 \\ 0.61 & 0.39 \\ 0.74 & 0.30 \end{bmatrix} \begin{bmatrix} 0.780 & 0.301 \\ -1.045 & 1.268 \end{bmatrix} = \begin{bmatrix} 0.93 & 0.02 \\ 0.91 & 0.04 \\ 0.75 & 0.10 \\ 0.71 & 0.06 \\ 0.61 & 0.09 \\ 0.01 & 0.87 \\ 0.21 & 0.62 \\ 0.00 & 0.71 \\ 0.07 & 0.68 \\ 0.26 & 0.60 \end{bmatrix} \tag{16.4}$$

If we wish to understand where the axes of the new oblique factors are in terms of the original orthogonal unrotated solution, we can refer to the inverse of the oblique $T$, which is:

$$T^{-1} = \begin{bmatrix} 0.780 & 0.301 \\ -1.045 & 1.268 \end{bmatrix}^{-1} = \begin{bmatrix} 0.973 & -0.231 \\ 0.801 & 0.598 \end{bmatrix} \tag{16.5}$$

As this is a rotation from an oblique to an orthonormal basis, we can again multiply this matrix by its transpose to find the degree of correlation between the factors of the oblique solution, as follows:

$$TT^{T} = \begin{bmatrix} 0.973 & -0.231 \\ 0.801 & 0.598 \end{bmatrix} \begin{bmatrix} 0.973 & 0.801 \\ -0.231 & 0.598 \end{bmatrix} = \begin{bmatrix} 1. & 0.64 \\ 0.64 & 1 \end{bmatrix} \tag{16.6}$$

The arccosine of 0.64 in degrees gives us the 50° angle between the oblique factors shown in Figure 16.5.

Although the rotations considered so far attempt to obtain a solution in which each variable loads on only one factor, rotations can be used to re-express an unrotated solution to any of the patterns of loadings talked about in Chapter 12. For example, if the items under analysis can be thought to be a type of mono-method multi-trait or multimethod mono-trait study, a rotation to the form suggested by a bifactor model may be more appropriate (as discussed in Chapter 12). In such models, all variables load on a single factor (corresponding to the single method or single trait), and subsets of items load on additional factors (the multiple traits or multiple methods).

Similarly, analogues to the hierarchical factor model in which a super-factor has, as its indicators, factors that are, in turn, related to the manifest variables are also possible.

## Simple Structure

So far, the discussion of an optimal rotation has focused only on finding a new set of axes in which each variable loads on only one variable. This ignores the fact that, in many research situations, the picture is slightly more complex. The idea of simple structure, first proposed by Thurstone (1947), is one way to think about characteristics of a rotated factor loading matrix more generally. Aspects of simple structure are desiderata and not a single mathematically operationalized criterion, which is why a variety of rotation techniques have been proposed to balance these various objectives differently. According to Thurstone, a proposed rotation demonstrates simple structure if:

1. Each variable has at least one zero loading on some factor.
2. Each factor has at least as many zero loadings as there are factors.
3. Each pair of factors has variables with significant loadings on one and zero loadings on the other.
4. Each pair of factors has a large proportion of zero loadings on both factors (if there are say four or more factors total).
5. Each pair of factors has only a few "complex" variables (variables which load on more than one factor).

To use an example, consider a matrix of eight manifest variables and rotation of a four-factor model. An unrotated factor matrix unrotated might look like the following if we use $X$'s to denote nonzero numbers and 0's to denote zeros:

$$\begin{bmatrix} X & X & 0 & 0 \\ X & X & X & 0 \\ X & X & 0 & X \\ X & X & X & 0 \\ X & 0 & X & X \\ X & X & 0 & 0 \\ X & X & X & 0 \\ X & 0 & X & 0 \end{bmatrix}$$

An unrotated factor pattern matrix could look like this because factor loadings are extracted in decreasing order of the available covariation. A simple structure rotated solution, on the other hand, could look something like this:

$$\begin{bmatrix} X & 0 & 0 & 0 \\ X & 0 & 0 & 0 \\ 0 & X & 0 & 0 \\ 0 & X & 0 & 0 \\ 0 & 0 & X & X \\ 0 & 0 & X & 0 \\ 0 & 0 & 0 & X \\ 0 & 0 & 0 & X \end{bmatrix}$$

Verify for yourself that the matrix just presented meets the five criteria mentioned by Thurstone and that the complex variable is indicated in the fifth row of the matrix.

An example of an interpretable factor loading matrix that does not correspond to simple structure could be a bifactor pattern such as this:

$$\begin{bmatrix} X & X & 0 & 0 \\ X & X & 0 & 0 \\ X & 0 & X & 0 \\ X & 0 & X & 0 \\ X & 0 & 0 & X \\ X & 0 & 0 & X \\ X & 0 & 0 & 0 \\ X & 0 & 0 & 0 \end{bmatrix}$$

As mentioned previously, if all items share a common method, if all individuals have pronounced response set (such as a five-point Likert scale), or if the individuals in the data set have a systematic inter-individual difference (such as, for example, body mass if you are asking items about the amount of alcohol needed to induce alcohol intoxication), a bifactor model would be a preferable conceptual alternative to simple structure. If the data to be factored are longitudinal, it may make sense to think of a general growth factor, with the remaining factors representing developmentally punctuated periods of covariation or environmentally delimited effects (such as young adults living in apartments or residence halls during their college years).

## Objective Rotations

### Global versus Locally Optimal Rotations

Objective rotations, like estimation of SEM parameters, are sometimes heuristic search procedures. It is therefore important to recognize that although the many rotations work very well in practice, there is often no guarantee that the computer program has recovered the globally optimal rotation according to a proposed rotation criterion. To some extent this limitation has been reduced given advances in computer programming and resources, making it easier for rotation programs to employ multiple random starts for rotation and to calculate rotations to greater precision.

A critical skepticism of the standard defaults for the tolerance and tuning parameters, however, is an issue that deserves attention in a thorough examination of a final rotation. Hattori, Zhang, and Preacher (2017), for example, investigated the role of the default values for a tuning parameter, $\varepsilon$, in Geomin rotation and found that the values used by the Mplus program performed acceptably. They and other methodologists have found, however, that the use of multiple random starts significantly improved the successful recovery of a global rotated solution rather than local rotated solutions.

Second, it is important to remember that more than one well-fitting rotation may exist. For example, it is often standard practice for rotation programs to rotate the pattern matrix so that loadings are all positive rather than all negative. Analytically, this is not a grave problem, but it should be kept in mind when interpreting the results of the program. For example, it has been repeatedly noted in studies of self-reports of distress or psychopathology that positive factor loadings are recovered, leading some to talk about the "karma" of multiple types of life stresses occurring to people or the fact that many psychopathologies co-occur. One can just as easily, however, think of interpreting the same general latent variable as consisting of negative loadings across these manifest variables, leading to the interpretation that individuals who do not have one type of life stress tend not to have another type of stress, or that individuals who do not suffer one psychopathology also tend not to suffer from other disorders as well.

In a similar vein, similar interpretations can be applied to measures of intellectual abilities that are often discussed as positively correlated abilities. There may be several reasons in common across intellectual tasks that cause individuals to fail to perform, such as participant fatigue, impulsivity, and reticence to engage with measurement instruments. Just as the letter G has been used to denote a general ability factor (which could be the result of several variables such as economic advantage, environment, or common method variance), one could just as easily speak of an F factor that is a proxy for any of the several reasons why one may not successfully complete a task.

This is a conceptual issue rather than a mathematical issue in rotation, but it should be kept in mind when thinking about the naming problem associated with interpreting the factor loadings of a rotated solution.

## Rotational Skepticism and Guidance

Many rotations are available in current statistics software. A quick count of the GPArotation package in R reveals 41 different possibilities, and the Mplus package lists 27. These counts do not include the different variations within rotation, which are possible using tuning parameters. This often leads a researcher to wonder which rotation is best for their data or, perhaps more naively, which rotation is "best." The answer, of course, is that there is no globally superior rotation. Fabrigar, Wegener, MacCallum, and Strahan (1999) note, though, that in many situations, oblique rotations are preferable to orthogonal ones given that most constructs are correlated.

Some provisional guidance regarding rotation choices does exist, however. Schmitt and Sass (2011) examined various simple and complex factor structures relative to eight rotation criteria which are discussed later in this chapter (Varimax, Promax, Geomin, Direct Quartimin/CF-Quartimax, CF-Varimax, CF-Equamax, CF-Parsimax, and CF-Pacfarsim). They recommended CF-Equamax and CF-Pacfarsim for initial exploratory rotation using manifest variables of unknown quality, but they recommended Geomin or CF-Quartimax for instruments where fewer and smaller cross-loadings are expected. Similarly, Asparouhov and Muthén (2009) report that oblique Geomin rotation appears to work well in situations where little is known about the true factor structure and the reference structure "contains simple or moderately complicated loading structures.". They note, though, that it fails in situations where three or more factors are present and significant numbers of variables have loadings on three or more variables.

These explorations all assumed that some kind of orthogonal or oblique factor structure underlies the data. Other parsimonious structures, such as random intercept or bifactor models, may be viable alternative models. The Bi-Geomin and Bi-CF-Quartimax rotations mentioned earlier correspond to such models and can be considered as alternatives. Conversely, however, it may also be that a well-fitting bifactor rotation of the data may overfit the data and the actual relationships in the data may consist only of correlated factors (Bonifay, Lane, & Reise, 2016; Snyder, Young, & Hankin, 2017). Similar concerns have been raised when evaluating the bifactor model relative to hierarchical factor models (e.g., Murray & Johnson, 2013). None of these papers, however, considered whether a random intercept factor exists in addition to or instead of the general factor.

Absent a strong conceptual reason for a particular rotation, consideration of a variety of rotations available seems advisable. The matrix algebra involved in rotation is literally that involved in looking at things from a different angle. Examining a variety of rotations may, in some situations, provide some comfort that a given rotation is not an idiosyncratic result. In other situations, it may provide unanticipated patterns that spark a new conjectural theory of the relationships. So long as the reporter of the research is transparent regarding how the discovery was made, the fledgling new model can be critically examined by assumption checking a proposed new model, by cross-validating the factor structure by analyzing split halves of the data or replicating the study on a different population.

Such alternative rotations can correspond to the narrative that a reasonable skeptic might propose as a response to an initially favored rotation, if the purpose of our analysis is to put a proposed theory to a hard test. Such exploration is part of thorough analysis and not post-hoc "Harking" (Hypothesizing After Results are Known), so long as it is in service of killing off a favored model.

## Types of Objective Rotations

To illustrate the general idea of the logic of objective rotations, three objective rotations will be discussed, the Quartimax, Varimax, and Promax rotations. These are presented in this order because of their simplicity and the fact that many statistical software packages use the Varimax and Promax rotations as their default rotations.

### *Quartimax Rotation*

Have you ever noticed that when you play with a calculator and square numbers that are less than one, they eventually become zero? You may have noticed that this happens relatively quickly if the number you start with is small, say 0.10, but that it takes longer if you start with a number close to 1, such as 0.95. This basic property underlies the Quartimax rotation. The Quartimax criterion is an orthogonal rotation that minimizes the fourth powers of a proposed rotation. Quartimax seeks to find the smallest sum across all variables and factors of the rotated factor pattern raised to the fourth power. Taking sums across $f$ factors and $v$ variables, we can write this criterion sum as:

$$L = \sum_{f} \sum_{v} p^4 \tag{16.7}$$

This formula is similar to that for the uncorrected sum of squares (USS) from basic statistics: USS = $\sum_{i=1}^{N} X^2$. Basically, this property of the Quartimax rotation leads it to orient the axes more toward variables that have larger communalities and also leads it to recover rotated solutions containing a single larger factor with other factors containing smaller numbers of variables.

### Varimax Rotation

The orthogonal Varimax rotation is the default rotation for many exploratory factor analysis programs and was first proposed by Harry Kaiser in his doctoral dissertation and subsequently published in 1958. It is similar to the Quartimax rotation in that it considers the fourth power of loadings, but it subtracts off the sum of the squared coefficients within each factor. The formula for the criterion, assuming $k$ manifest variables and $f$ factors, is:

$$L = \sum_{f}\sum_{v} p^4 - \frac{1}{v}\sum_{f}\left(\sum_{v} p^2\right)^2 \qquad (16.8)$$

Readers making connections between this equation and their other classes in statistics will recognize the similarity between this criterion and the formula for a variance:

$$\sigma_X^2 = \frac{\Sigma\left(X-\bar{X}\right)^2}{N} = \frac{\Sigma X^2}{N} - \left(\frac{\Sigma X}{N}\right)^2 \qquad (16.9)$$

The original Varimax rotation is roughly analogous to maximizing the variance of the squared loadings. We could, for example, divide the first term of Eq. (16.8) by the number of variables and still maximize the same quantity. In practice it has been observed that Varimax rotations are more prone to generally apportion items relatively equally across factors.

### Promax Rotation

The Promax rotation (Henderson & White, 1964) is an oblique rotation that is, by design, likely to produce a factor loading matrix that closely resembles that obtained by Varimax, due to the fact that it begins with a Varimax rotation as a starting point but then adjusts the axes to allow for factors to be correlated. It accomplishes this by considering a matrix consisting of the

Varimax rotated factor loadings raised to some power (as of this writing, for example, Proc Factor in SAS raises this matrix to the third power, while *fa()* in R raises this matrix to the fourth power and restores the original sign of the loading). As in the Quartimax rotation description, this results in smaller loadings more quickly approximating zero than larger loadings. This matrix is then rescaled so that the loadings are expressed as a proportion of the largest loading in the column by dividing each loading by the largest loading in the column. This matrix is referred to as the Procrustes Target matrix in honor of the hotelier from Greek mythology who fit travelers to his guest bed by stretching or lopping them as needed.

Using the data in the oblique rotation example earlier, the Varimax loadings, loadings raised to the third power, and Procrustes Target Matrix are shown in Table 16.3.

Assuming $P_0$ represents the Varimax rotated factor pattern and $P_t$ represents the target Procrustes matrix, calculation of the oblique solution then proceeds by first calculating the following initial translation matrix which for these data are:

$$T_i = \left(P_0^T P_0\right)^{-1} P_0^T P_t = \begin{bmatrix} 1.10 & -0.40 \\ -0.44 & 1.11 \end{bmatrix} \tag{16.10}$$

A reader wanting to make connections across the material in the class will recognize that the second term in Eq. (16.10) is simply the matrix formula for the regression weights discussed in Chapter 2 and that we are simply predicting the Procrustes target matrix from the Varimax rotated factor loadings.

**TABLE 16.3**

Varimax, Varimax$^3$, and Procrustes Target Matrices for Promax Rotation

| Item | Varimax | | Varimax$^3$ | | Procrustes Target | |
|---|---|---|---|---|---|---|
| Variable | *F1* | *F2* | *F1* | *F2* | *F1* | *F2* |
| C gatherings more fun | 0.88 | 0.34 | 0.68 | 0.04 | 1.00 | 0.07 |
| D improves parties | 0.87 | 0.35 | 0.66 | 0.04 | 0.97 | 0.08 |
| A enjoy party | 0.74 | 0.35 | 0.40 | 0.04 | 0.59 | 0.08 |
| B be sociable | 0.69 | 0.30 | 0.32 | 0.03 | 0.47 | 0.05 |
| E to celebrate | 0.60 | 0.30 | 0.22 | 0.03 | 0.32 | 0.05 |
| Q pleasant feeling | 0.30 | 0.82 | 0.03 | 0.55 | 0.04 | 1.00 |
| K like the feeling | 0.40 | 0.66 | 0.07 | 0.29 | 0.10 | 0.52 |
| P to get high | 0.24 | 0.68 | 0.01 | 0.31 | 0.02 | 0.56 |
| O it's exciting | 0.29 | 0.66 | 0.02 | 0.29 | 0.04 | 0.53 |
| R it's fun | 0.45 | 0.66 | 0.09 | 0.28 | 0.13 | 0.51 |

This resulting matrix, though, is not the desired translation matrix because the Procrustes target matrix does not necessarily transform to the same unit length. To rescale $T_i$, we first need to calculate the inverse of $\left(T_i^T\right)T_i$, which for these data is:

$$\left(\left(T_i^T\right)T_i\right)^{-1} = \begin{bmatrix} 1.26 & 0.84 \\ 0.84 & 1.28 \end{bmatrix} \tag{16.11}$$

The square root of the diagonal elements of this matrix are post-multiplied by $T_i$ to accomplish the rescaling:

$$T_i\sqrt{\text{Diag}\left(\left(T_i^T\right)T_i\right)^{-1}} = \begin{bmatrix} 1.26 & 0.84 \\ 0.84 & 1.28 \end{bmatrix}\begin{bmatrix} \sqrt{1.26} & 0 \\ 0 & \sqrt{1.28} \end{bmatrix} = \begin{bmatrix} 1.24 & -0.45 \\ -0.49 & 1.26 \end{bmatrix} \tag{16.12}$$

This then gives the desired Promax rotation:

$$\begin{bmatrix} 0.88 & 0.34 \\ 0.87 & 0.35 \\ 0.74 & 0.35 \\ 0.69 & 0.30 \\ 0.60 & 0.30 \\ 0.30 & 0.82 \\ 0.40 & 0.66 \\ 0.24 & 0.68 \\ 0.29 & 0.66 \\ 0.45 & 0.66 \end{bmatrix}\begin{bmatrix} 1.24 & -0.45 \\ -0.49 & 1.26 \end{bmatrix} = \begin{bmatrix} 0.93 & 0.02 \\ 0.91 & 0.04 \\ 0.75 & 0.10 \\ 0.71 & 0.06 \\ 0.61 & 0.09 \\ 0.01 & 0.87 \\ 0.21 & 0.62 \\ 0.00 & 0.71 \\ 0.07 & 0.68 \\ 0.26 & 0.60 \end{bmatrix} \tag{16.13}$$

Most computer programs reporting Promax rotation provide the translation matrix that directly transforms the unrotated solution to the oblique Promax rotation. This is accomplished by multiplying the $T$ matrix for the Varimax rotation ($T_V$) by the $T$ matrix for the Promax solution ($T_P$), or in equation form:

$$T = T_V T_P = \begin{bmatrix} 0.83 & 0.55 \\ -0.55 & 0.83 \end{bmatrix}\begin{bmatrix} 1.24 & -0.45 \\ -0.49 & 1.26 \end{bmatrix} = \begin{bmatrix} 0.78 & 0.30 \\ -1.04 & 1.27 \end{bmatrix} \tag{16.14}$$

### *Geomin Rotation*

The Geomin rotation (Browne, 2001; Yates, 1987) is an increasingly popular rotation technique mainly because it appears to function well when substantial loadings exist on more than one factor, that is, there are a number of complex variables in the data (refer to Thurstone's fifth criterion mentioned earlier) (Asparouhov & Muthén, 2009; Browne, 2001). It is available in both orthogonal and oblique versions.

The basic idea of the Geomin rotation uses the idea of the geometric mean. In an equation form, the geometric mean can be calculated as follows:

$$\left[\prod_{j=1}^{m}(X_j)\right]^{\frac{1}{m}} \tag{16.15}$$

The $\Pi$ in this case denotes the product of the terms rather than the $\Sigma$ used in calculating the sum. The geometric mean of 1.2, 1.5, and 1.9, for example, is $\sqrt[3]{1.2+1.5+1.9} = \sqrt[3]{4.6} \cong 1.507$. The idea of the geometric mean is a bit novel for many people, so a short introduction as to what it is may be useful. It is often used in the real world when the data under consideration deal with growth or interest problems. It's useful to notice that the geometric mean exists only for positive numbers and is always going to be less than the arithmetic mean (unless all the terms are the same, in which case the arithmetic and geometric mean are the same).

The criterion associated with the Geomin rotation is the geometric mean of the squared loadings summed across the factors. It can be expressed as follows assuming $p$ factors and $m$ variables:

$$\sum_{i=1}^{p}\left[\prod_{j=1}^{m}\left(\lambda_{ij}^2+\varepsilon\right)\right]^{\frac{1}{m}} \tag{16.16}$$

The term $\varepsilon$, was proposed by Browne (2001) because he noted that the geometric mean of the $m$ squared factor loadings would be zero if one of the $m$ loadings was exactly zero. This would create an indeterminacy problem because other nonzero loadings could then no longer contribute to the geometric mean used to penalize cross loadings or emphasize large loadings. As a result, a variety of equally well-fitting rotated solutions could be found.

There was some question as to what a good value for $\varepsilon$ would be, however. Hattori, Zhang, and Preacher (2017) examined a variety of $\varepsilon$ values for simulated data and note the trade-off between the number of local solutions, the number of random starts, and the chosen $\varepsilon$ value. Specifically, small $\varepsilon$ values are more prone to large numbers of local rotations and, if a global rotation exists, require more random starts to find it. Choice of $\varepsilon$ appears to be different depending on the number of factors under consideration. Defaults for the present version of Mplus, for example, use a value of 0.0001 for two factors, 0.001 for three factors, and 0.01 for four or more factors. The *fa()* program in R, by contrast, uses a value of 0.00001 for all rotations and advises the user to specify additional random starts using the Random.Start function. Table 16.4 shows the Geomin rotation for this example.

## Factor Congruence

Given that many different types of rotations can (and should) be entertained, some assessment of the degree of similarity between the rotated solutions is of interest. Mathematically, the congruence coefficient is simply the unadjusted cosine between the factor vectors. Although Burt (1948) is credited with the original proposal of the coefficient, Tucker's (1951) use of the

**TABLE 16.4**

Geomin Rotation, Factor Correlations, and Translation Matrix

| **Variable** | ***F1*** | ***F2*** |
|---|---|---|
| | Factor Loadings | |
| C gatherings more fun | 0.96 | −0.03 |
| D improves parties | 0.94 | 0.00 |
| A enjoy party | 0.77 | 0.07 |
| B be sociable | 0.73 | 0.02 |
| E to celebrate | 0.63 | 0.07 |
| Q pleasant feeling | −0.04 | 0.90 |
| K like the feeling | 0.18 | 0.64 |
| P to get high | −0.04 | 0.74 |
| O it's exciting | 0.04 | 0.70 |
| R it's fun | 0.24 | 0.61 |
| | Factor Correlations | |
| *F1* | 1.00 | 0.69 |
| *F2* | 0.69 | 1.00 |
| | Translation Matrix *T* | |
| *F1* | 0.79 | 0.27 |
| *F2* | −1.14 | 1.36 |

coefficient resulted in the coefficient being often referred to as Tucker's congruence coefficient. The formula for the congruence between two factors is:

$$\phi = \frac{\sum_m \lambda_1 \lambda_2}{\sqrt{\sum_m \lambda_1^2 \sum_m \lambda_2^2}} \tag{16.17}$$

To see how this is different from a simple correlation, we can compare this with the formula for the correlation coefficient and substitute the definitions of the standard deviations involved to show what the correlation between the factor loadings would be:

$$\rho = \frac{\sum_m (\lambda_1 - \bar{\lambda}_1)(\lambda_2 - \bar{\lambda}_2)}{\sqrt{\sum_m (\lambda_1 - \bar{\lambda}_1)^2 \sum_m (\lambda_2 - \bar{\lambda}_2)^2}} = \frac{\sigma_{1,2}}{\sigma_1 \sigma_2} \tag{16.18}$$

From this we can see that the only difference between the two coefficients is that the correlation coefficient between sets of factors subtracts out the average loading within each respective factor. Whereas examination of the correlation between factors assesses the similarity across factors, the congruence coefficient also assesses the relative magnitude of the recovered factor loadings as well.

Although these coefficients are easy to calculate, the R package factor.congruence program will calculate them automatically. The Varimax, Promax, and Geomin(oblique) rotations for the Social and Enhancement items are shown in Table 16.5. As can be seen in the table, the correspondence between the Geomin and Promax solutions is effectively identical, and the correspondence between the first and second factors of the Varimax rotation with the oblique solutions is pronounced but not identical, and the marked

**TABLE 16.5**

Congruence Coefficients for Varimax, Geomin, and Promax Rotations

| Rotation | | Varimax | | Geomin | | Promax | |
|---|---|---|---|---|---|---|---|
| | | *F1* | *F2* | *F1* | *F2* | *F1* | *F2* |
| Varimax | *F1* | 1.00 | | | | | |
| | *F2* | 0.75 | 1.00 | | | | |
| Geomin | *F1* | 0.95 | 0.49 | 1.00 | | | |
| | *F2* | 0.42 | 0.92 | 0.10 | 1.00 | | |
| Promax | *F1* | 0.94 | 0.47 | 1.00 | 0.08 | 1.00 | |
| | *F2* | 0.41 | 0.91 | 0.09 | 1.00 | 0.07 | 1.00 |

cross-correlations across factors argues for the view that the Varimax rotated factors represent a different rotation from the oblique rotations considered.

The congruence coefficient is also used within the context of investigating the correspondence between factor loadings based on analyses of different data sets. In these situations, the researcher has a choice between examining the congruence resulting from unrotated or rotated factor solutions. Congruence based on unrotated factor loadings can provide some insights into the relative patterns of variability and selection effects present in the data using the sequential and orthogonal comparisons implicit in unrotated solutions, whereas comparison based on rotated solutions provides a general assessment of both the sample differences and the correspondence of the final rotated solution for each data set. The use of target rotations described later in this section is a better approach for that situation, however.

## Additional Rotations

### *Bifactor Rotations*

As mentioned in "Bifactor Models" in Chapter 12, the bifactor model is a parsimonious model in which a single general factor with loadings across all variables is thought to exist as well as a number of smaller factors that have loadings on subsets of items. Although a number of bifactor rotations exist, Table 16.6 shows the BI-CF-Quartimax and Bi-Geomin rotations for the Reasons for Drinking measure. For both rotations, if a general factor underlies the data, it evidently deals with items dealing with social and enhancement reasons for drinking. For both rotations, an additional factor consisting of only the social items is also recovered (Factor 4 in the Bi-CF-Quartimax rotation and Factor 2 in the Bi-Geomin rotation), as well as items dealing with coping: Factor 2 of the Bi-CF-Quartimax and Factor 3 of the Bi-Geomin rotation. Factor 3 of the Bi-CF-Quartimax and Factor 4 of the Bi-Geomin represent conformity reasons. The factor correlations between the subfactors also appear similar. The correlation between the coping items (e.g., "helps me forget my worries") and the group acceptance items (e.g., "to get in with a group I like") is similar to the correlation observed for the oblique Geomin rotation shown in Table 16.4.

The congruence coefficients calculated across the two rotations further reinforce the informal observations regarding the correspondence of the factors, with high correspondence between $F1$, $F2$, $F3$, and $F4$ of the Bi-CF-Quartimax and $F1$, $F3$, $F4$, and $F2$ of the Bi-Geomin rotation, respectively (Table 16.7).

Taken together, consideration of the rotations for the instrument suggests that a four-factor oblique solution fits the data well. If, however, a single general factor underlies the items, it is likely composed only of the items dealing with social and enhancement reasons for drinking. It is possible that future

**TABLE 16.6**

Bi-CF-Quartimax and Bi-Geomin Rotations for Reasons for Drinking Survey

| **Items** | | Bi-CF-Quartimax | | | | Bi-Geomin | | | |
|---|---|---|---|---|---|---|---|---|---|
| | | Factor | | | | Factor | | | |
| | | 1.00 | 2.00 | 3.00 | 4.00 | 1.00 | 2.00 | 3.00 | 4.00 |
| C because it makes social gatherings more fun | | 0.85 | −0.02 | −0.02 | 0.41 | 0.80 | 0.49 | −0.03 | −0.02 |
| D because it improves parties | | 0.85 | −0.01 | −0.02 | 0.39 | 0.81 | 0.48 | −0.02 | −0.03 |
| A because it helps me enjoy party | | 0.76 | 0.02 | 0.05 | 0.31 | 0.73 | 0.38 | 0.02 | 0.05 |
| B to be sociable | | 0.69 | 0.02 | 0.06 | 0.31 | 0.66 | 0.37 | 0.02 | 0.06 |
| E to celebrate | | 0.63 | 0.02 | −0.07 | 0.24 | 0.60 | 0.30 | 0.01 | −0.07 |
| G because it helps me when I feel depressed or nervous | | 0.34 | 0.83 | 0.02 | −0.02 | 0.37 | −0.03 | 0.82 | 0.02 |
| J because it helps me forget about my problems | | 0.36 | 0.81 | −0.01 | −0.01 | 0.38 | −0.01 | 0.80 | 0.00 |
| F because it helps me forget worries | | 0.38 | 0.80 | −0.02 | 0.03 | 0.40 | 0.03 | 0.79 | −0.02 |
| H to cheer myself up | | 0.42 | 0.76 | −0.01 | −0.02 | 0.44 | −0.01 | 0.75 | −0.01 |
| I because it helps me to feel more self-confident | | 0.52 | 0.40 | 0.08 | 0.05 | 0.53 | 0.08 | 0.39 | 0.08 |
| Q because it gives me a pleasant feeling | | 0.80 | −0.05 | −0.04 | −0.34 | 0.83 | −0.26 | −0.07 | −0.04 |
| K because I like the feeling | | 0.76 | 0.08 | −0.12 | −0.16 | 0.77 | −0.09 | 0.06 | −0.12 |
| P to get high | | 0.66 | 0.04 | 0.06 | −0.29 | 0.69 | −0.23 | 0.02 | 0.06 |
| O because it's exciting | | 0.69 | 0.00 | 0.05 | −0.24 | 0.71 | −0.17 | −0.02 | 0.05 |
| R because it's fun | | 0.78 | −0.09 | −0.09 | −0.13 | 0.79 | −0.04 | −0.11 | −0.09 |
| V to be liked | | 0.20 | 0.01 | 0.90 | 0.00 | 0.21 | 0.00 | 0.02 | 0.89 |
| U to get in with a group I like | | 0.19 | 0.00 | 0.90 | 0.00 | 0.20 | −0.01 | 0.01 | 0.90 |
| T so others won't kid me | | 0.12 | 0.01 | 0.84 | 0.00 | 0.13 | −0.01 | 0.01 | 0.84 |
| W so I won't feel left out | | 0.21 | −0.02 | 0.80 | 0.03 | 0.22 | 0.04 | −0.02 | 0.79 |
| S because my friends pressure me | | 0.19 | 0.00 | 0.75 | −0.03 | 0.20 | −0.03 | 0.00 | 0.75 |
| | | Factor Correlations | | | | Factor Correlations | | | |
| | 1.00 | 1.00 | | | | 1.00 | | | |
| Factors | 2.00 | 0.00 | 1.00 | | | 0.00 | 1.00 | | |
| | 3.00 | 0.00 | 0.33 | 1.00 | | 0.00 | 0.00 | 1.00 | |
| | 4.00 | 0.00 | −0.03 | −0.07 | 1.00 | 0.00 | −0.05 | 0.31 | 1.00 |

**TABLE 16.7**

Congruence Coefficients for Bi-CF-Quartimax and Bi-Geomin Rotations

| | Bi-CF-Quartimax | | | | Bi-Geomin | | | |
|---|---|---|---|---|---|---|---|---|
| *F* | 1 | 2 | 3 | 4 | 1 | 2 | 3 | 4 |
| 1 | 1.00 | | | | | | | |
| 2 | 0.33 | 1.00 | | | | | | |
| 3 | 0.14 | 0.01 | 1.00 | | | | | |
| 4 | 0.19 | 0.02 | 0.01 | 1.00 | | | | |
| 1 | 1.00 | 0.35 | 0.15 | 0.15 | 1.00 | | | |
| 2 | 0.39 | 0.01 | 0.00 | 0.97 | 0.36 | 1.00 | | |
| 3 | 0.30 | 1.00 | 0.02 | 0.02 | 0.32 | 0.01 | 1.00 | |
| 4 | 0.14 | 0.01 | 1.00 | 0.01 | 0.15 | 0.00 | 0.02 | 1.00 |

research could investigate the differential prediction of these constructs or differences in invariance of these items across groups as a strategy for judging their relative merits.

### *Target Rotation*

Target rotation (Browne, 2001) performs rotation relative to a subset of loadings to known numeric values. Often these values of zero are chosen to identify which manifest variables should not load on particular factors. As such, target rotation can be thought of as a compromise between the solution resulting from a hand rotation, confirmatory factor model, and the mechanical (objective) rotations described earlier.

Target rotations can be either orthogonal or oblique. A minimum number of targets must be specified, which depends on the number of factors involved. For oblique solutions with $m$ factors, the number of targets must be at least $m(m-1)$. For orthogonal rotations, the number must be $\frac{m(m-1)}{2}$.

The specified target values will not necessarily be recovered in the rotation – i.e., a value other than zero may result given an initial target value of zero. This may indicate a poor match of the factor structure to the data, but it is not necessarily indicative of a failure to arrive at a final solution.

The rotation criterion for target rotation can be expressed as follows for $m$ factors and $p$ variables

$$\prod_{i=1}^{p}\sum_{j=1}^{m} a_{ij}\left(\lambda_{ij}-b_{ij}\right)^2, \tag{16.19}$$

where $b_{ij}$ refers to the target values for loadings, $a_{ij}=1$ if the loading corresponds to a target loading and $a_{ij}=0$ otherwise.

Asparouhov and Muthén (2009) provide a discussion and Monte Carlo studies comparing the relative merits of target rotation relative to Geomin and ESEM (Exploratory Structural Equation Modeling) approaches to the issue of rotation. Given that target rotations are now relatively easy to implement with available software, such explorations seem like an attractive alternative, especially in those situations when the researcher wants to directly compare the factor structure of a data set relative to a previously published rotation. One way to proceed in such a situation is to freely estimate the highest-loading item in each factor of the previous solution (assuming it is not a complex item), and then specify zero loading targets for the remaining factors.

*Example Target Rotation*

For example, consider a target rotation in which the first items of each subtest are rotated to a target value of zero within each factor, providing the necessary 4(3) = 12 targets to identify the rotation. The resulting target rotation is shown in Table 16.8. Although the recovered factor pattern matrix appears quite similar to that recovered from the oblique Geomin rotation shown in Table 16.4, the correlations between Factors 1, 2, and 3 appear slightly lower than the Geomin rotation, suggesting a lower amount of factor complexity for those factors. Item I ("helps me feel more self-confident") cross-loads slightly more with *F*1, suggesting that feeling more self-confident may be more associated with social reasons for drinking.

## CF Rotational Criteria

Crawford and Ferguson (1970) proposed a family of rotations based on differential weighting of row and column complexity in rotations. The general formula for rotations in this class is:

$$(1-\kappa)\sum_{\substack{i=1\\m}}^{p}\sum_{j=1}^{m}\sum_{l\neq j,l=1}^{m}\lambda_{ij}^{2}\lambda_{il}^{2}+\kappa\sum_{\substack{i=1\\m}}^{p}\sum_{j=1}^{m}\sum_{l\neq j,l=1}^{m}\lambda_{ij}^{2}\lambda_{lj}^{2} \qquad (16.20)$$

for $m$ factors and $p$ variables. The first term of Eq. (16.20) represents the row complexity, while the second row represents column complexity. Row complexity is minimized when there is only one nonzero loading in the rows of the factor loading matrix, and column complexity is minimized when there is a single nonzero loading in each column of the factor loading matrix. Rotations within the CF family represent different choices of the weighting parameter, $\kappa$.

**TABLE 16.8**

Target Rotation for Reasons for Drinking Survey

| | Target Rotation Factors | | | |
|---|---|---|---|---|
| **Items** | *F1* | *F2* | *F3* | *F4* |
| C makes social gatherings more fun | 0.82 | 0.00 | 0.00 | 0.00 |
| D improves parties | 0.83 | 0.01 | 0.02 | 0.00 |
| A helps me enjoy party | 0.67 | 0.04 | 0.05 | 0.06 |
| B to be sociable | 0.61 | 0.04 | 0.01 | 0.07 |
| E to celebrate | 0.50 | 0.04 | 0.07 | −0.05 |
| G helps me when I feel depressed or nervous | 0.00 | 0.90 | 0.00 | 0.00 |
| J helps me forget about my problems | 0.03 | 0.90 | 0.01 | −0.03 |
| F helps me forget worries | 0.10 | 0.90 | −0.03 | −0.04 |
| H to cheer myself up | 0.06 | 0.84 | 0.06 | −0.02 |
| I helps me to feel more self-confident | 0.26 | 0.47 | 0.13 | 0.09 |
| Q gives me a pleasant feeling | 0.00 | 0.00 | 0.82 | 0.00 |
| K because I like the feeling | 0.18 | 0.12 | 0.59 | −0.09 |
| P to get high | −0.03 | 0.09 | 0.77 | 0.10 |
| O it's exciting | 0.05 | 0.04 | 0.66 | 0.08 |
| R because it's fun | 0.23 | −0.05 | 0.55 | −0.05 |
| V to be liked | 0.00 | 0.00 | 0.00 | 0.68 |
| U to get in with a group I like | −0.01 | −0.01 | 0.00 | 0.67 |
| T so others won't kid me | −0.02 | −0.01 | −0.03 | 0.61 |
| W so I won't feel left out | 0.05 | −0.03 | −0.01 | 0.71 |
| S my friends pressure me | −0.02 | −0.01 | 0.04 | 0.61 |
| | Factor Correlations | | | |
| *F1* | 1.00 | | | |
| *F2* | 0.29 | 1.00 | | |
| *F3* | 0.66 | 0.30 | 1.00 | |
| *F4* | 0.13 | 0.41 | 0.16 | 1.00 |

*Direct Quartimin/CF-Quartimax*

Direct Quartimin (also known as CF-Quartimax) results when $\kappa$ is chosen to be 0. This means that this rotation seeks to identify a solution in which each variable loads on only one factor. So if, for example, working with a previous well-developed instrument in which each variable loads on only one factor, this rotation may be an attractive option. The drawback to use of this rotation, however, is that it can result in overinflated estimates of factor correlations (Carroll, 1953; Gorsuch, 1983; Yates, 1987).

*CF-Varimax*

When $\kappa$ is chosen to be $1/p$ (the reciprocal of the number of variables), the result is orthogonal CF-Varimax. When the factors underlying the data are orthogonal, Varimax and CF-Varimax provide similar answers. In situations, however, where the factors are highly correlated, it has been observed that Varimax tends to provide factors with near-zero loadings across all items (called "factor collapse"), and this is less often encountered with CF-Varimax (Browne, 2001; Crawford, 1975).

*CF-Equamax*

When $\kappa$ is chosen to be $m/2p$ (i.e., the number of factors divided by twice the number of variables), the resulting CF-Equamax rotation tends to spread variances more equally across factors (Gorsuch, 1983). However, one should be aware that the relative weighting of row and column complexity depends on the number of factors and items under analysis.

*CF-Facparsim*

CF-Facparsim results when $\kappa = 1$ – i.e., all weight is given to reducing row complexity and, as a result, complex relationships between factors may be estimated. While this may not be a practical final rotation, examining it may provide a "worst case" scenario of factor complexity.

## ESEM: Multigroup EFA Rotation

The ideas of factor rotation and multigroup invariance models in Chapter 14 can be combined, although the fundamental questions of the dimensionality and comparability of measurement model probably argue for a group-by-group exploration, using a combination of the extraction and rotation approaches. Estimation of multigroup exploratory factor models for examining rotation and invariance properties are a way to combine invariance and rotation simultaneously (Asparouhov & Muthén, 2009, 2014). Another advantage of the ESEM approach is that it allows the researcher to specify part of the model as an exploratory factor model while estimating other parts of the model as a CFA. For example, one could specify an additional confirmatory factor involving alcohol use that is predicted by the ESEM factors associated with reasons for drinking. Specification of models with mean factor effects for all latent variables (i.e., complex growth curve models) is also possible allowing the researcher to explore unanticipated patterns of trajectories.

As an example of how one might organize a multigroup exploratory factor analysis, consider the items from the four subscales of the Reasons for Drinking scale within a multigroup analysis of men and women. This data set consists of 1,294 women and 865 men. The sample sizes and the relative strength of associations between the variables observed earlier would appear adequate to assess whether the dimensionality of drinking motives is the same across men and women. To that end, a parallel analysis and Velicer's MAP test for each group was conducted (not shown here in the interest of space), which revealed that four factors appeared to account for the data in each group. The process of ESEM allows the researcher to both rotate an EFA solution and employ the multigroup comparisons discussed in Chapter 14 to reflect the measurement assumptions and possible group differences in the data.

### Model A: ESEM Factor Loadings, Errors, and Intercepts Variant

Under this model, the number of factors is assumed to be the same across groups. However, the pattern of loadings may be completely different across groups. Some items that load on one factor are allowed to load on a completely different factor in the other group, and the degree to which factors correlate within group may be completely different as well. Given that the latent variables may correspond to entirely different constructs, it doesn't really make sense to ask whether the means of the factors are different across groups. For this reason, allowing the intercepts to vary across groups is a parsimonious way to express differences in mean elevation of scores across all variables. As with the earlier multigroup models, factor variances were constrained to one within group for identification purposes. In this analysis an oblique Geomin rotation was used, although one could explore other rotations such as a target rotation as well. Model fit statistics for the model are given in the first row of Table 16.9.

### Model B: ESEM Factor Loadings Invariant, Errors, and Intercepts Variant

Model B makes only the assumption that factor loadings were the same across groups, but allows for the possibility that the variances of the factors might be greater, say, for men than for women in their reasons for drinking and that perhaps the degree to which the factors are correlated for women might be different from those for men. This model does not address the question of whether the overall level of responses was different across men and women, however. For that reason, factor means across the groups are set to zero for purposes of mathematical identification, and the intercepts are freely estimated for each group.

**TABLE 16.9**

Selected ESEM Multigroup Model Comparisons

| | Parameters | | | | | | Fit Measures | | | | | |
|---|---|---|---|---|---|---|---|---|---|---|---|---|
| **Model** | **Loadings** | **Intercepts** | **Factor Variances** | **Factor Covariances** | **Factor Means** | **Error Variances** | $\chi^2(df)$ | **BIC** | **TLI** | **CFI** | **RMSEA** | **AIC** |
| A | N | N | 1 | N | 0 | N | 1,349.64 (232) | 80,654.74 | 0.95 | 0.97 | 0.07 (0.06–0.07) | 79,360.29 |
| B | Y | N | N | N | 0 | N | 1,481.27 (296) | 80,295.02 | 0.96 | 0.97 | 0.06 (0.06–0.06) | 79,363.93 |
| C | Y | Y | N | N | 0 | N | 1,636.81 (316) | 80,297.01 | 0.95 | 0.96 | 0.06 (0.06–0.06) | 79,479.46 |
| D | Y | Y | 1 | Y | N | N | 1,664.10 (322) | 80,278.23 | 0.95 | 0.96 | 0.06 (0.06–0.06) | 79,494.75 |
| E | Y | Y | 1 | Y | 0 | N | 1,714.28 (326) | 80,297.70 | 0.95 | 0.96 | 0.06 (0.06–0.07) | 79,536.93 |
| F | Y | Y | 1 | Y | N | Y | 1,890.02 (342) | 80,350.60 | 0.95 | 0.96 | 0.06 (0.06–0.07) | 79,680.67 |
| G | Y | N | 1 | Y | 0 | Y | 1,555.51 (306) | 80,292.48 | 0.95 | 0.96 | 0.06 (0.06–0.06) | 79,418.16 |

### Model C: ESEM Factor Loadings, Errors, and Intercepts Variant

If we assume that the loadings and manifest variable intercepts are equal across groups, it then becomes possible to summarize group differences at the latent variable in terms of both the means and variances of the latent variables. This is the default invariance model for Mplus. If this model fits well, one advantage it has is that it allows discussion of group differences in elevation (e.g., "Men appear to endorse higher levels of reasons for drinking") as well as variability (e.g., "Men appear to be more variable in their reports of whether drinking to conform to a group is a reason to drink alcohol").

### Model D: ESEM Factor Loadings, Intercepts, Factor Variances and Covariances Invariant, Errors and Factor Means Variant

Model D is appropriate if it is assumed that the latent variables in the EFA represent exactly the same latent variables in terms of loadings, factor variability, and covariances between factors, but with the difference that mean differences exist across the groups and that the measurement errors are different across groups.

### Model E, F, and G

Model E assumes that a common EFA underlies both groups, except that the factor means differ across groups. Model F assumes that factor loadings, intercepts, factor variances and covariances, and factor means are invariant, but that measurement errors vary across groups. Model G represents a completely invariant model in which all parameters of the EFA are identical across the groups.

### Multigroup Model Comparison

Table 16.9 shows the invariance constraints as well as the fit statistics for the various multigroup models. As can be seen, the pattern of $\chi^2$ statistics consistently decrease the more parameters are in the model; however, these differences in model fit can be attributed to the large sample sizes used. The AIC also prefers the completely variant model A as well. If we use the BIC as the basis for model comparison, however, Model D appears to have the best fit, suggesting that mean differences exist between men and women in their overall levels of reporting reasons for drinking, but that the factors appear invariant across men and women. Given that the TLI, CRI, and RMSEA statistics appear almost identical across the models, a case can be made that Model D appears the most appropriate for the data.

The unstandardized solution for Model D is shown in Table 16.10. When we compare it to the exploratory factor analysis based on the data as a single

**TABLE 16.10**

Estimated Parameters Model D: Factor Loadings, Variances, Correlations, and Intercepts Invariant

| Items | Factor Loadings Λ | | | | Intercepts | Error Variances Female | Error Variances Male |
|---|---|---|---|---|---|---|---|
| | *F1* | *F2* | *F3* | *F4* | | | |
| C because it makes social gatherings more fun | 0.83 | −0.01 | −0.01 | −0.01 | 3.05 | 0.07 | 0.10 |
| D because it improves parties | 0.84 | 0.00 | 0.00 | −0.01 | 2.98 | 0.10 | 0.10 |
| A because it helps me enjoy party | 0.68 | 0.03 | 0.05 | 0.05 | 2.91 | 0.26 | 0.25 |
| B to be sociable | 0.62 | 0.03 | 0.00 | 0.06 | 2.96 | 0.31 | 0.32 |
| E to celebrate | 0.51 | 0.02 | 0.06 | −0.05 | 3.10 | 0.38 | 0.35 |
| G because it helps me when I feel depressed or nervous | −0.05 | 0.92 | −0.01 | 0.02 | 1.88 | 0.18 | 0.16 |
| J because it helps me forget about my problems | −0.02 | 0.91 | 0.00 | −0.01 | 1.91 | 0.21 | 0.23 |
| F because it helps me forget worries | 0.05 | 0.91 | −0.04 | −0.02 | 2.00 | 0.23 | 0.24 |
| H to cheer myself up | 0.02 | 0.85 | 0.05 | −0.01 | 2.00 | 0.26 | 0.24 |
| I because it helps me to feel more self-confident | 0.24 | 0.46 | 0.12 | 0.09 | 2.23 | 0.57 | 0.60 |
| Q because it gives me a pleasant feeling | −0.01 | −0.04 | 0.84 | 0.00 | 2.80 | 0.25 | 0.15 |
| K because I like the feeling | 0.17 | 0.09 | 0.60 | −0.09 | 2.92 | 0.36 | 0.35 |
| P to get high | −0.05 | 0.06 | 0.78 | 0.10 | 2.24 | 0.56 | 0.60 |
| O because it's exciting | 0.04 | 0.01 | 0.67 | 0.08 | 2.49 | 0.45 | 0.50 |
| R because it's fun | 0.22 | −0.08 | 0.57 | −0.05 | 3.07 | 0.31 | 0.21 |
| V to be liked | 0.01 | 0.02 | 0.00 | 0.66 | 1.30 | 0.08 | 0.09 |
| U to get in with a group I like | 0.00 | 0.01 | 0.01 | 0.65 | 1.27 | 0.06 | 0.11 |
| T so others won't kid me | −0.02 | 0.01 | −0.02 | 0.60 | 1.28 | 0.11 | 0.19 |
| W so I won't feel left out | 0.06 | −0.01 | −0.02 | 0.69 | 1.45 | 0.29 | 0.19 |
| S because my friends pressure me | −0.02 | 0.01 | 0.05 | 0.60 | 1.41 | 0.21 | 0.29 |

| | Factor Variances and Correlations | | | |
|---|---|---|---|---|
| | *F*1 | *F*2 | *F*3 | *F*4 |
| *F*1 (Social) | 1.00 | 0.36 | 0.68 | 0.12 |
| *F*2 (Coping) | | 1.00 | 0.38 | 0.37 |
| *F*3 (Enhancement) | | | 1.00 | 0.14 |
| *F*4 (Conformity) | | | | 1.00 |
| | Factor Means | | | |
| | *F*1 | *F*2 | *F*3 | *F*4 |
| Females | 0.00 | 0.00 | 0.00 | 0.00 |
| Males | 0.10 | 0.06 | 0.20 | 0.26 |
| Z-Score Percentile Male Mean | 0.54 | 0.52 | 0.58 | 0.60 |

*Note*: Underlined values denote fixed parameter values.

group shown in Table 15.1, you can see that the estimated factor loadings from the EFA based on the entire sample for Factors 1, 3, and 4 are uniformly higher than for the ESEM multigroup EFA, due to the fact that sex differences associated with these factors contribute to greater observed variability in the manifest variables in the single group analysis. Loadings for Factor 2 appear similar across the two tables. The patterns of cross-loadings appear slightly different, with the item "I because it helps me to feel more self-confident" now loading on the *F*1, which deals with social reasons for drinking. Because the factor variances are set equal (and fixed to unity for mathematical identification purposes), the covariances between factors are correlations. These seem roughly similar to the single group factor analysis as well. The freely estimated factor means for the men can be interpreted as *z*-scores, and the percentile associated with the means for men relative to women is shown in the last row of Table 16.10, showing that sex differences are most pronounced for Factors 3 and 4, which deal with enhancement and conformity reasons for drinking. Although significant differences in estimated error variances were found (comparing the BIC values for Model D and F), no evident pattern in the magnitude of estimated error is evident from the table.

## Concluding Comments

Much ground has been covered concerning rotation possibilities. For those inclined to do a deeper dive into the rationale, theory, and mathematics of factor rotation to maintain their understanding of unfolding developments, Browne's (2001) survey of rotations and Jennrich's (2001, 2002) gradient projection method to orthogonal and oblique rotations repay study. The initial comments on rotation choices give only general guidance on how these rotations have behaved in simulated data or in real-world data in which item pools consist of orthogonal or oblique factor structures. To some extent, the avenues of exploration can be reduced when all researchers in the area might agree on the dimensionality of the model, but in other situations exploring dimensions in proximity to that suggested by parallel analysis is a reasonable alternative, provided it is remembered that parallel analysis provides information about the plausibility of orthogonally identifiable factors. Rotations then proceed to exploration of rotations to examine the degree to which they correspond to the parsimonious models described in Chapter 12. Examination of a replication of a study to either a new data set or across groups within a single data set is a second scenario. Given examination of the congruence coefficients resulting from rotations of datasets, it seems more reasonable to employ the target rotation or ESEM approaches for a more direct comparison to original research.

Under both scenarios, it is important to entertain the counterarguments of a reasonable skeptic to the analysis, including assumption checks of the model that appear to have been met. The identification and final results of factor analyses are not expected to replicate across studies that vary widely in terms of sample size or the number and quality of the variables. The field of exploratory factor analysis continues to evolve as computer software and machinery make more complex models available. As advances in computing power and artificial intelligence make widespread exploration of data more feasible, it is important that researchers be explicit about their process of exploring and evaluating models.

# 17

# *SEM Assumption Checking*

The results of an SEM involve not only the parameters associated with relationships between manifest variables. Assumption checks regarding the overall fit of an SEM, discrepancies between observed and predicted values of manifest variables, and evaluation of the linearity of parameters involving latent variables are a few of the possible avenues for assumption checking the claims associated with an SEM. As with the regression model assumption checking described in Chapter 5, the influence of a small single or small number of observations is unlikely to affect an SEM if the sample size is large. However, regardless of sample size, the exploration of other aspects of the model, such as the assumption of linearity of path coefficients, is worthwhile regardless of sample size. In this chapter, assumption checks will be provided for the orthogonal latent basis model of the McArdle dataset analyzed in Chapter 13.

## Assumption Checking for Large and Small Sample Sizes

### LOESS Regressions of Factor Scores

To calculate LOESS regression lines involving the factor scores, it is necessary to first save the factor scores and then specify the LOESS regressions predicting each of the manifest variables from the factor scores. Bivariate regression plots are shown in Figure 17.1. From these plots it appears reasonable to assume that the relationships between the factors and the manifest variables is largely linear with the possible exception of the fourth measurement occasion, where it seems that lower intercept factor scores are associated with a slight flattening of the predicted Time 4 scores based on individuals with a lower level of the intercept factor. The assumption of linear relationships between the latent variables with their respective manifest variables appears reasonable based on the plot.


DOI: 10.1201/9781003595267-17

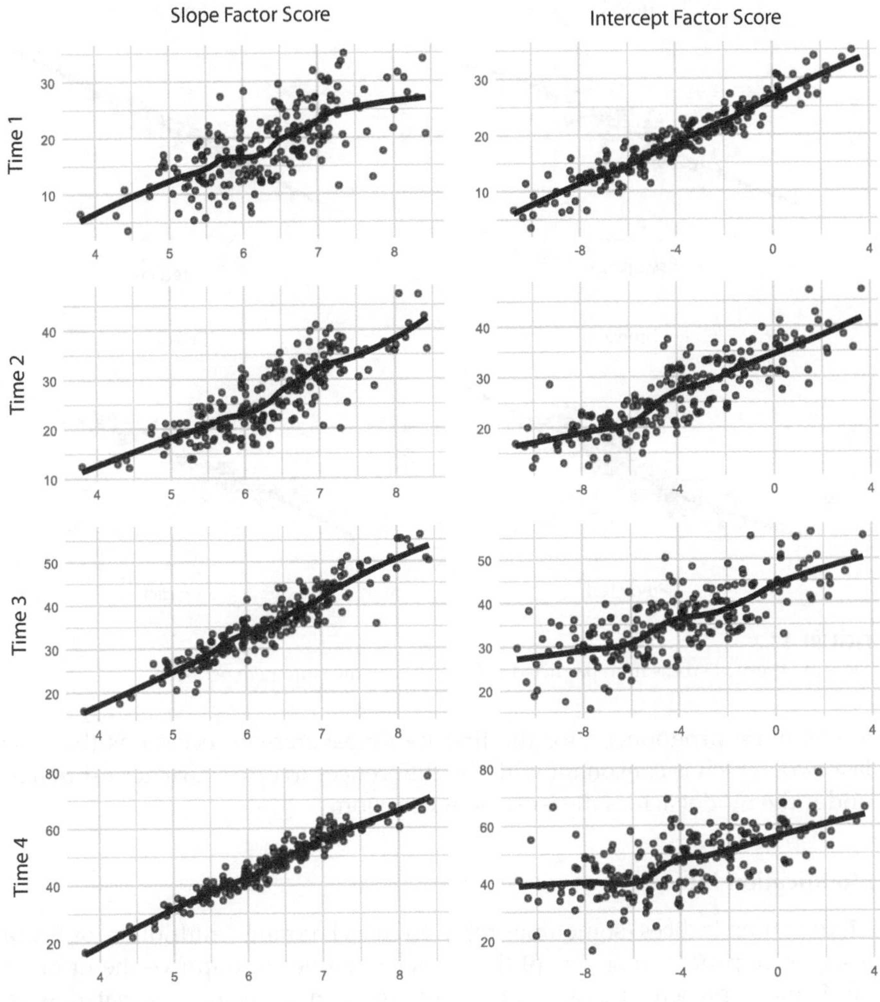


**FIGURE 17.1**
LOESS regressions predicting times 1 through 4 from slope and intercept factor scores.

## Residual Plots

Although the LOESS plots are a usual way to investigate the linearity associated with factor loadings, they do not inform the researcher concerning the overall accuracy of the model for each of the variables. Residual plots, in which the actual manifest variables are plotted as a function of the predicted scores for each variable, are a useful check to this end. These are shown in Figure 17.2, and show that there is a relatively homoscedastic but linear relationship between the predicted and observed scores. Variability in prediction

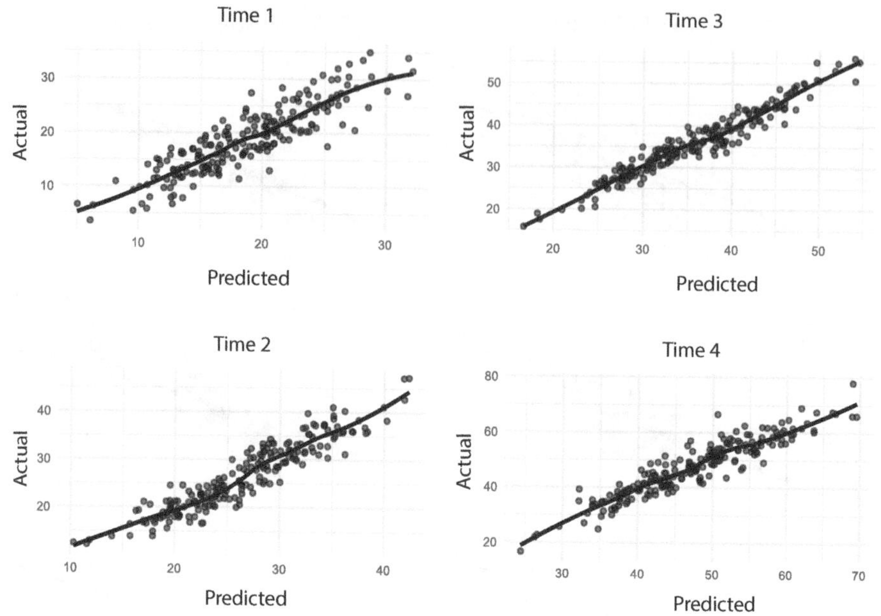


**FIGURE 17.2**
Plots of observed scores from predicted values for each measurement occasion.

seems more pronounced for the first two measurement occasions than the last two, which is consonant with the differences in error variance estimated under the model across measurement occasion.

## Modification Indices

Modification indices (sometimes referred to as Lagrange multipliers or Score tests) were proposed as one of the earlier remedies to improve the fit of an SEM and represent a single degree of freedom $\chi^2$ value associated with an additional parameter that could be added to the model. Its calculation involves the Hessian matrix of second derivatives (mentioned in Chapter 8's explanation of the iterative solution of SEMs using Newton's method) and the first derivative of the log-likelihood for the model. These additions to a model may involve the relaxation of an equality constraint or the addition of a path coefficient, loading, or covariance to the model. (The addition of a new factor, while logically possible, is not the addition of a single new parameter.) Modification indices are usually ranked and the largest coefficients considered for the largest values of added parameters. There is general consensus that new parameters should be added to the model only if a compelling conceptual case can be made for their addition, because significant modification indices can occur due to the experiment-wise error associated with the large number of potential parameter additions. Saris, Satorra, and Van der Veld (2009) recommend examining the modification index along with the Expected Parameter Change (EPC) as a way to judge the addition of a potential parameter. Mulaik (2009a) also cautions that modifications using the modification index approach will only work if the "true" model has the same structure and that use of modification indices to an incorrect model may be totally misleading (p. 438).

These cautions being acknowledged, it is sometimes useful to examine modification indices for clues as to whether a single additional parameter to the SEM can have an outsized effect on the resulting parameter estimates or even invalidate the conclusions of a model favored to that point. For the McArdle dataset under consideration, modification indices for the latent basis model are probably inadvisable, given that the overall model $\chi^2$ is not significant. However, if they are calculated, the intercept for Time 3 appears large (MI = 7.92, EPC = 6.35), suggesting that a change in level for WISC vocabulary scores occurred at that occasion. Other modifications involving Times 3 and 4 also were significant (the intercept loading for Time MI = 6.22, EPC = –0.182; the intercept for Time 4, MI = 5.85, EPC = –9.728; and intercept loading for Time 4, MI = 5.54, EPC = 0.32). If the Time 3 intercept is freed, however, no modification indices are significant, highlighting the fact that modification indices are not independent of each other. Again, however, given that the initial model $\chi^2$ is not significant, this modification to the model is probably ill-advised.

## Assumption Checking for Smaller Data Sets

Analogous to assumption checking for traditional linear models, a single observation or small set of observations is unlikely to have an outsized effect on a large data set. Individual observations for smaller data sets can have substantial effects on the fit and parameters of SEMs.

### $\Delta\chi^2$

$\Delta\chi^2$ represents the difference between the estimated model $\chi^2$ and the $\chi^2$ when a given observation is removed from the data (Pek & MacCallum, 2011). This is an often-unexplored option for improving model fit, given that focus has often been on changing the model by including additional parameters. The resulting plot of $\Delta\chi^2$ is inspected to identify individuals who unduly affect the general fit of the model. For the data example, the resulting

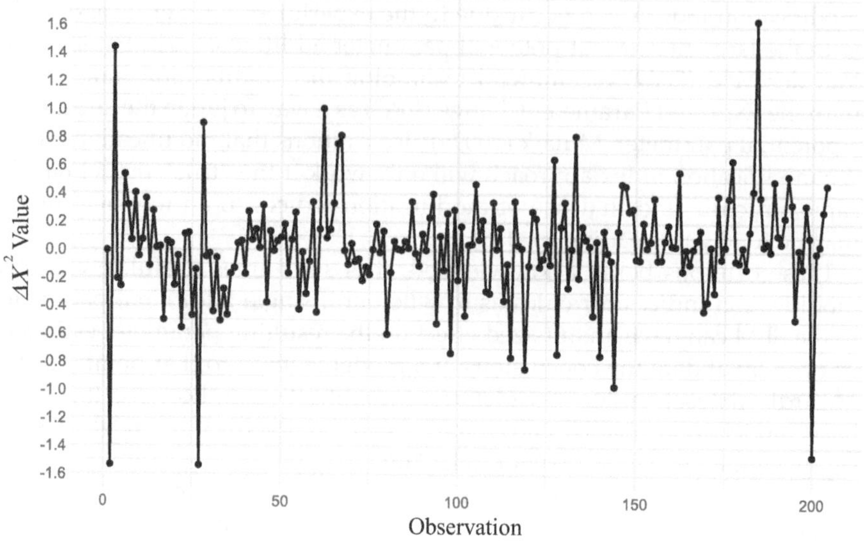


**FIGURE 17.3**
$\Delta\chi^2$ values for McArdle data.

plot is shown in Figure 17.3. Positive values of $\Delta\chi^2$ indicate that removing the observation would improve model fit, while negative values indicate that removing the observation would decrease model fit. There is no statistical cutoff value for significant improvement, and the procedure is used to flag observations for further inspection.

## Likelihood Distance

The likelihood distance (LD) is calculated as twice the difference between the log-likelihood of the original model and the log-likelihood of the model with the case removed. Larger LD values indicate a greater influence of the deleted case on the model's fit. It is similar information to that obtained from the $\Delta\chi^2$, and for that reason the two statistics are often plotted together to identify observations that seem influential under both criteria as shown in Figure 17.4. As can be seen from the graph, the information in the plot is not redundant. In particular, the LD value for observation 120 in the dataset (0.925) is substantially higher than for the remaining observations. The scores of this observation are somewhat unusual in that they go from an extremely low initial assessment (12.95, a score at about the 10th percentile)

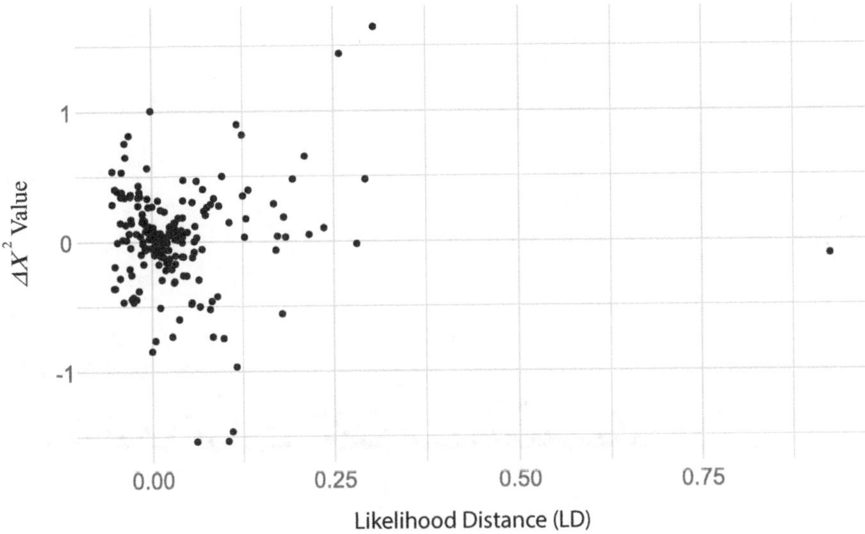


**FIGURE 17.4**
Plot of $\Delta\chi^2$ by LD.

to a high vocabulary score at the Time 4 assessment (66.62), a value above the 90th percentile of scores.

## Generalized Cook's Distance (gCD)

gCD (Pek & MacCallum, 2011) is a version of Cook's distance standardized across all estimated parameters of the SEM, although it can be calculated on a subset of parameters if desired. Recall that Cook's distance is a measure intended to weight an atypical observation both in terms of its status as an outlying or influential observation. The plot of the gCP is shown in Figure 17.5. Again, observation 120 appears to show the highest gCD value of 0.99 for these data.

## Parameter Influence (ParInfluence)

It is also possible to investigate the effect of individual observations on the estimated parameters of the SEM. Figure 17.6 shows the results of investigation of the loadings for the four measurement occasions ($\lambda$) and factor means ($\alpha$) for the intercept and slope factors of the orthogonal FCSI model. While several individual observations are associated with changes in the

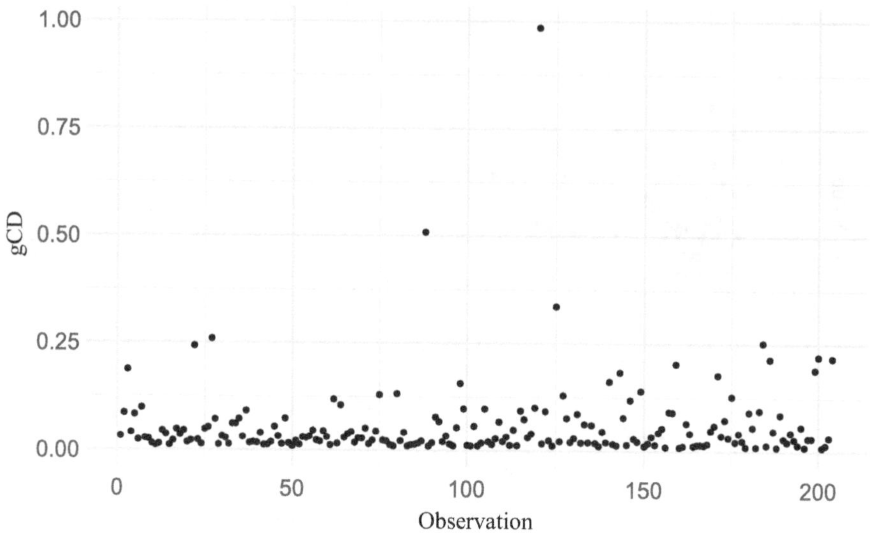


**FIGURE 17.5**
Generalized Cook's distance (gCD) for McArdle data.

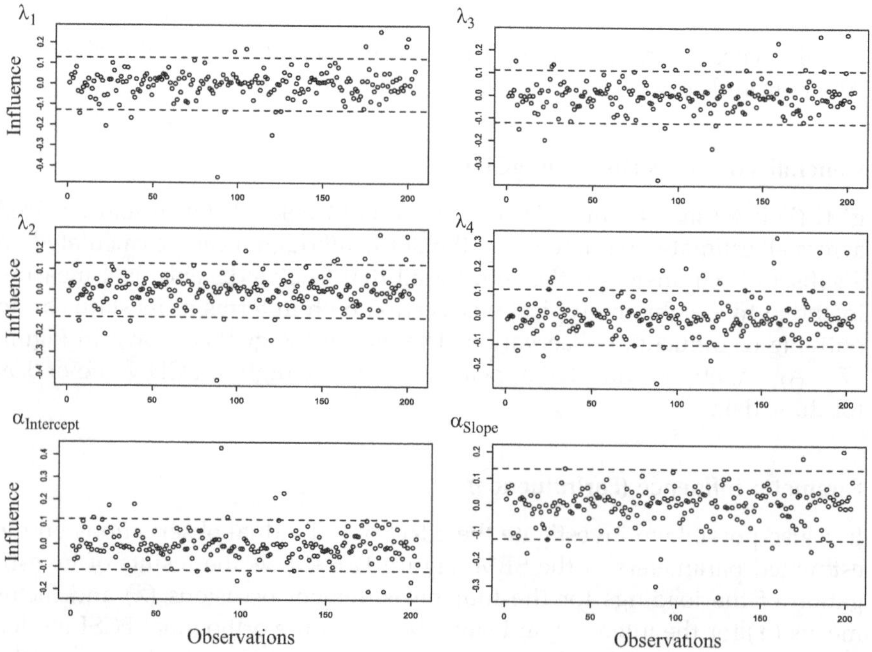


**FIGURE 17.6**
Individual observation influence on parameter estimates.

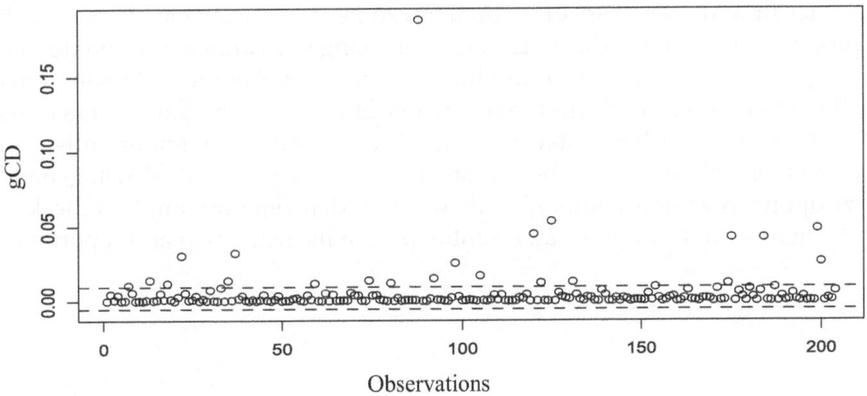


**FIGURE 17.7**
Generalized Cook's distance (gCD) for $\alpha_1$, estimate of intercept factor mean ($a_{\text{Intercept}}$)

parameters, they appear to be somewhat small, with the possible exception of subject 88, who scored at about the 10th percentile across all measurement occasions.

It is also possible to examine gCD relative to individual parameters of the model. For these data, the change in parameter values was somewhat unremarkable except again for the 88th subject, as shown in Figure 17.7.

## Conclusion

Although evaluation of a favored model often focuses on discussion of modification indices to improve model fit, this chapter has discussed how a variety of tools exist for critical examination of a model. Evaluation of the LOESS curves associated with indicator variables of factors can lead to adjustments of the scaling, such as collapsing values associated with evident ceiling or floor effects in the LOESS plot. Residual plots can inform the analyst of a ceiling or floor effects associated with the manifest variable, which can indicate problems with the use of the manifest variable as a measuring tool, indicate the presence of unanticipated subgroups of individuals in the data, or, for smaller data sets, the existence of outlying observations. For larger data sets, this examination may point to the need to represent some of the manifest variables in the model as ordered categorical rather than continuous measures, as discussed in the next chapter.

The diagnostics of the model fit of smaller data sets (the likelihood distance and $\Delta\chi^2$) inform the researcher of individual observations that may be responsible for lack of model fit and are an alternative to the modification index searches often reported. More importantly, however, they are diagnostic of factors that are responsible for an overly optimistic assessment of model fit and form part of a counterargument a reasonable skeptic might propose for the data. The assessments of change in parameter estimates and the gCD (conducted at the individual parameter level or more globally across all parameters) can inform the researcher of reasons why parameters of the model failed to achieve statistical significance; but, more importantly, they provide insight as to whether parameters under consideration will generalize upon replication. Although somewhat burdensome, assumption checking of a final model composes an essential part of its evaluation and reporting.

# 18

# *Categorical Variable Dependent Variables*

> The division of the perceived university into parts and whole is convenient and may be necessary, but no necessity determines how it shall be done
>
> **– Gregory Bateson (1979)**

The chapter emphasizes that modeling categorical dependent variables requires careful attention to assumptions, sample size, and model identification strategies. When done appropriately, these models allow researchers to better respect the ordinal nature of many psychological and behavioral measures, yielding more valid inferences about latent processes.

## Latent Response Variables

### Why a Categorical Dependent Variable Is a Problem

To this point, manifest variables have been represented as continuous measures or as independent categorical variables such as dummy, effect, and orthogonally coded variables. It is inappropriate to use a binary or categorical variable as a dependent variable, though, as this would not meet the assumptions of regression. There are at least three straightforward reasons why a binary variable, $Y$, cannot be simply used in a traditional regression model. First, the distribution of errors of prediction from a regression model are not normally distributed. Errors of prediction can be only either $1-\hat{Y}$ or $0-\hat{Y}$, which is clearly not normally distributed.

Second, error variances in $Y$ as a function of a predictor variable are not homoscedastic as a function of a predictor variable. The variance of a binary variable, for example, is $\pi(1-\pi)$ where $\pi$ is the proportion of 1's in the population. For example, if the proportion of 1's in the population is 0.5, the variance is 0.25 (which is the largest variance that can occur in a binary variable). However, if the proportion is 0.1, the variance is 0.09.

To carry this fact over to the case of a single predictor regression of a dependent categorical variable from $X$, it can be shown (Kutner, Nachtsheim, Neter & Li, 2005, pp. 557–558), that the population error variance $\sigma^2_{\varepsilon_i}$ for the $i$th individual is equal to:

DOI: 10.1201/9781003595267-18

$$\sigma_{\varepsilon_i}^2 = (\beta_0 + \beta_1 X_i)(1 - \beta_0 - \beta_1 X_i), \quad (18.1)$$

which depends on $X_i$. Therefore, the error variance associated with a binary variable will be different at different levels of $X$. So, for example, if the regression constant $\beta_0$ is 0.2 and the regression weight $\beta_1$ is 0.3, the error variance for a regression equation with $X_i$ at zero will be $(0.2)(0.8) = 0.16$, while the error variance when $X_i$ is equal to 1 will be $(0.5)(0.5) = 0.25$. Finally, when the dependent variable takes on only two values, the predicted value of the variable must fall on the interval $[0,1]$, which is a type of constraint not often made in linear modeling or structural equation software.

## Latent Response Variables: Definitions and Terminology

A mathematical model that allows the researcher to specify a linear regression relationship between an independent variable and a categorical variable proceeds by introducing a new continuous latent variable into the proposed structural model containing only observations from a dependent variable measured in discrete categories. This new type of latent variable is termed a latent response variable and is often represented as $y^*$. The distributions of the two most common latent response variables are the normal and logistic distributions, which are symmetric. It is worth noting that possible latent response variables need not be symmetric. A third distribution, the Gumbel or log-log response function, may be more appropriate if the underlying distribution is thought to be asymmetric. As most SEM software does not allow for these latent response variables, and fitting SEMs to them would take the discussion into the larger topic of generalized linear models, they will not be covered in this chapter.

Most statistical software packages readily accommodate categorical variables that represent some type of ordinal continuum. Doing so requires estimation of additional parameters in the model and, for this reason, specifying a dependent variable as categorical may be more appropriate in larger samples where the additional parameters can be accurately estimated, especially if some of the categories have relatively low base rates.

### *Normal Density Functions*

The form of the normal distribution with zero mean and unit variance is well known:

$$f(Y) = \frac{e^{-\frac{1}{2}(Y)^2}}{\sqrt{2\pi}} \quad (18.2)$$

where the variable $Y$ is continuous and ranges from $-\infty$ to $\infty$. In most situations such as cross-sectional data, this form of the normal distribution is used, as it assumes that the mean $(\mu)$ and standard deviation $(\sigma)$ are in the usual standard form ($\mu = 0$ and $\sigma = 1$). This is the normal distribution form used to calculate z-scores. In other situations, however, such as data containing repeated measurements, the mean and/or variance of the latent response variable associated with later measurement occasions may be assumed to vary over time. In these cases, the latent response variable has a mean and standard deviation denoted in as $\mu$ and $\sigma$, respectively as shown in Eq. (18.3).

$$f(Y) = \frac{e^{-\frac{1}{2}\left(\frac{Y-\mu}{\sigma}\right)^2}}{\sqrt{2\pi}\sigma} \tag{18.3}$$

The probability density function associated with the normal distribution is shown in the solid line in Figure 18.1.

### *Logistic Density Function*

The cumulative logistic density function in standard form has a simpler form than the normal density function:

$$F(Y) = \frac{e^{-Y}}{\left(1+e^{-Y}\right)^2} \tag{18.4}$$

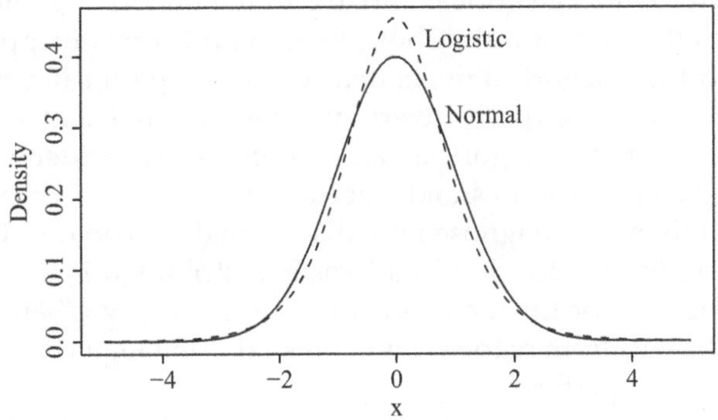


**FIGURE 18.1**
Probability density functions of the logistic and normal distributions.

It has a mean of zero and standard deviation of $\frac{\pi}{\sqrt{3}}$. In order to make the general form of the distribution similar to the normal, the distribution is often standardized by dividing by the standard deviation. The probability distribution for the logistic curve is shown in the dashed lines of Figure 18.1. As can be seen, the resulting distribution is slightly heavier in the tails than in the normal distribution and slightly more leptokurtic; but generally speaking, the two distributions are quite similar. Because of its simplicity, the logistic response variable is often used because it's a bit easier to work with mathematically. In addition, the logistic latent response variable can be used to calculate a statistic called the odds ratio, an estimate of the relative risk effect of a predictor variable on the categorical outcome, which is useful in situations when the categorical variable is a measure of a diagnosis or categorical outcome such as school graduation.

Just as the case for the normal latent response variable, it is also possible to consider the cumulative logistic distribution with nonstandard mean $(\mu)$ and standard deviation $(\sigma)$, as shown in Eq. (18.5).

$$F(Y) = \frac{e^{-\left(\frac{Y-\mu}{\sigma}\right)}}{\sigma\left(1+e^{-\left(\frac{Y-\mu}{\sigma}\right)}\right)^2} \tag{18.5}$$

### *Latent Response Variable Threshold*

The relationship of a categorical variable to the underlying latent response variable uses the idea of a threshold to accomplish this mapping between the SEM and the categorical dependent variable. Specifically, the threshold determines a cut score that converts the continuous latent response variable into the observed categorical values. It helps to consider an example to understand the idea of a threshold. For example, consider a data set in which 38% of the individuals diagnose for a disease and 62% do not. The cut-point corresponding to this division in a logistic distribution is 0.5, as shown in Figure 18.2. Choice of this link function is termed "logit," and, as we will see, the coefficients from regressions expressed using logits have some mathematically nice properties.

Alternatively, if we were to think of a normally distributed link function, the corresponding threshold value would be 0.3, as shown in Figure 18.3. A threshold estimated using the normal distribution is referred to as a "probit."

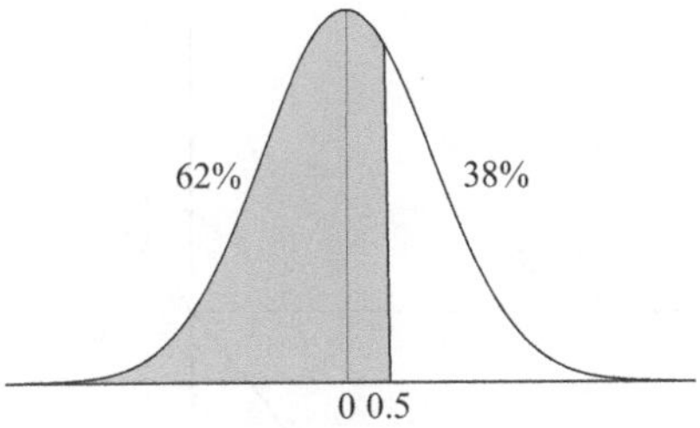


**FIGURE 18.2**
Logit threshold example.

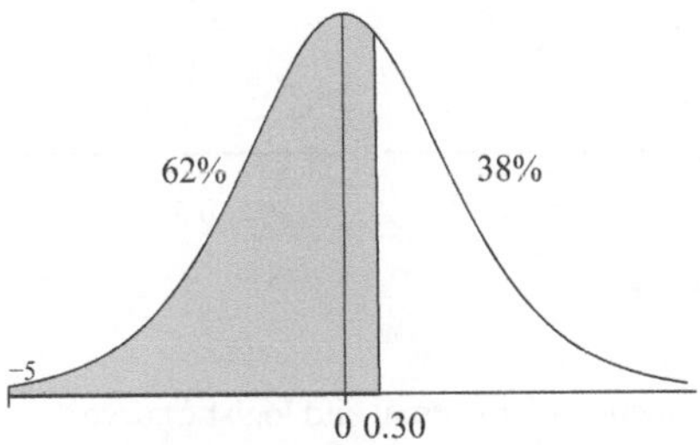


**FIGURE 18.3**
Probit threshold example.

### *Latent Response Variable Regression*

*Logit and Probit Link Estimated Parameters*. It is easiest to formally introduce the idea of the link function involving a categorical dependent variable when that variable is binary. Although one might think that the probit link function would be easiest to understand because it involves the normal distribution, the logit link function actually has some desirable mathematical properties, which makes it a bit easier to understand. Eqs. (18.4) and (18.5) on pages 341 and 342 give the equation for the shape of the percentile points of standard logistic and nonstandard logistic distributions, respectively, and are called "cumulative density functions" for that reason. The cumulative probability density functions for the normal and logistic distributions are given in Figure 18.4. As can be seen from the reference lines of 0.3 and 0.5, the threshold values correspond to a proportion value of 0.62.

It might help to look first at the cumulative distribution function of the logit in Eq. (18.6) given a particular value of $Y$ because it is a simple mathematical expression. Note that if $Y = 0 e^0 = 1$, making the proportion in the first

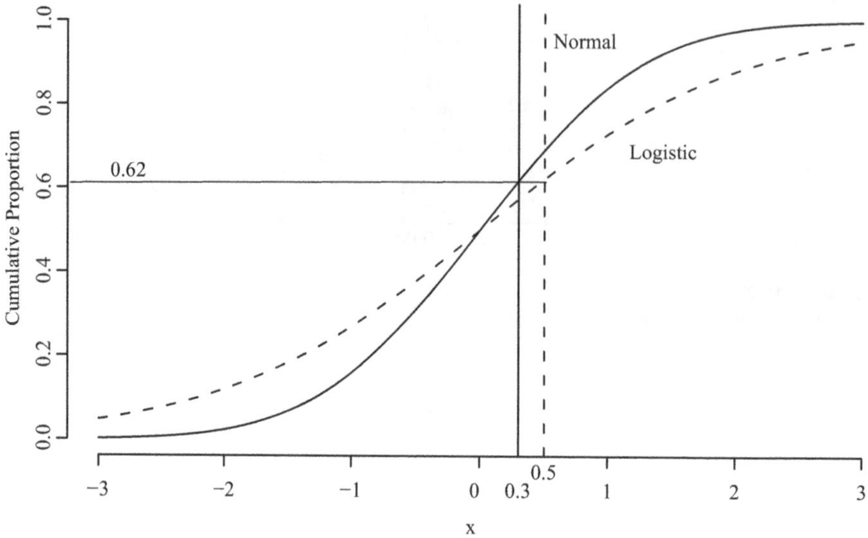


**FIGURE 18.4**
Cumulative distribution functions for normal and logistic curves.

category 0.5. Values above 0 indicate more than half of the observations in the first category and values below 0 indicate proportions less than 0.5.

$$p\,|\,Y = \frac{1}{1+e^{-Y}} \tag{18.6}$$

The cumulative distribution function for the normal distribution in Figure 18.4 is also *S*-shaped, but its mathematical form involves an integral as shown here:

$$F^{-1}(Y) = \frac{1}{\sqrt{2\pi}} \int_{-\infty}^{Y} e^{-\frac{1}{2}\left(\frac{Y-\mu}{\sigma}\right)^2} dY, \tag{18.7}$$

which, for the case of a standard normal (*Z*-score) distribution, reduces to:

$$F^{-1}(Z) = \frac{1}{\sqrt{2\pi}} \int_{-\infty}^{Z} e^{-\frac{1}{2}Z^2} dX \tag{18.8}$$

Evaluating this integral is difficult to do directly, and so numerical approximations such as Taylor's series expansion are used to estimate it.

In most applications, such as factor analysis of individual categorical items, it is simplest to use the standard normal expression of Eq. (18.8) for the data. In longitudinal data, however, it may be useful to think of the mean or standard deviation of the normal distribution as changing across subsequent measurement occasions. Although analogous expressions for the logistic function involve a mean and value analogous to the standard deviation, these are not frequently employed.

## Regression Equations with the Logit Link Function

Linear regression using the logistic link function is the same as for regular linear regression, except that the dependent variable takes the form of the natural logarithm of the odds ratio. Specifically, if X is the adjoined raw score matrix, the predicted value of the dependent variable is:

$$\hat{Y} = \ln\left(\frac{p}{1-p}\right) = X\beta \tag{18.9}$$

This means that the regression predicts the natural logarithm of the odds ratio, i.e., the probability that $Y = 1$ divided by the probability that $Y = 0$. A little bit of math to calculate the probability that $Y = 1$ gives:

$$\text{Prob}\left(Y = 1 | X\right) = \frac{1}{1 + e^{X\beta}} \tag{18.10}$$

Consider the simplest example. If no regression parameters were estimated other than the regression intercept $\beta_0$, a threshold of 0.5 as shown in Figure 18.2 will simply return the base-rate proportion in the data.

When predictor variables are included in the model, the interpretation of the regression intercept $\beta_0$ in a regression model is analogous to that of regular regression regression in that it represents the natural logarithm of the odds ratio associated with the scenario where all predictors are zero – i.e., the $Y$ intercept. Often, the $Y$ intercept based on raw data is not a value that has any useful interpretation for the researchers because predictor variables are positive values and the $Y$ intercept does not correspond to a set of values that could naturally occur in the data. For this reason, rather than use the adjoined matrix based on raw data values, the column of ones is adjoined to the deviation score version of the data, meaning that the intercept corresponds to the natural logarithm of the odds ratio associated with the centroid of the predictors – i.e., the value associated with the mean of all of the predictors. As is true for regression more generally, centering predictor values is also valuable when interaction terms are to be included in the model in order to reduce multicollinearity.

In the more general case where the logistic regression contains one or more continuous predictors, the regression weight associated with the continuous variable represents the increase or decrease in the odds ratio associated with a one-point increase in the predictor variable.

### Regression Equations with the Probit Link Function

The same general pattern is used to interpret regression weights associated with the probit link function, except that the interpretation is made relative to the normal distribution. For example, a stand-alone intercept is interpreted as an estimate of the *z*-score associated with the proportion of individuals in the base category. Regression weights associated with the probit link function are interpreted as the expected gain in the *z*-score associated with a one-point gain in the respective predictor variable. In equation form, this is:

$$\mathrm{Prob}\left(Y = 1 \middle| X\right) = \frac{1}{\sqrt{2\pi}} \int_{-\infty}^{X\beta} e^{-\frac{1}{2}X\beta^2}\, dX \tag{18.11}$$

### Estimation Methods

There are multiple ways to estimate SEMs with categorical variables. Developments in this area are certain to change given the attention such models receive. The following is a description of the approaches often used and a discussion of their relative merits. To begin this discussion with the "bottom line" of what is recommended, the maximum likelihood estimation with robust standard errors (MLR) approach is often most preferred, provided that the assumptions made about the latent response variables in your model are appropriate for your research question. Bayesian estimation (not covered here) has these same advantages and, although less widely known, is increasingly easier to implement given advances in software. Weighted least square with mean and variance adjusted estimation (WLSMV) is less computationally intensive and currently the default for programs like Mplus and lavaan. It also has the advantages that it can accommodate latent response variables in non-standard form. Finally, pairwise tetrachoric correlations are even less computationally demanding and adequate for large samples. I present these alternatives here in reverse order as they are in increasing order of complexity and highlight the advantages and limitations of their use.

## Pairwise Correlation Estimates

As noted earlier, one strategy for creating an adjusted correlation matrix for use in SEM is to generate a correlation matrix in which an observed correlation coefficient is replaced with an estimate of the correlation that would result if the underlying normally distributed form of the categorical variables in the data were available. These estimates can vary depending on whether one or both variables are categorical and, if categorical, if two or more categories are present in the variable. To be fair, these adjustments are calculated automatically by statistical software; however, some points about these adjustments result from examination of their formulae, which will be summarized at the end of this section. As an illustration, the drinking data from Chapter 8 will be used with a dichotomous variable of student report of whether or not they consumed five or more drinks in a sitting during the fall and spring semesters and, as a continuous variable, scores on the Enhancement scale of the Reasons for Drinking measure.

### Tetrachoric Correlations

One approach to estimating a correlation matrix from two categorical variables is to use some estimate of the correlation under the assumption that the observed categories reflect some underlying continuum. The tetrachoric correlation has a rather straightforward formula based on the observed frequencies $f$ of the $2 \times 2$ cross-tabulation can be expressed as in Table 18.1

The tetrachoric correlation, $\rho$, can be calculated as:

$$\rho = \frac{\gamma - 1}{\gamma + 1} \tag{18.12}$$

**TABLE 18.1**

Contingency Table for Two Dichotomous Variables

| Frequencies | Variable *A* | |
|---|---|---|
| Variable *B* | 0.00 | 1.00 |
| 0.00 | $f_{00}$ | $f_{01}$ |
| 1.00 | $f_{10}$ | $f_{11}$ |

where

$$\gamma = \left(\frac{f_{00}f_{11}}{f_{01}f_{10}}\right)^{\frac{\pi}{4}} \tag{18.13}$$

provided that all cells of the table are nonzero. If $f_{01}$ or $f_{10}$ are zero, the correlation is 1, and if either $f_{00}$ or $f_{11}$ are zero, the correlation is −1.

Another formula sometimes used is similar. If $\delta$ is defined as follows:

$$\delta = 1 + \sqrt{\frac{f_{00}f_{11}}{f_{10}f_{01}}} \tag{18.14}$$

then the tetrachoric correlation, $\rho$, is

$$\rho = \cos\frac{\pi}{\delta}, \tag{18.15}$$

with the cases for zero cells defined as before. The particular derivations for these formulae need not concern us, except to note that in Eqs. (18.13) and (18.14), the tetrachoric correlation is a function of a ratio involving the product of agreements across the two variables to the product of disagreements and, as shown in Eq. (18.15), the correlation represents the cosine of an angle, as discussed in "Correlation between Variables" in Chapter 6.

As an example, consider the case of a dichotomous variable indicating whether a person has drunk five or more drinks in a sitting measured during the fall and spring semesters for the college sample we have been using. The cross-tabulation for the data is shown in Table 18.2. Multiplying the frequencies of agreement across the two variables and dividing by the product of the disagreements yield a value of 19.26, producing a $\delta$ of 5.39, a $\gamma$ of 10.21, and a tetrachoric correlation of 0.83 using either formula.

**TABLE 18.2**

Cross-Tabulation of Reports of Drinking Five or More Drinks in a Sitting During Freshman Year

| Frequencies | Spring | |
|---|---|---|
| Fall | 0 (No) | 1 (Yes) |
| 0 (No) | 804 | 218 |
| 1 (Yes) | 171 | 893 |

## Polychoric Correlations

The idea of the tetrachoric correlation can be extended to the case of more than two categories. This approach, applied to only two variables at a time, with the number of thresholds being one less than the number of categories of the variable. (For example, if a variable has three levels, two thresholds are necessary.) It becomes necessary to use maximum likelihood estimation to estimate these thresholds, and therefore no simple formula can be used.

## Biserial Correlations

The biserial correlation is used when attempting to estimate the latent correlation between two variables when one variable is observed as a dichotomous variable but the other is observed as a continuous variable. For a continuous variable $Y$ and a dichotomous variable $X$, the biserial correlation can be expressed in either of the following two forms:

$$r_b = \frac{\left(\bar{Y}_1 - \bar{Y}_0\right)}{s_Y}\sqrt{\frac{p_0\left(1-p_0\right)}{h}} = \frac{r_{XY}\sqrt{p_0\left(1-p_0\right)}}{h} \tag{18.16}$$

where $\bar{Y}_1$ and $\bar{Y}_0$ represent the means of $Y$ for the observations scoring 1 and 0 on $X$, respectively, and $s_Y$ the standard deviation of $Y$.$p_0$ represents the proportion of scores of one level of the categorical variable, and $h$ represents the normal probability density function of $X$ associated with the observed proportion of the categorical variable $X$. In the first example, we can see that the biserial correlation is affected by the magnitude of difference in the continuous variable. Note that the $z$-score difference on the continuous variable between the groups is contained in the first formula for the coefficient. Second, note that the raw correlation between the categorical and continuous variables is corrected by the proportions on the categorical variable, being largest when the groups are evenly split and smaller for all other scenarios.

For example, if we were to consider the case of drinking five or more drinks and the enhancement reasons for drinking scale during the fall freshman semester in our data example, we can observe that 45.72% of students did not consume five or more drinks in a sitting while 54.28% did. The 45.72% number corresponds to a $z$-score of −0.11, and the height of the normal PDF corresponding to this (via Eq. (18.3)) is 0.40, which gives us a biserial correlation of 0.50.

## Polyserial Correlations

Analogous to the situation with polychoric correlations, polyserial correlations can also be calculated between a categorical variable with more than two categories and a continuous variable. As with polychoric correlations, maximum likelihood estimation must be used to estimate the correlation.

## Generation of the Adjusted Correlation Matrix Using Pairwise Adjustments

The tedious task of specifying each type of correlation to generate a correlation matrix for their data across multiple variables is handled by computer software. This is often then used to generate a correlation matrix to be fed into either an exploratory factor analysis or SEM program. Use of standard Pearson correlations when the underlying variables are ordinal in nature results in underestimates (in absolute value) of the association between two variables. Some cautions are in order, however, when using these adjustments. First, as can be seen from the formulae for tetrachoric correlations, empty cells in the cross-tabulation of dichotomous variables can automatically result in assignments of perfect correlations between the variables. As a result, smaller sample sizes are likely to result in spurious estimates of perfect associations between variables due to sampling errors in which empty cells might occur. Such cells are also likely to occur when proportions of responses are markedly unequal. Second, there is no guarantee that the resulting matrix made by such pairwise adjustments is of full rank, meaning that it might not be invertible or provide consistent estimates. In practice, it may be necessary to group sparse categories together or with more frequently occurring values to provide a useful analysis. In small-sample situations with several variables, it may be necessary to simply admit that the adjustments described here are not possible and that treating the variables as continuous is the only possible analysis until larger data sets become available. However, in situations where large samples are employed and none of the individual categories are rare events, the various adjustments described earlier may work well in practice to provide a useful analysis.

## Diagonally Weighted Least Squares Alternatives

Diagonally weighted least squares (DWLS; see Chapter 8) is often proposed as an alternative to the serial pairwise adjustments just described. A robust version of this, weighted least squares with mean and variance adjustments (WLSMV), is the default estimation approach in the Mplus and lavaan programs. In WLSMV, only the probit link function is used in both the Mplus and lavaan programs.[1] In plain language, this estimation technique is used when an Mplus program is written as for any structural model, with the addition of a statement identifying which variables are ordered categorical instead of continuous. In Mplus, specifying WLSMV results in the generation of robust (Satorra-Bentler) standard errors and a mean- and variance-adjusted

$\chi^2$ statistic that has been found to perform fairly well even using small sample sizes (e.g., $N > 200$; Forero, Maydeu-Olivares, & Gallardo-Pujol, 2009). It is important to realize that when doing this, one cannot directly compare the $\chi^2$ statistics from printouts to assess relative model fit among competing models. Use of this adjusted $\chi^2$ statistic, however, means that model comparisons save the derivatives from the base model, as is done when using the DIFFTEST option in Mplus.

### Maximum Likelihood with Robust Standard Errors

MLR (ML estimation with a sandwich estimator) represents an alternative approach that is preferable in many situations, although the resulting model fit, parameter estimates, and parameter significance are often largely similar to the values obtained under WLSMV. The method of estimating the standard errors is called the "Huber-White" sandwich estimator. ML estimation, because it is a full information estimation technique, is likely to have a performance edge when the usual missing assumptions of the data are not met because it uses all available information for each case during estimation. The ML approach uses numeric integration to estimate polychoric correlations instead of the raw covariance matrix. For this reason, using MLR estimation is somewhat more computationally intensive. An additional limitation of this approach is that the variance of the latent response variable is fixed to the value of the standard link function and cannot be freely estimated. As such, if it is desired to estimate a model in which the latent response variable variation increases or decreases (as may occur in repeated measures or multigroup models), some other approach like WLSMV must be used.

The MLR approach has some advantages. It is more efficient than WLSMV (meaning that the estimated parameters do a better job of estimating their respective population values as sample size increases). Second, because it is a simultaneous estimation approach, it is more appropriate when the data are missing at random (MAR). The robust ML method appears to perform slightly better when the latent variable distribution is skewed (DeMars, 2012). As for WLSMV, robust (Satorra-Bentler) standard errors work well for sample sizes larger than about 250, and these perform better than the unadjusted full-information ML estimates in simulation studies (Yang-Wallentin & Jöreskog, 2001).

Fitting a single marginal threshold for each variable that is then used to estimate the SEM associated with the continuous latent response variables is similar to Olsson's (1979) two-step approach, in which a single threshold is estimated for the categorical variable adjustment, which is then used to generate the matrix used to calculate the SEM with continuous latent response variable (see the discussion on weighted least squares in Chapter 8).

## Some General Guidance Regarding Categorical Dependent Variables

It is helpful to note that, as mentioned in the discussion of pairwise correlation estimates, the ability to successfully estimate thresholds or adjust correlations depends to some extent on the sample size and relative magnitude of the proportions being estimated. These adjustments work best in practice when more data are available and when the proportions of response are moderate, meaning that they are neither too rare nor too commonplace, or, in the case of more than one category, thresholds are not too closely spaced together. In practice, when multiple thresholds are estimated for a single variable, as occurs in variables with more than two categories, categories may be combined to improve estimation, for example. It may not be possible to estimate thresholds for data sets that are too small, and analysis of the data in their continuous forms may be all that can be done.

Table 18.3, adapted from Muthén, Muthén, and Asparouhov (2015), provides a summary of the relative merits of the three estimation methods for categorical dependent variables. Note that weighted least squares is preferred when the model under consideration has a large number of factors, and a large number of subjects is being analyzed. This is because under these conditions, efficiency of estimation is not a huge consideration, and the weighted bivariate adjustment of weighted least squares is a simple approach. When testing or exploration of the latent response variable (LRV)

**TABLE 18.3**

Relative Advantages for Latent Dependent Categorical SEM's

| Model Characteristics | Weighted Least Squares | Maximum Likelihood | Bayes |
|---|---|---|---|
| Large number of factors | + | – | + |
| Large number of variables | – | + | + |
| Large number of subjects | + | – | – |
| Small number of subjects | – | + | + |
| Statistical efficiency | – | + | + |
| Missing data handling | – | + | + |
| Test of LRV structure | + | – | + |
| Ordered polytomous | | | |
| Ordered polytomous variables | + | – | – |
| Heywood cases | – | – | + |
| Zero cells | – | + | + |
| Residual correlations | + | – | ± |

and/or multiple ordered categorical variables are under consideration, weighted least squares is also a good idea because this approach can allow the researcher to have variable scaling factors. So, for example, a researcher interested in looking at a longitudinal factor model of multiple Likert-format items across measurement occasions would probably prefer weighted least squares, given that the flexibility of the scaling factor approach allows specification of more general models than ML estimation, which is limited to unit scaling factors. It is also quite easy for the researcher to specify correlated measurement errors in the data. Notice, though, that zero cells in the cross-tabulation of categorical variables are a problem because, as noted before, these are estimated as +1 or −1 correlations.

By contrast, if relatively few observations are being analyzed and efficiency of estimation is a consideration, maximum likelihood may be preferred. The maximum likelihood approach, because it estimates thresholds marginally across all the cells associated with a variable, has the additional advantage that it provides good estimates if the data are missing at random (MAR). Maximum likelihood is at somewhat of a disadvantage if a large number of factors or individuals is being analyzed due to the estimation properties of these models. A large number of resource intensive integration points are often necessary to successfully estimate the model.

Notice that the right-most column, Bayesian estimation, has several advantages in that it can deal with a large number of factors, is efficient, provides unbiased estimates if data are missing at random, permits a test of the latent response variable structure, and, by its mathematical assumptions, prevents any negative estimates of variance components (and, for that matter, other mathematically improper solutions such as estimated correlations greater than unity). Like maximum likelihood, it is also appropriate if zero cells are observed in the cross-tabulations of variables and permits estimation of residual correlations. Although Bayesian estimation is becoming increasingly easy to program with available software, details of the logic and estimation procedures require an extended discussion that cannot be included here in the interests of space constraints.

## Variance of the Latent Response Variable, $y^*$

The WLSMV estimator comes in two different versions, depending on scaling choices of the latent response variable within the SEM. In other words, the variance of the latent response variable in a given SEM is the sum of the variance of the variable using the regular tracing rules plus the assumed variance of the underlying latent response variable. Writing this out in general matrix notation is straightforward, but the equations are probably a bit distracting for the new reader in this area. They are, however, shown in Muthén and Muthén (1998–2017).

For this reason, it helps to consider a simple example, the case of a categorical dependent variable that is an indicator of a factor. The total variance of the latent response variable under a logit link function would be

$$\sigma^2_{y^*} = \lambda^2\psi + \frac{\pi^2}{3} \tag{18.17}$$

because, as noted earlier, the standard deviation of the logistic distribution is $\frac{\pi}{\sqrt{3}}$. Similarly, for the probit link function, the standard normal distribution assumed with a unit variance, making the variance of $y^*$ making the example of the variance of an indicator of a factor model equal to:

$$\sigma^2_{y^*} = \lambda^2\psi + 1 \tag{18.18}$$

Under these formulae, it is not possible to specify different values for the variances of the latent response variable, as might occur, for example, if a factor analysis of repeated measures is constructed.

## Theta and Delta Parameterization

Under WLSMV estimation, there are two ways to estimate the latent response variable, termed delta and theta parameterization. These two methods, though different in their mathematical forms, can be thought of as equivalent in that the overall model fit, parameter significance, and have standardized coefficients that are identical under either parameterization (Finney & DeStefano, 2013).

In most SEM programs delta parameterization is the default because its numerical estimation is slightly better behaved (Muthén & Asparouhov, 2002). On the other hand, theta parameterization is used when the researcher wants to estimate an explicit error variance as might occur, for example, if a researcher wanted to estimate error covariances between variables that are indicators in a factor model, or if equality constraints on error variances across items are to be considered. Theta parameterization, however, is the required approach if it is the case that the dependent categorical variable is both influenced by and influences either another observed dependent variable or a latent variable (Muthén & Muthén, 1998–2017, p. 675).

Delta parameterization (which is referred to as "marginal parameterization" by some) constrains the residual variance to one (because the underlying normal distribution is thought to have unit variance) and estimates the variance of $y^*$ (the latent response distribution). By default, then, this will result in the same predicted variance as found in Eq. (18.17) and Eq. (18.18) for the logit and probit link functions, respectively.

By contrast, under theta parameterization, the variance of $y^*$ is fixed to one and the error variance is freely estimated. Because variance of $y^*$ is now not one, it is necessary to use a scaling adjustment in order to talk about the thresholds and their associated proportions of $y^*$ in terms of the standard normal cumulative distribution function shown in Figure 18.4. In equation form, this scaling factor is the reciprocal of Eq. (18.18):

$$\Delta_j = \frac{1}{\sqrt{\lambda_j^2 \Psi + 1}} \tag{18.19}$$

This scaling factor is used by software such as Mplus to calculate standardized estimates.

## Categorical Dependent Variable Examples

Given that a lot of ground has been covered at this point, it is helpful to start with a simple example of a computer program that recovers these proportions.

### Single Predictor Logistic Regression

#### *LOESS Regression*

Suppose that a researcher was interested in determining whether scores on the Enhancement Reasons for Drinking scale could be used to predict whether an individual consumed five or more drinks in a sitting. Just as

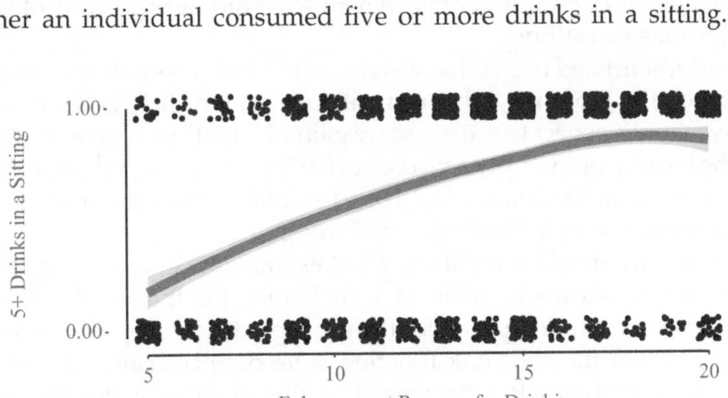


**FIGURE 18.5**
LOESS regression for dependent categorical variable.

with regression with continuous dependent variables, the first step in fitting such a model begins with assumption checking. Figure 18.5 shows a LOESS regression for these data. As can be seen from the figure, there is a slight bend at the upper levels of the predictor variable, suggesting a leveling off of predicted scores for values 17 and above. Accordingly, a slight rescaling of the predictor variable may be in order in which values larger than 17 are simply set to 17, thereby capturing the general linear trend in the data. In some situations the shape of the resulting LOESS curve may suggest fitting a quadratic or other curve to the data; but, given that no a priori theory exists for such a relationship, making a minor adjustment to the scale appears appropriate in this setting.

### *Fitting the Logistic Model*

After this adjustment, the logistic regression model was fit to the data, and influence diagnostics for the data were examined. In the case of this data set, rerunning the analysis excluding influential observations did not appreciably affect the results, and this is probably due to the fact that the sample size of 1,969 was rather large. In smaller data sets, however, such observations may substantially affect parameter estimates and model fit (shown in Table 18.4).

The general fit statistics are given in the first three rows of the table and could be used to compare with other non-nested models. In this data set, there were 618 individuals who did not consume five or more drinks in a sitting and 1,351 individuals who did (i.e., 31.4% and 68.6%, respectively). As we can see from this, the standardized probit link function solution value of -.53 is the *z*-score associated with 30% of the distribution. This intercept tells us that approximately 30% of the individuals who score at the average of the Reasons for Drinking Enhancement scale did not report drinking five or more drinks in a sitting.

The unstandardized regression weight of 0.15 indicates that given a gain of one point on the modified enhancement scale (in which 17 is the maximum score), we would expect to see a *z*-score gain of 0.15; thus, a gain of 10 points above the mean would imply a *z*-score of $0.97\left(\left(0.15^{*}10\right)-0.53\right)$, meaning that roughly 83% of individuals who score 10 points above the mean are predicted to drink five or more drinks in a sitting.

Similar results are obtained if the MLR estimation is done using the logit link function, as shown in Table 18.5. As before, the threshold obtained in the standardized solution produces the same predicted proportions as in the same, except that the logit link function is used. Specifically, the predicted proportion of individuals who scored at the average of the Reasons for Drinking Enhancement scale who consumed five or more drinks (that is, the proportion in the second category) is equal to:

**TABLE 18.4**

Probit MLR Regression Output Excerpt

| | | | | | |
|---|---|---|---|---|---|
| | | **Selected Output** | | | |
| Information Criteria | Akaike (AIC) | 2150.15 | | | |
| | Bayesian (BIC) | 2161.81 | | | |
| | Sample-Size Adjusted BIC | 2155.45 | | | |
| Model Results | | Estimate | S.E. | Est./S.E. | P-Value |
| Plus5 on Enhance | | 0.15 | 0 | 16.92 | 0 |
| Thresholds | Plus5$1 | −0.53 | 0 | −16.92 | 0 |
| | | **Proportions Plus5** | | | |
| | Category 1 | 0.31 | | | |
| | Category 2 | 0.69 | | | |
| | | **STDYX Standardization** | | | |
| Plus5 on Enhance | | 0.47 | 0 | 20.89 | 0 |
| Thresholds | Plus5$1 | −0.47 | 0 | −16.98 | 0 |
| R-Square | Plus5 | 0.22 | 0 | 9.77 | 0 |

$$\text{prob}(x=1) = \frac{1}{1-e^{\beta_0}} = \frac{1}{1-e^{-0.88}} = 0.71 \quad (18.20)$$

and, from this, that the probability of being in the first category is 0.29, which is somewhat similar to the 0.30 value obtained under the probit link function.

It is worth noting at this point that the parameters of logistic regression are slightly different from the parameters of linear regression. Recall that in linear regression that the fitted regression line always passes through the "centroid," or value corresponding to the estimated mean of all predictors and criterion. In logistic regression, the proportion of each category predicted by the model is the average over all observations. This average of the probabilities is not the same as the probability of the average (of $x$) due to the latent response variable function being nonlinear in its argument. It is this average that is shown in both Table 18.5 and 18.6 under the row labeled "Results in Probability Scale."

### *Standardized Results*

Both Tables 18.5 and 18.6 show standardized coefficients as well. The standardized threshold becomes a $z$-score for the continuous latent response variable $Y^*$ given that this variable has unit variance when standardized. The standardization divides the threshold by the model-estimated SD of $Y^*$. The $r^2$ for the model represents the square of the standardized estimate of 0.43 or, in the context of multiple predictors, the proportion of variance in the

**TABLE 18.5**

Logit MLR Regression Output Except

| Selected Output | | | | | |
|---|---|---|---|---|---|
| Information Criteria | Akaike (AIC) | 2,150.15 | | | |
| | Bayesian (BIC) | 2,161.81 | | | |
| | Sample-Size Adjusted BIC | 2,154.97 | | | |
| Model Results | | Estimate | S.E. | Est./S.E. | P-Value |
| Plus5 on Enhance | | 0.25 | 0.02 | 16.39 | 0 |
| Thresholds | Plus5$1 | −0.88 | 0.05 | −16.33 | 0 |
| **Results in Probability Scale** | | | | | |
| Plus5 | Category 1 | 0.31 | | | |
| | Category 2 | 0.69 | | | |
| **Logistic Regression Odds Ratio** | | | | | |
| Plus5 on Enhance | | 1.28 | 0.02 | 14.5 | 0 |
| **STDYX Standardization** | | | | | |
| Plus5 on Enhance | | 0.43 | 0.02 | 19.53 | 0 |
| Thresholds | Plus5$1 | −0.44 | 0.03 | −16.66 | 0 |
| R-Square | Plus5 | 0.19 | 0.02 | 9.77 | 0 |

**TABLE 18.6**

Observed Frequencies and Predicted Proportions under Rasch Model

| Measurement Occasion | Consumed 5+ Drinks? | | Predicted Rasch Proportions | | | |
|---|---|---|---|---|---|---|
| | No | Yes | No | Yes | Threshold ($\tau_i$) | Difficulty ($\tau_i/\lambda$) where $\lambda = 3.23$ |
| Pre-College | 56% | 44% | 56% | 44% | 0.59 | 0.18 |
| | 1264 | 981 | | | | |
| Freshman | 47% | 53% | 45% | 56% | −0.51 | −0.16 |
| Fall | 728 | 824 | | | | |
| Freshman | 47% | 53% | 43% | 57% | −0.61 | −0.19 |
| Spring | 700 | 801 | | | | |
| Sophomore | 46% | 54% | 42% | 58% | −0.76 | −0.24 |
| Fall | 617 | 726 | | | | |
| Sophomore | 46% | 54% | 43% | 57% | −0.65 | −0.2 |
| Spring | 676 | 800 | | | | |
| Junior | 41% | 59% | 38% | 62% | −1.11 | −0.34 |
| Fall | 579 | 839 | | | | |
| Junior | 42% | 58% | 39% | 61% | −1.07 | −0.33 |
| Spring | 593 | 820 | | | | |
| Senior | 40% | 60% | 36% | 64% | −1.32 | −0.41 |
| Fall | 565 | 861 | | | | |
| Senior | 41% | 59% | 37% | 63% | −1.22 | −0.38 |
| Spring | 544 | 788 | | | | |

dependent variable expressed by the tracing rules involving the predictor variables as done, for example in the standardized two-predictor regression example of Chapter 4.

## Item Response Theory

Latent response variables can also be used within any structural model. This is another way that structural equation modeling can be used to understand models in item response theory, and here it becomes possible to explore new models using the structural models we introduced so far. As we noted in Chapter 7, within psychometrics we view manifest variables as indicators of a latent variable. The focus of our inquiry, though, is on the properties of the manifest variable's job as a measuring tool in terms of the level of the latent variable. Two measurement properties of items are most often the focus, termed difficulty (sometimes also called the "location," for reasons described shortly) and discrimination. An item's difficulty is familiar to anyone with experience with classroom tests: Some items are much less likely to be correctly answered by individuals in the class than others. Those items that are correctly answered by only a small proportion of individuals in the class are said to be more difficult than items that a larger proportion can correctly answer. If we think about a plot of the probability of correctly answering the question as a dependent variable as a function of values of the latent variable items, the difficulty associated with the *i*th item can be expressed as:

$$b_i = \frac{\tau_i}{\lambda_i} \tag{18.21}$$

where $\tau_i$ is the threshold associated with the *i*th item and $\lambda_i$ is the factor loading for the item.

### Rasch Models

Rasch models assume that items differ from each other only in terms of their difficulty. The structural equation model underlying the Rasch model is the tau equivalent factor model as presented in the tau equivalent measures model discussed in Chapter 10 – a model in which the factor variance is fixed to one and all factor loadings are constrained to equality. The Rasch model is also sometimes referred to as the one-parameter item response model.

## Longitudinal Example

As a simple example of the Rasch model in action, consider the fact that the college drinking study we have been looking at contains an assessment done in the summer before enrollment as well as the fall and spring semesters for the next four years, making nine "items" that could be used to describe a continuum of severity of heavy alcohol use.

Frequently, we are interested in looking at how endorsing these items is related to the underlying latent variable. From the data of our example here, for example, reporting drinking 5+ drinks is more common during the junior and senior assessments when individuals can legally consume alcohol than in the Pre-College assessment when students are more likely to be living in their parents' home and it is illegal for them to purchase alcohol. An individual who endorses consuming 5+ drinks at the Pre-College assessment is more indicative of a more extreme level of problematic alcohol consumption than someone who reports doing so as a junior or freshman. The predicted proportions of reports of drinking 5+ drinks for the model are shown in the middle two columns. The estimated thresholds from the Rasch/Tau equivalent factor model with freely estimated thresholds are shown in the column labeled "Threshold," and the estimated difficulties for the nine dichotomous items, using Eq. (18.21), are estimated as shown in the right-most column of Table 18.6. Note that the estimated thresholds of the model will not correspond numerically to the predicted proportions because, as explained earlier, these proportions are arrived at by taking the expected value across all observations in the data set, but they are, largely, close.

## ICC Curves

Item characteristic curves (ICCs) can be thought of as a type of cumulative distribution function (CDF) to describe the probability of endorsing an item as a function of the underlying trait. (Recall that a CDF was shown in Figure 18.4.) To make a plot of the relationship between endorsement of a particular item and the underlying distribution of ability, the assumed link function and the estimated value of the factor loading are used to describe the steepness of the curve as a function of level of the latent variable. Within item response theory, this steepness is termed the "discrimination" and is defined for the *i*th item as:

$$a_i = \frac{1}{\sqrt{\lambda_i^2 - 1}} \tag{18.22}$$

The Rasch model, as we noted, assumes that the factor loadings for all items are the same and that, therefore, the steepness of the item curves equal. The item response theory model presented here is more specifically known as the two-parameter model because items differ, at most, in terms of their difficulty and discrimination.

It is possible to draw an item characteristic curve as a function of the level of the latent variable, usually denoted by the symbol $\theta$, the item discrimination, $a_i$, and the difficulty, $b_i$, using the following formula:

$$\text{Prob}\left(Y_i = 1 | \theta\right) = \frac{1}{1 + e^{-a_i(\theta - b_i)}} \quad (18.23)$$

As an example, for a latent value of 0.6, the predicted probability of endorsing the +5 item in the Pre-College assessment is

$$\text{Prob}\left(Y_i = 1 | 0.6\right) = \frac{1}{1 + e^{-3.23(0.6-0.18)}} = 0.79,$$

whereas the corresponding probability for the latent value of $\theta = 0.18$ is 0.50. This latter value helps us to interpret the difficulty coefficient as that point on the latent variable scale that corresponds to exactly half of the individuals endorsing the item or correctly answering the question.

We can use these predicted probabilities to construct a graph of how item difficulties are related to the odds of endorsing or correctly answering an item as a function of level on the latent variable. Figure 18.6 shows the ICCs across the nine waves of the Plus 5 drinking variable. As can be seen from

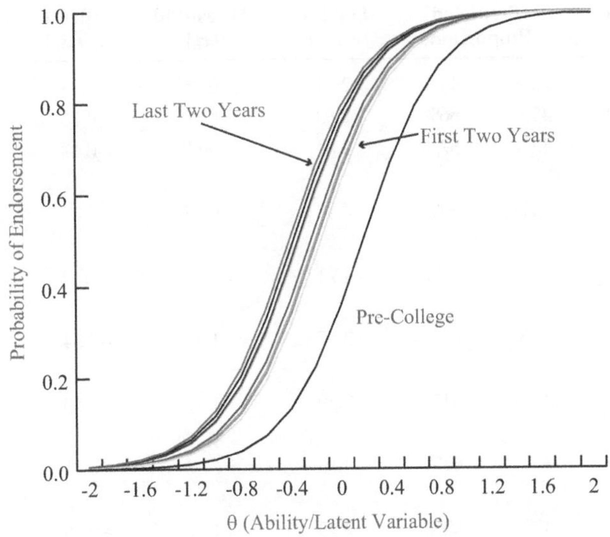


**FIGURE 18.6**
ICC's for Rasch model data.

the figure, the Pre-College drinking variable exists to the right of the other curves, indicating that those individuals who have higher scores on the problematic drinking factor are more likely to endorse this item. The second band of curves associated with the four measurement occasions gathered during the first two years of college are associated with lower levels of problematic alcohol consumption, and the four measurement occasions associated with the junior and senior years of college are associated with lower levels of the latent variable. This pattern, of course, should not be taken to mean that juniors and seniors who consume five or more drinks are not at risk for severe consequences of consumption, but rather that students who report consuming five or more drinks during the Pre-College assessment are likely more heavily alcohol-involved than those who report such consumption at later waves.

## Two-Parameter Item Response Model

Extension of this model, to allow for the fact that some items may better discriminate between individuals than other items, amounts to fitting the data to the standard congeneric factor model mentioned in Chapter 10 – i.e., a single factor model with freely estimated loadings (Table 18.7).

**TABLE 18.7**

Observed Frequencies and Predicted Proportions Under Two-Parameter Model

| Measurement Occasion | Two-Parameter Predicted Proportions | | Loadings ($\lambda_i$) | Threshold ($\tau_i$) | Difficulty ($\tau_i / \lambda_i$) | Discrimination Eq. (18.22) |
|---|---|---|---|---|---|---|
| Pre-College | 56% | 44% | 2.31 | 0.45 | 0.18 | 2.31 |
| Freshman Fall | 45% | 56% | 2.83 | −0.46 | −0.16 | 2.83 |
| Freshman Spring | 43% | 57% | 3.64 | −0.7 | −0.19 | 3.64 |
| Sophomore Fall | 42% | 59% | 4.19 | −0.98 | −0.24 | 4.19 |
| Sophomore Spring | 43% | 57% | 4.71 | −0.94 | −0.2 | 4.71 |
| Junior Fall | 39% | 62% | 3.67 | −1.23 | −0.34 | 3.67 |
| Junior Spring | 39% | 61% | 3.59 | −1.17 | −0.33 | 3.59 |
| Senior Fall | 36% | 64% | 3.16 | −1.3 | −0.41 | 3.16 |
| Senior Spring | 37% | 63% | 2.93 | −1.12 | −0.38 | 2.93 |

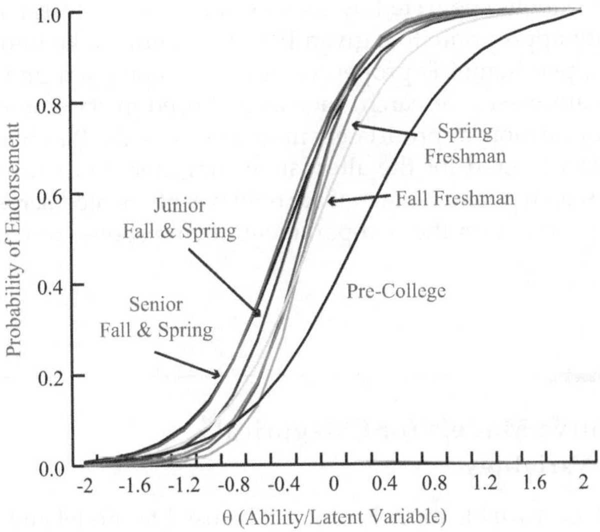


**FIGURE 18.7**
ICC's for two-parameter model.

In similar fashion, the individual item difficulties and discrimination can now be used to generate ICCs for this model, as shown in Figure 18.7. As can be seen from the figure, the Pre-College report of drinking five or more drinks still discriminates individuals at higher levels of problematic alcohol consumption, but the slope of this line is not as steep as the remaining items. The curves for the senior and junior assessments also seem a little more distinct, with individuals at the lower levels of the latent variable being slightly more likely to report consuming five or more drinks as a senior than as a junior.

## Summary of Item Response Theory Models

Using IRT models to assess the properties of items has many useful applications. Within the general area of the assessment of abilities, for example, it helps us to select items for assessment that accomplish specific goals we may have. For example, we may select those assessment items that allow us not only to identify individuals who are high in ability, but to identify those items that may more clearly discriminate between individuals at a high level (using items with not only a high difficulty, but also higher discrimination values). Use of these models within other areas of psychology would allow us to select items which identify individuals at specific levels of the traits of interest.

As with any structural models, best practice in specifying and fitting such models also involves assumption checking and the side analyses and model specifications suggested by such examination. In addition, the degree of complexity appropriate to a given IRT model must take into account the data and the psychometric properties of the phenomenon under investigation. Small-sample-size research, such as that used in classroom assessment contexts, may be more appropriately modeled using the Rasch model, as one may be underpowered for the alternative congeneric measurement model. Similarly, research phenomena that are only weakly related across items may not be well-powered for the two-parameter item response model.

## Growth Curve Models for Categorical Observed Variables

Latent response variable models can also be used to model any of the structural models discussed so far in this book. The case of binary dichotomous items is somewhat limiting, as it is not possible to develop separate estimates for the threshold and scaling factor. This means that, for binary items, one can freely estimate only thresholds, meaning that increases in the variability of the latent response variable must be held to equality.

Growth curve models involving latent response variables require some care in specification in order to specify factor mean-level effects for the growth and/or random intercept factors. Often this takes the form of fixing some threshold of the latent response variable to a known value, such as zero. Similarly, if it is desired to specify that the scaling factors of the assumed latent response variable vary across measurement occasions, it is necessary to fix the scaling factor associated with one of the measurement occasions. It should be noted that these decisions allow the researcher to specify a mathematically identified model, but that the scale and mean level of the unstandardized solutions for such growth models will be different depending on how the model is mathematically identified. Standardized solutions expressing the proportion of variation in the latent response variables should be similar across identification strategies, however.

### Plus 5 Drinking Growth Curve Example

The original question involving drinking five or more drinks in a sitting had more than two categories. The response categories are shown in Table 18.8. The original format of the item included response options for saying every day and more than once every day, but these were collapsed into the 5+-per-week group to produce sufficient frequencies to increase the accuracy of the estimated thresholds in the model.

**TABLE 18.8**

Proportions and Frequencies of Reports of Drinking Five or More Drinks in a Single Sitting

| Times in Past 30 Days Consumed 5+ Drinks in a Sitting? | | Freshman | | Sophomore | | Junior | | Senior | |
|---|---|---|---|---|---|---|---|---|---|
| | Pre-College | Fall | Spring | Fall | Spring | Fall | Spring | Fall | Spring |
| | | | | Proportions | | | | | |
| 0 Times | 0.56 | 0.47 | 0.47 | 0.46 | 0.46 | 0.41 | 0.42 | 0.4 | 0.41 |
| 1 Time | 0.11 | 0.14 | 0.14 | 0.14 | 0.12 | 0.17 | 0.15 | 0.16 | 0.16 |
| 2–3 times | 0.14 | 0.17 | 0.17 | 0.15 | 0.17 | 0.19 | 0.18 | 0.18 | 0.18 |
| 1 or 2 a week | 0.1 | 0.16 | 0.16 | 0.17 | 0.18 | 0.18 | 0.18 | 0.2 | 0.19 |
| 3–4 a week | 0.06 | 0.05 | 0.06 | 0.06 | 0.06 | 0.05 | 0.06 | 0.06 | 0.05 |
| 5+ a week | 0.03 | 0.02 | 0.01 | 0.01 | 0.01 | 0.01 | 0.01 | 0.01 | 0.01 |
| | | | | Frequencies | | | | | |
| Did Not | 1,264 | 728 | 700 | 617 | 676 | 579 | 593 | 565 | 544 |
| Once | 256 | 217 | 209 | 193 | 179 | 237 | 206 | 234 | 219 |
| 2–3 times | 306 | 261 | 254 | 206 | 256 | 266 | 254 | 249 | 238 |
| 1 or 2 a week | 215 | 252 | 235 | 234 | 262 | 256 | 257 | 280 | 246 |
| 3–4 a week | 131 | 71 | 86 | 80 | 90 | 67 | 89 | 81 | 69 |
| 5+ a week | 73 | 23 | 17 | 13 | 13 | 13 | 14 | 17 | 16 |

**TABLE 18.9**

Parameter estimates for Free Curve model with categorical indicators.

| | | Intercept | Growth | | |
|---|---|---|---|---|---|
| | | Loadings | | Scaling Factors | Thresholds |
| Pre-College | | 1 | −0.61 | 1 | t1 = 0.00 |
| Freshman | Fall | 1 | −0.42 | 1.12 | t2 = 0.29 |
| | Spring | 1 | −0.34 | 1.22 | t3 = 0.67 |
| Sophomore | Fall | 1 | −0.26 | 1.26 | t4 = 1.27 |
| | Spring | 1 | −0.19 | 1.32 | t5 = 1.85 |
| Junior | Fall | 1 | −0.07 | 1.39 | |
| | Spring | 1 | −0.03 | 1.37 | |
| Senior | Fall | 1 | 0.06 | 1.4 | |
| | Spring | 1 | 0.03 | 1.36 | |
| | Means | 0.18 | 0.46 | | |
| | Variances | 0.41 | 1 | | |

An orthogonal free curve model was fit to the data. For purposes of mathematical identification, the first threshold is fixed to zero across all measurement occasions and the remaining thresholds fixed to equality across their corresponding measurement occasion. The scaling factor for the Pre-College assessment is fixed to 1, and the remaining scaling factors are freely estimated.

The resulting model appeared to fit relatively well. Although the $\chi^2$ fit of the model was statistically significant ($\chi^2_{57} = 216.83, p < 0.001$), the remaining fit measures suggest an adequate fit to the data (RMSEA = 0.04, 95% CI = [0.03–0.04], CFI = TLI = 1.00). The resulting parameter estimates are shown in Table 18.9. As can be seen in the model, factor loadings for the growth factor are negative at earlier measurement occasions and are positive during the senior year, meaning that the aperture under a linear model occurs somewhere between the junior and senior years. This also implies that the correlation between the intercept and slope would be observed to be negative if an oblique model were fit relative to earlier measurement occasions but would be close to zero if set relative to the senior assessments. (For these data, the correlation between the intercept and growth factors if set to the first time of measurement is −0.69 and is −0.09 if the senior fall semester is set as the reference measurement occasion with identical fit across all models.)

Just as the estimated factor loadings and means can be used to generate a predicted curve (as was done for the Free Curve model in Chapter 13), the predicted shape of the estimated curve can be calculated as the sum of the random intercept mean and the product of the respective estimated growth loading and growth factor mean. This is shown for the Free Curve estimates

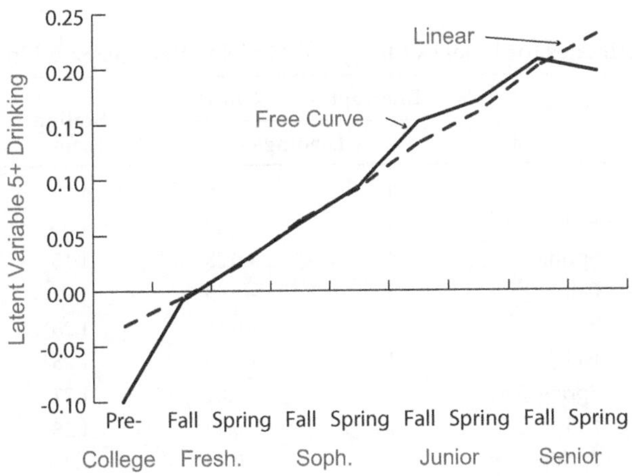


**FIGURE 18.8**
Estimated FCSI and linear curves for categorical 5+ drinking.

as the solid line in Figure 18.8. As can be seen, the general trend of growth in the latent variable appears roughly linear across measurement occasions, with a slight downward dip associated with the Pre-College assessment.

An orthogonal linear growth model can be estimated as a parsimonious conjecture for the data. This was discussed as the Orthogonal Linear SI model in Chapter 13. Doing so requires the specification of nonlinear constraints on the intermediary loadings between the first and last measurement occasion as a weighted composite of the first estimated loading, $\lambda_0$, and the last loading, $\lambda_8$. For this example, the corresponding constraints are necessary: $\lambda_1 = 0.10^* \lambda_8 + 0.90^* \lambda_0$; $\lambda_2 = 0.21^* \lambda_8 + \left(0.79^* \lambda_0\right)$; $\lambda_3 = 0.36^* \lambda_8 + \left(0.64^* \lambda_0\right)$; $\lambda_4 = 0.47^* \lambda_8 + \left(0.53^* \lambda_0\right)$; $\lambda_5 = 0.63^* \lambda_8 + \left(0.37^* \lambda_0\right)$; $\lambda_6 = 0.73^* \lambda_8 + \left(0.27^* \lambda_0\right)$; $\lambda_7 = 0.89^* \lambda_8 + \left(0.11^* \lambda_0\right)$.

The resulting model fits about as well as the Free Curve model, but with fewer parameters ($\chi^2_{64}$, $p < 0.001$, RMSEA = 0.04, 95%CI[0.03–0.04), TLI = CFI = 1.00). Parameter estimates for the resulting model are shown in Table 18.10. As can be seen from the table, the estimated thresholds are roughly the same as those in Table 18.8, and the estimated means and random intercept variance are also approximately the same. The only substantive improvement of the Free Curve model over the Linear model is in the predicted proportion of individuals who did not consume five or more drinks, with the Free Curve model's prediction of 54% of individuals not consuming five or more drinks being closer to the observed proportion of 56%, and the Linear model predicting only 51%. Although it would be attractive to formally compare the fit of these two models, it is not possible to do so using Mplus for the orthogonal Free Curve model because of the use of model constraints.

**TABLE 18.10**

Parameter Estimates for Linear Orthogonal Model with Categorical Indicators

| | | Intercept | Growth | Scaling Factors | Thresholds |
|---|---|---|---|---|---|
| | | Loadings | | | |
| Pre-College | | 1 | −0.48 | 1 | $t1 = 0.00$ |
| Freshman | Fall | 1 | −0.42 | 1.06 | $t2 = 0.31$ |
| | Spring | 1 | −0.35 | 1.15 | $t3 = 0.71$ |
| Sophomore | Fall | 1 | −0.25 | 1.19 | $t4 = 1.34$ |
| | Spring | 1 | −0.18 | 1.23 | $t5 = 1.96$ |
| Junior | Fall | 1 | −0.08 | 1.28 | |
| | Spring | 1 | −0.01 | 1.27 | |
| Senior | Fall | 1 | 0.09 | 1.28 | |
| | Spring | 1 | 0.16 | 1.28 | |
| | Means | 0.17 | 0.41 | | |
| | Variances | 0.48 | 1 | | |

It should be noted more generally that the assumption of equal scaling factors across items may be more reasonable in analysis of such things as Likert-format questionnaire items; if this is the case, MLR estimation, which requires that scaling factors be equal across manifest items, may be a preferable approach to the data.

## Conclusion

As discussed at the beginning, the problem of categorical dependent variables involves the assumption of some type of latent response variable to enable the machinery of estimation as an SEM. Advances in statistical theory and software have made estimation of these models much more straightforward than in the past. Models that assume ordered categorical assessment involve a more realistic assumption about the measurement of many manifest variables beyond the continuous measurement assumptions of the previous chapters. Assumption checking of such models is an important aspect of conscientious analysis. Relatively larger sample sizes may be necessary for their successful implementation if the number of ordered categories becomes large, and ad hoc adjustments to the categorization of variables may be necessary based on both the base rates of the categories and the departures from nonlinearity that may be present.

## Note

1 There are many other ways to estimate an asymptotic covariance matrix, however. Finney and DeStefano (2013) has a review of the approaches and more will no doubt be developed in the future. EQS, for example, uses an approach called "partitioned maximum likelihood" (PML), in which different submodels are estimated depending on whether only categorical, only continuous, or categorical and continuous variables are being considered. After computation, the AGLS estimator is used to produce the asymptotic covariance matrix.

# 19

# *Postscript*

Structural equation modeling (SEM) is both a way to understand the limitations and advantages of basic linear statistical models and the wider variety of structural equation models (SEMs). The advantages of SEMs more adequately address some of the fundamental problems with earlier models, namely the problem of measurement error in exogenous manifest variables, the need for parsimonious explanations for covariation between variables, and the variety of psychometric models available for such parsimony. Exploration of the external validity of these SEMs across groups in terms of the components of an SEM is an additional advantage. Although SEMs are a powerful framework for testing complex theoretical relationships among latent and observed variables, its effective use hinges critically on aligning the model's complexity with the sample size and selection of measurement models that appropriately reflect the data's structure and limitations. Overfitting due to an excessive number of free parameters relative to sample size can produce failure to produce estimates, unstable estimates, spurious significance, and poor generalizability. Conversely, overly restrictive models may mask important relationships or generate biased parameter estimates.

## Right-Sizing

Right-sizing an SEM involves a balance of theoretical fidelity and being given the available data. Across the chapters of this book, when a variety of latent variable models appear plausible, the three-step approach of first examining the dimensionality underlying the data, evaluating a parsimonious measurement model for the data, and, where appropriate, examining how the SEM can be used to summarize mean level effects appears to assist in navigating the large number of conceptually plausible candidate models.


DOI: 10.1201/9781003595267-19

## Assumption Checking

When examining a favored model or candidate alternatives, assumption checking is essential, both to the interpretation of results and potential replicability of findings. This checking occurs at both the level of the model as a whole as well as more focused attention to individual observations. LOESS plots, modification indices, scatterplots, alternative measurement models, and, where appropriate, alternate rotations are all part of a model-level evaluation of linearity, conditional normality, homoscedasticity, the existence of ceiling/floor effects, and alternative latent variable structures. Individual observation-level checking includes examination of scatterplots, $\Delta\chi^2$, the generalized Cook's distance, and changes in parameter estimates and likelihood distance. Such diagnostics are particularly important in smaller samples where the impact of outliers is magnified.

## Summary

Together, right-sizing and assumption checking enhance the validity and replicability of SEM findings. They shift the focus from mechanical fit indices to thoughtful model construction and data-model alignment. Incorporating these practices strengthens inferences drawn from SEM and improves the likelihood that conclusions will hold in future applications or replications. SEM, when right-sized and critically examined, becomes not only a confirmatory tool but also a robust instrument for scientific insight.

It is important to recall the observation made in the Introduction that scientific models are oversimplifications of the real world. There is no true model, and the researcher is not absolved of the problem of competing models by pre-specifying their SEM before data collection. The favored model, is, in a sense, the one "left standing" after competing alternatives have been eliminated. As a result, contrary to many discussions, the interpretation of statistical models is not a simple decision between acceptance and rejection of a model. A third interpretation of a statistical model is that the data at hand simply do not permit examination of an alternative theory due to limitations of the observations, number of variables, and available psychometric models for the data.

As advances in computing power, the ability of software to easily specify multiple models and, yes, artificial intelligence unfold, it is hoped a more rigorous analysis of SEMs will provide sounder evaluation of the strengths, limitations, and generalizability of our models as explanations of the data.

# *References*


Akaike, H. (1973). Information theory and an extension of the maximum likelihood principle. In B. N. Petrov & F. Csáki (Eds.), *2nd International Symposium on Information Theory*, Tsahkadsor, Armenia, USSR, September 2–8, 1971 (pp. 267–281), Budapest: Akadémiai Kiadó.

Alwin, D. F., & Jackson, D. J. (1981). Applications of simultaneous factor analysis to issues of factorial invariance. In D. J. Jackson & E. F. Borgatta (Eds.), *Factor analysis and measurement in sociological research* (pp. 249–279). Beverly Hills, CA: Sage.

American Psychological Association. (2001). *Publication manual of the American Psychological Association* (5th ed.). Washington, DC: Author.

Anscombe, F. J. (1973). Graphs in statistical analysis. *The American Statistician, 27*(1), 17–21. https://doi.org/10.1080/00031305.1973.10478966

Asparouhov, T., & Muthén, B. (2009). Exploratory structural equation modeling. *Structural Equation Modeling: A Multidisciplinary Journal, 16*(3), 397–438. https://doi.org/10.1080/10705510903008204

Asparouhov, T., & Muthén, B. (2014). Multiple-group factor analysis alignment. *Structural Equation Modeling, 21*(4), 495–508. https://doi.org/10.1080/10705511.2014.919210

Bacon, R. (ca. 1267). *Opus Majus*.

Bagozzi, R. P., & Edwards, J. R. (1998). A general approach for representing constructs in organizational research. *Organizational Research Methods, 1*(1), 45–87. https://doi.org/10.1177/109442819800100104

Banks, R. B. (2013). *Growth and diffusion phenomena: Mathematical frameworks and applications* (Vol. 14). Berlin, Germany: Springer Science & Business Media. https://doi.org/10.1007/978-3-662-03052-3

Baron, R. M., & Kenny, D. A. (1986). The moderator-mediator variable distinction in social psychological research: Conceptual, strategic and statistical considerations. *Journal of Personality and Social Psychology, 51*(6), 1173–1182. https://doi.org/10.1037/0022-3514.51.6.1173

Bastiaansen, J. A., Kunkels, Y. K., Blaauw, F. J., Boker, S. M., Ceulemans, E., Chen, M., Chow, S-M., de Jonge, P., Emerencia, A. C., Epskamp, S., Fisher, A. J., Hamaker, E. L., Kuppens, P., Lutz, W., Meyer, M. J., Moulder, R., Oravecz, Z., Riese, H., Rubel, J., … Bringmann, L. F. (2020). Time to get personal? The impact of researchers choices on the selection of treatment targets using the experience sampling methodology. *Journal of Psychosomatic Research, 137*, 110211. https://doi.org/10.1016/j.jpsychores.2020.110211

Bateson, G. (1979). *Mind and nature: A necessary unity*. London, UK: Wildwood House.

Bentler, P. M. (1990). Comparative fit indexes in structural models. *Psychological Bulletin, 107*(2), 238–246. https://doi.org/10.1037/0033-2909.107.2.238

Bishop, J., Geiser, C., & Cole, D. A. (2015). Modeling latent growth with multiple indicators: A comparison of three approaches. *Psychological Methods, 20*(1), 43–62. https://doi.org/10.1037/met0000018

Bollen, K. A. (1989). *Structural equations with latent variables*. New York, NY: John Wiley & Sons. https://doi.org/10.1002/9781118619179

Bollen, K. A., & Long, J. S. (Eds.). (1993). *Testing structural equation models*. Newbury Park, CA: Sage.

Bonifay, W., Lane, S. P., & Reise, S. P. (2016). Three concerns with applying a bifactor model as a structure of psychopathology. *Clinical Psychological Science, 5*(1), 184–186. https://doi.org/10.1177/2167702616657069

Borsboom, D. (2006). The attack of the psychometricians. *Psychometrika, 71*(3), 425–440. https://doi.org/10.1007/s11336-006-1447-6

Breznau, N., Rinke, E. M., Wuttke, A., Nguyen, H. H. V., Adem, M., Adriaans, J., Alvarez-Benjumea, A., Andersen, H. K., Auer, D., Azevedo, F., Bahnsen, O., Balzer, D., Bauer, G., Bauer, P. C., Baumann, M., Baute, S., Benoit, V., Bernauer, J., Berning, C., Berthold, A., … Zóltak, T. (2022). Observing many researchers using the same data and hypothesis reveals a hidden universe of uncertainty. *Proceedings of the National Academy of Sciences of the United States of America, 119*(44), e2203150119. https://doi.org/10.1073/pnas.2203150119

Browne, M. W. (1982). Covariance structures. In D. M. Hawkins (Ed.), *Topics in applied multivariate analysis* (pp. 72–141). Cambridge, UK: Cambridge University Press

Browne, M. W. (1984). Asymptotically distribution-free methods for the analysis of covariance structures. *British Journal of Mathematical and Statistical Psychology, 37*(1), 62–83. https://doi.org/10.1111/j.2044-8317.1984.tb00789.x

Browne, M. W. (2001). An overview of analytic rotation in exploratory factor analysis. *Multivariate Behavioral Research, 36*(10), 111–150. https://doi.org/10.1207/S15327906MBR3601_05

Browne, M. W., & Mels, G. (1990). *RAMONA user's guide*. Unpublished report, Department of Psychology, Ohio State University.

Burnham, K. P., & Anderson, D. R. (2004). Multimodel inference: Understanding AIC and BIC in model selection. *Sociological Methods & Research, 33*(2), 261–304. https://doi.org/10.1177/0049124104268644

Burt, C. (1948). The factorial study of temperamental traits. *British Journal of Statistical Psychology, 1*(3), 178–203. https://doi.org/10.1111/j.2044-8317.1948.tb00236.x

Byrne, B. M. (2010). *Structural equation modeling with AMOS: Basic concepts, applications, and programming* (2nd ed.). New York, NY: Routledge/Taylor & Francis Group.

Byrne, B. M. (2012). Multivariate applications series. *Structural equation modeling with Mplus: Basic concepts, applications, and programming*. New York, NY: Routledge/Taylor & Francis Group.

Byrne, B. M., Shavelson, R. J., & Muthén, B. O. (1989). Testing for the equivalence of factor covariance and mean structures: The issue of partial measurement invariance. *Psychological Bulletin, 105*, 456–466. https://doi.org/10.1037/0033-2909.105.3.456

Campbell, D. T. (1974). Evolutionary epistemology. In: P. A. Schilpp (Ed.), *The philosophy of Karl R. Popper* (pp. 412–463). LaSalle: Open Court.

Campbell, D. T., & Fiske, D. W. (1959). Convergent and discriminant validation by the multitrait-multimethod matrix. *Psychological Bulletin, 56*(2), 81–105. https://doi.org/10.1037/h0046016

Carroll, J. B. (1953). An analytic solution for approximating simple structure in factor analysis. *Psychometrika, 18*(1), 23–38. https://doi.org/10.1007/BF02289025

Cattell, R. B. (1988). Psychological theory and scientific method. In J. R. Nesselroade & R. B. Cattell (Eds.), *Handbook of multivariate experimental psychology* (2nd ed., pp. 3–20). New York, NY: Plenum Press. https://doi.org/10.1007/978-1-4613-0893-5_1

Cattell, R. B., & Vogelmann, S. (1977). A comprehensive trial of the scree and KG criteria for determining the number of factors. *Multivariate Behavioral Research, 12*(3), 289–325. https://doi.org/10.1207/s15327906mbr1203_2

Chen, F., Curran, P. J., Bollen, K., Kirby, J., & Paxton, P. (2008). An empirical evaluation of the use of fixed cutoff points in the RMSEA test statistic in structural equation models. *Sociological Methods and Research, 36*, 462–494. https://doi.org/10.1177/0049124108314720

Cheung, G. W., & Rensvold, R. B. (2002). Evaluating goodness-of-fit indexes for testing measurement invariance. *Structural Equation Modeling, 9*(2), 233–255. https://doi.org/10.1207/S15328007SEM0902_5

Cleveland, W. S. (1979). Robust locally weighted regression and smoothing scatterplots. *Journal of the American Statistical Association, 74*(368), 829–836. https://doi.org/10.1080/01621459.1979.10481038

Cleveland, W. S., & Devlin, S. J. (1988). Locally weighted regression: An approach to regression analysis by local fitting. *Journal of the American Statistical Association, 83*(403), 596–610.

Cole, D. A., & Maxwell, S. E. (2003). Testing mediational models with longitudinal data: Questions and tips in the use of structural equation modeling. *Journal of Abnormal Psychology, 112*(4), 558–577. https://doi.org/10.1037/0021-843X.112.4.558

Colibazzi, T., Zhu, H., Bansal, R., Schultz, R. T., Wang, Z., & Peterson, B. S. (2008). Latent volumetric structure of the human brain: Exploratory factor analysis and structural equation modeling of gray matter volumes in healthy children and adults. *Human Brain Mapping, 29*(11), 1302–1312. https://doi.org/10.1002/hbm.20466

Cooper, M. L. (1994). Motivations for alcohol use among adolescents: Development and validation of a four-factor model. *Psychological Assessment, 6*(2), 117–128. https://doi.org/10.1037/1040-3590.6.2.117

Corley, J., Conte, F., Harris, S. E., Taylor, A. M., Redmond, P., Russ, T. C., Deary, I. J., & Cox, S. R. (2023). Predictors of longitudinal cognitive ageing from age 70 to 82 including APOE e4 status, early-life and lifestyle factors: The Lothian Birth Cohort 1936. *Molecular Psychiatry, 28*(3), 1256–1271. https://doi.org/10.1038/s41380-022-01900-4

Crawford, C. B. (1975). A rotation criterion with weighting for row and column complexity. *Psychometrika, 40*(2), 253–259. https://doi.org/10.1007/BF02291557

Crawford, C. B., & Ferguson, G. A. (1970). A general rotation criterion and its use in orthogonal rotation. *Psychometrika, 35*(3), 321–332. https://doi.org/10.1007/BF02310792

Cronbach, L. J., & Furby, L. (1970). How we should measure "change": Or should we? *Psychological Bulletin, 74*(1), 68–80. https://doi.org/10.1037/h0029382

Curran, P. J., West, S. G., & Finch, J. F. (1996). The robustness of test statistics to nonnormality and specification error in confirmatory factor analysis. *Psychological Methods, 1*(1), 16–29. https://doi.org/10.1037/1082-989X.1.1.16

Davis, S. C., & Boundy, R. G. (2018). *Transportation energy data book: Edition 37*. Oak Ridge, TN: Oak Ridge National Laboratory. Retrieved from http://cta.ornl.gov/data

DeMars, C. E. (2012). A comparison of limited-information and full-information methods in Mplus for estimating item response theory parameters for nonnormal populations. *Structural Equation Modeling, 19*(4), 610–632. https://doi.org/10.1080/10705511.2012.713272

Duncan, O. D., Haller, A. O., & Portes, A. (1968). Peer influences on aspirations: A reinterpretation. *American Journal of Sociology, 74*(2), 119–137.

Eid, M., & Diener, E. (1999). Intraindividual variability in affect: Reliability, validity, and personality correlates. *Journal of Personality and Social Psychology, 76*(4), 662–676. https://doi.org/10.1037/0022-3514.76.4.662

Eid, M., Nussbeck, F. W., Geiser, C., Cole, D. A., Gollwitzer, M., & Lischetzke, T. (2008). Structural equation modeling of multitrait-multimethod data: Different models for different types of methods. *Psychological Methods, 13*(3), 230–253. https://doi.org/10.1037/a0013219

Einstein, A. (1931). The world as I see it. Philosophical Library. https://history.aip.org/exhibits/einstein/essay.htm

Everitt, B. S. (1987). *Introduction to optimization methods and their application in statistics.* Dordrecht: Springer.

Fabrigar, L. R., Wegener, D. T., MacCallum, R. C., & Strahan, E. J. (1999). Evaluating the use of exploratory factor analysis in psychological research. *Psychological Methods, 4*(3), 272–299. https://doi.org/10.1037/1082-989X.4.3.272

Fan, X., & Sivo, S. A. (2005). Sensitivity of fit indexes to misspecified structural or measurement model components: Rationale of two-index strategy revisited. *Structural Equation Modeling, 12*(3), 343–367. https://doi.org/10.1207/s15328007sem1203_1

Ferrer, E., Hamagami, F., & McArdle, J. J. (2004). Modeling latent growth curves with incomplete data using different types of structural equation modeling and multilevel software. *Structural Equation Modeling, 11*(3), 452–483. https://doi.org/10.1207/s15328007sem1103_8

Finney, S. J., & DeStefano, C. (2013). Nonnormal and categorical data in structural equation modeling. In G. R. Hancock & R. O. Mueller (Eds.), *Structural equation modeling: A second course* (2nd ed., pp. 439–492). Charlotte, NC: Information Age Publishing.

Fisher, R. A. (1966). *The design of experiments* (8th ed.). New York, NY: Hafner Publishing.

Flora, D. B. (2008). Specifying piecewise latent trajectory models for longitudinal data. *Structural Equation Modeling, 15*(3), 513–533. https://doi.org/10.1080/10705510802154349

Forero, C. G., Maydeu-Olivares, A., & Gallardo-Pujol, D. (2009). Factor analysis with ordinal indicators: A Monte Carlo study comparing DWLS and ULS estimation. *Structural Equation Modeling, 16*(4), 625–641. https://doi.org/10.1080/10705510903203573

Friendly, M. (2002). Corrgrams: Exploratory displays for correlation matrices, *The American Statistician, 56*(4), 316–324. https://doi.org/10.1198/000313002533

Galilei, G. (1623/1957). *The assayer* (S. Drake, Trans.). Berkeley, CA: University of California Press.

Galton, F. (1886). Regression towards mediocrity in hereditary stature. *The Journal of the Anthropological Institute of Great Britain and Ireland. 15*, 246–263. https://doi.org/10.2307/2841583. JSTOR 2841583.

Geiser, C. (2023). Structural equation modeling with the Mplus and lavaan programs. In R. H. Hoyle (Ed.), *Handbook of structural equation modeling* (pp. 241–258). New York, NY: Guilford Press.

Ghisletta, P., & McArdle, J. J. (2012). Latent curve models and latent change score models estimated in R. *Structural Equation Modeling, 19*(4), 651–682. https://doi.org/10.1080/10705511.2012.713275

Gonzalez, R., & Griffin, D. (2001). Testing parameters in structural equation modeling: Every "one" matters. *Psychological Methods, 6*(3), 258–269. https://doi.org/10.1037/1082-989X.6.3.258

Gorsuch, R. L. (1983). *Factor analysis* (2nd ed.). Hillsdale, NJ: Erlbaum.

Gorsuch, R., & Nelson, J. (1981, November). *CNG scree test: An objective procedure for determining the number of factors*. Annual Meeting of the Society for Multivariate Experimental Psychology, Vancouver, Canada.

Grimm, K. J., Ram, N., & Estabrook, R. (2017). *Growth modeling: Structural equation and multilevel modeling approaches*. New York, NY: Guilford Press.

Grimm, K. J., Zhang, Z., Hamagami, F., & Mazzocco, M. M. M. (2013). Modeling nonlinear change via latent change and latent acceleration frameworks: Examining velocity and acceleration of growth trajectories. *Multivariate Behavioral Research, 48*(1), 117–143. https://doi.org/10.1080/00273171.2012.755111

Guilford, J. P. (1956). The structure of intellect. *Psychological Bulletin, 53*(4), 267–293. https://doi.org/10.1037/h0040755

Hall, H. S., & Knight, S. R. (1899). *Higher algebra*. London, UK: Macmillan.

Hamagami, F., & McArdle, J. J. (2007). Dynamic extensions of latent difference score models. In S. M. Boker & M. J. Wenger (Eds.), *Data analytic techniques for dynamical systems* (pp. 47–85). Mahwah, NJ: Lawrence Erlbaum Associates Publishers.

Hancock, G. R., & Choi, J. (2006). A vernacular for linear latent growth models. *Structural Equation Modeling, 13*(3), 352–377. https://doi.org/10.1207/s15328007sem1303_2

Harmon, H. H. (1976). *Modern factor analysis* (3rd ed., revised). Chicago, IL: University of Chicago Press.

Hattori, M., Zhang, G., & Preacher, K. J. (2017). Multiple local solutions and geomin rotation. *Multivariate Behavioral Research, 52*(6), 720–731. https://doi.org/10.1080/00273171.2017.1361312

Henderson, A. E., & White, P. D. (1964). Promax: A quick method for rotation to obtain oblique simple structure. *British Journal of Mathematical and Statistical Psychology, 17*(1), 65–70. https://doi.org/10.1111/j.2044-8317.1964.tb00244.x

Horn, J. L. (1965). A rationale and test for the number of factors in factor analysis. *Psychometrika 30*, 179–185. https://doi.org/10.1007/BF02289447

Horn, J. L., & McArdle, J. J. (1992). A practical and theoretical guide to measurement invariance in aging research. *Experimental Aging Research, 18*(3), 117–144. https://doi.org/10.1080/03610739208253916

Hu, L. T., & Bentler, P. M. (1999). Cutoff criteria for fit indexes in covariance structure analysis: Conventional criteria versus new alternatives. *Structural Equation Modeling, 6*(1), 1–55. https://doi.org/10.1080/10705519909540118

Hu, L. T., Bentler, P. M., & Kano, Y. (1992). Can test statistics in covariance structure analysis be trusted?. *Psychological Bulletin, 112*(2), 351–362. https://doi.org/10.1037/0033-2909.112.2.351

Hurley, C. (2004). Clustering visualizations of multidimensional data. *Journal of Computational and Graphical Statistics, 13*(4), 788–806. https://doi.org/10.1198/106186004X12425

Jennrich, R. I. (2001). A simple general procedure for orthogonal rotation. *Psychometrika, 66*(2), 289–306. https://doi.org/10.1007/BF02294840

Jennrich, R. I. (2002). A simple general method for oblique rotation. *Psychometrika, 67*(1), 7–19. https://doi.org/10.1007/BF02294706

Johnston, L. D., O'Malley, P. M., & Bachman, J. G. (1995). *National survey results on drug use from the Monitoring the Future study, 1975-1994. Volume I: Secondary school students.* (NIH Publication No. 95-4026). Rockville, MD: National Institute on Drug Abuse.

Jolliffe, I. T. (1986). The use of principal components in regression. *Journal of the Royal Statistical Society: Series C (Applied Statistics), 25*(3), 318–329.

Jolliffe, I. T. (2002). *Principal component analysis* (2nd ed.). Springer. https://doi.org/10.1007/b98835

Jöreskog, K. G. (1971). Simultaneous factor analysis in several populations. *Psychometrika, 36*(4), 409–426.

Kaiser, H. F. (1958). The varimax criterion for analytic rotation in factor analysis. *Psychometrika* 23(3), 187–200.

Kaiser, H. F., & Rice, J. (1974). Little Jiffy, Mark IV. *Educational and Psychological Measurement, 34*(1), 111–117. https://doi.org/10.1177/001316447403400115

Kaplan, D. (2008). *Structural equation modeling: Foundations and extensions* (Vol. 10). Los Angeles, CA: SAGE Publications.

Kass, R., & Raftery, A. (1995). Bayes factors. *Journal of the American Statistical Association, 90*(430), 773–795. https://doi.org/10.1080/01621459.1995.10476572

Kass, R. E., & Wasserman, L. (1995). A reference Bayesian test for nested hypotheses and its relationship to the Schwarz criterion. *Journal of the American Statistical Association, 90*(431), 928–934. https://doi.org/10.2307/2291327

Kenny, D. A., Kaniskan, B., & McCoach, D. B. (2014). The performance of RMSEA in models with small degrees of freedom. *Sociological Methods & Research*, 44(3), 486–507. https://doi.org/10.1177/0049124114543236

Kerr, N. L. (1998). HARKing: Hypothesizing after the results are known. *Personality and Social Psychology Review*, 2(3), 196–217. https://doi.org/10.1207/s15327957pspr0203_4

Kievit, R. A., Brandmaier, A. M., Ziegler, G., van Harmelen, A.-L., de Mooij, S. M. M., Moutoussis, M., Goodyer, I. M., Bullmore, E., Jones, P. B., Fonagy, P., Lindenberger, U., & Dolan, R. J. (2018). Developmental cognitive neuroscience using latent change score models: A tutorial and applications. *Developmental Cognitive Neuroscience, 33*, 99–117. https://doi.org/10.1016/j.dcn.2017.11.007

Klein, A., & Moosbrugger, H. (2000). Maximum likelihood estimation of latent interaction effects with the LMS method. *Psychometrika, 65*(4), 457–474. https://doi.org/10.1007/BF02296338

Kuder, G. F., & Richardson, M. W. (1937). The theory of the estimation of test reliability. *Psychometrika*, 2(3), 151–160. https://doi.org/10.1007/BF02288391

Kutner, M. H., Nachtsheim, C. J., Neter, J., & Li, W. (2005). *Applied linear statistical models* (5th ed.). Boston, MA: McGraw-Hill Irwin.

Langer, D. A., Wood, J. J., Bergman, R. L., & Piacentini, J. C. (2010). A multitrait–multimethod analysis of the construct validity of child anxiety disorders in a clinical sample. *Child Psychiatry and Human Development*, *41*(5), 549–561. https://doi.org/10.1007/s10578-010-0187-0

Lawley, D. N. (1943). A note on Karl Pearson's selection formulae. *Proceedings of the Royal Society of Edinburgh*, *2*, 28–30.

Lim, S., & Jahng, S. (2019). Determining the number of factors using parallel analysis and its recent variants. *Psychological Methods*, *24*(4), 452–467. https://doi.org/10.1037/met0000230

Lin, L. C., Huang, P. H., & Weng, L. J. (2017). Selecting path models in SEM: A comparison of model selection criteria. *Structural Equation Modeling*, *24*(6), 855–869. https://doi.org/10.1080/10705511.2017.1363652

Little, T. D., Slegers, D. W., & Card, N. A. (2006). A non-arbitrary method of ying and scaling latent variables in SEM and MACS models. *Structural Equation Modeling*, *13*(1), 59–72. https://doi.org/10.1207/s15328007sem1301_3

Littlefield, A. K., Sher, K. J., & Wood, P. K. (2010). Do changes in drinking motives mediate the relation between personality change and "maturing out" of problem drinking? *Journal of Abnormal Psychology*, *119*(1), 93–105. https://doi.org/10.1037/a0017512

Liu, J., & Perera, R. A. (2022). Estimating knots and their association in parallel bilinear spline growth curve models in the framework of individual measurement occasions. *Psychological Methods*, *27*(5), 703–729. https://doi.org/10.1037/met0000309

Loader, C. (1999). *Local regression and likelihood*. New York, NY: Springer. https://doi.org/10.1007/b98858

Loehlin, J. C. (2004). *Latent variable models: An introduction to factor, path, and structural equation analysis* (4th ed.). Mahwah, NJ: Lawrence Erlbaum Associates.

Loehlin, J. C., & Beaujean, A. A. (2017). *Latent variable models: An introduction to factor, path, and structural equation analysis* (5th ed.). New York, NY: Routledge.

Lord, F. M., Novick, M. R., & Birnbaum, A. (1968). *Statistical theories of mental test scores*. Oxford, England: Addison-Wesley.

MacCallum, R. C., Browne, M. W., & Sugawara, H. M. (1996). Power analysis and determination of sample size for covariance structure modeling. *Psychological Methods*, *1*(2), 130–149. https://doi.org/10.1037/1082-989X.1.2.130

MacCallum, R. C., Zhang, S., Preacher, K. J., & Rucker, D. D. (2002). On the practice of dichotomization of quantitative variables. *Psychological Methods*, *7*(1), 19–40. https://doi.org/10.1037/1082-989X.7.1.19

Marsh, H. W., Balla, J. R., & McDonald, R. P. (1988). Goodness-of-fit indexes in confirmatory factor analysis: The effect of sample size. *Psychological Bulletin*, *103*(3), 391–410. https://doi.org/10.1037/0033-2909.103.3.391

Mason, S. J. (1956). Feedback theory – Further properties of signal flow graphs. *Proceedings of the IRE*, *44*(7), 920–926. https://doi.org/10.1109/jrproc.1956.275147

Maxwell, S. E., Delaney, H. D., & Kelley, K. (2017). *Designing experiments and analyzing data: A model comparison perspective*. New York, NY: Routledge.

Maydeu-Olivares, A., & Coffman, D. L. (2006). Random intercept item factor analysis. *Psychological Methods*, *11*(4), 344–362. https://doi.org/10.1037/1082-989X.11.4.344

McArdle, J. J. (2009). Latent variable modeling of differences and changes with longitudinal data. *Annual Review of Psychology, 60*(1), 577–605. https://doi.org/10.1146/annurev.psych.60.110707.163612

McArdle, J. J., & Epstein, D. (1987). Latent growth curves within developmental structural equation models. *Child Development, 58*(1), 110–133. https://doi.org/10.2307/1130295

McArdle, J. J., & McDonald, R. P. (1984). Some algebraic properties of the Reticular Action Model for moment structures. *The British Journal of Mathematical and Statistical Psychology, 37*(Pt 2), 234–251. https://doi.org/10.1111/j.2044-8317.1984.tb00802.x

McDonald, R. P. (1967). *Nonlinear factor analysis* (Psychometrika Monograph No. 15). Richmond, VA: Psychometric Corporation.

McDonald, R. P., & Ho, M. H. R. (2002). Principles and practice in reporting structural equation analyses. *Psychological Methods, 7*(1), 64–82. https://doi.org/10.1037/1082-989X.7.1.64

McDonald, R. P., & Marsh, H. W. (1990). Choosing a multivariate model: Noncentrality and goodness of fit. *Psychological Bulletin, 107*(2), 247–255. https://doi.org/10.1037/0033-2909.107.2.247

McGraw, K. O., & Wong, S. P. (1996a). Forming inferences about some intraclass correlation coefficients. *Psychological Methods, 1*(1), 30–46. https://doi.org/10.1037/1082-989X.1.1.30

McGraw, K. O., & Wong, S. P. (1996b). "Forming inferences about some intraclass correlations coefficients": Correction. *Psychological Methods, 1*(4), 390. https://doi.org/10.1037/1082-989X.1.1.30

Meredith, W. (1964). Notes on factorial invariance. *Psychometrika, 29*(2), 177–185. https://doi.org/10.1007/BF02289699

Meredith, W. (1993). Measurement invariance, factor analysis and factorial invariance. *Psychometrika, 58*(4), 525–543. https://doi.org/10.1007/BF02294825

Meredith, W., & Tisak, J. (1990). Latent curve analysis. *Psychometrika, 55*(1), 107–122. https://doi.org/10.1007/BF02294746

Merkle, E. C., & Rosseel, Y. (2018). blavaan: Bayesian structural equation models via parameter expansion [Preprint]. arXiv. https://arxiv.org/abs/1809.05263

Millsap, R. E. (2011). *Statistical approaches to measurement invariance.* London: Routledge/Taylor & Francis Group.

Morey, R. Psychonomic Society. (January 16, 2019). [April 2019]. *You must tug that thread: Why treating preregistration as a gold standard might incentivize poor behavior.* https://featuredcontent.psychonomic.org/you-must-tug-that-thread-why-treating-preregistration-as-a-gold-standard-might-incentivize-poor-behavior

Mulaik, S. A. (2009a). *Linear causal modeling with structural equations.* Boca Raton, FL: CRC Press.

Mulaik, S. A. (2009b). *Foundations of factor analysis* (2nd ed.). Boca Raton, FL: CRC Press.

Murray, A. L., & Johnson, W. (2013). The limitations of model fit in comparing the bi-factor versus higher-order models of human cognitive ability structure. *Intelligence, 41*(5), 407–422. https://doi.org/10.1016/j.intell.2013.06.004

Muthén, B., & Asparouhov, T. (2002). Latent variable analysis with categorical outcomes: Multiple-group and growth modeling in Mplus. Mplus Web Notes: No. 4. www.statmodel.com

Muthén, B., & Muthén, L. K. (2000). Integrating person-centered and variable-centered analyses: growth mixture modeling with latent trajectory classes. *Alcoholism, Clinical and Experimental Research, 24*(6), 882–891. https://doi.org/10.1097/00000374-200006000-00020

Muthén, B. O. (1993). Goodness of fit with categorical and other non-normal variables. In K. A. Bollen& J. S. Long (Eds.), *Testing structural equation models* (pp. 205–243). Newbury Park, CA: Sage.

Muthén, B. O., & Kaplan, D. (1985). A comparison of some methodologies for the factor analysis of non-normal Likert variables. *British Journal of Mathematical and Statistical Psychology, 38*(2), 171–189. https://doi.org/10.1111/j.2044-8317.1985.tb00832.x

Muthén, B. O., Muthén, L. K., & Asparouhov, T. (2015, March 24). *Estimator choices with categorical outcomes* (Technical note). https://www.statmodel.com

Muthén, L. K., & Muthén, B. O. (1998–2017). *Mplus user's guide* (8th ed.). Muthén & Muthén. https://www.statmodel.com

Muthén, L. K., & Muthén, B. O. (2002). How to use a Monte Carlo study to decide on sample size and determine power. *Structural Equation Modeling, 9*(4), 599–620. https://doi.org/10.1207/S15328007SEM0904_8

Nash, L. (1963). *The nature of the natural sciences.* Boston, MA: Little, Brown.

Neale, M. C., & Cardon, L. R. (1992). *NATO ASI series D: Behavioural and social sciences, Vol. 67. Methodology for genetic studies of twins and families.* New York, NY: Kluwer/Plenum. https://doi.org/10.1007/978-94-015-8018-2

Neale, M. C., & Miller, M. B. (1997). The use of likelihood-based confidence intervals in genetic models. *Behavior Genetics, 27*(2), 113–120. https://doi.org/10.1023/A:1025681223921

Nesselroade, J. R., & Cattell, R. B. (Eds.). (1988). *Handbook of multivariate experimental psychology* (2nd ed.). New York, NY: Plenum. https://doi.org/10.1007/978-1-4613-0893-5

Newsom, J. T., Hachem, Z. A., Granger, A. M., & Denning, E. C. (2023). Where did I go wrong with my model? Ten tips for getting results in SEM. *Structural Equation Modeling, 30*(3), 491–500. https://doi.org/10.1080/10705511.2022.2101113

Nubé, M., & Van Den Boom, G. J. (2003). Gender and adult undernutrition in developing countries. *Annals of Human Biology, 30*(5), 520–537. https://doi.org/10.1080/0301446031000119601

Olsson, U. (1979). Maximum likelihood estimation of the polychoric correlation coefficient. *Psychometrika, 44*(4), 443–460. https://doi.org/10.1007/BF02296207

Osborne, R. T., & Lindsey, J. M. (1967). A longitudinal investigation of change in the factorial composition of intelligence with age in young school children. *Journal of Genetic Psychology, 110*(1), 49–58.

Osborne, R. T., & Suddick, D. E. (1972). A longitudinal investigation of the intellectual differentiation hypothesis. *Journal of Genetic Psychology, 121*(1), 83–89. https://doi.org/10.1080/00221325.1972.10533131

Panik, M. J. (2014). *Growth curve modeling: Theory and applications.* Hoboken, NJ: Wiley.

Pedhazur, E. J. (1997). *Multiple regression in behavioral research* (3rd ed.). Fort Worth, TX: Harcourt Brace.

Pek, J., & MacCallum, R. C. (2011). Sensitivity analysis in structural equation models: Cases and their influence. *Multivariate Behavioral Research, 46*, 202–228.

Popper, K. R. (1992). *The open universe: An argument for indeterminism* (W. W. Bartley III, Ed.). New York, NY: Routledge.

Popper, K. R. (2002). *The logic of scientific discovery*. New York, NY: Routledge.

Raykov, T., Tomer, A., & Nesselroade, J. R. (1991). Reporting structural equation modeling results in *Psychology and Aging*: Some proposed guidelines. *Psychology and Aging*, *6*(4), 499–503. https://doi.org/10.1037//0882-7974.6.4.499

Reddy, M. K., Anderson, B. J., Liebschutz, J., & Stein, M. D. (2013). Factor structure of PTSD symptoms in opioid-dependent patients rating their overall trauma history. *Drug and Alcohol Dependence*, *132*(3), 597–602.

Rhemtulla, M., Brosseau-Liard, P. É., & Savalei, V. (2012). When can categorical variables be treated as continuous? A comparison of robust continuous and categorical SEM estimation methods under suboptimal conditions. *Psychological Methods*, *17*(3), 354–373. https://doi.org/10.1037/a0029315

Riegel, K. F. (1977). The dialectics of time. In N. Datan & H. W. Reese (Eds.), *Life-span developmental psychology: Dialectical perspectives on experimental research* (pp. 4–46). New York, NY: Academic Press.

Rigdon, E. E. (1998). Structural equation modeling. In G. A. Marcoulides (Ed.), *Methodology for business and management. Modern methods for business research* (pp. 251–294). Mahwah, NJ: Erlbaum.

Rindskopf, D. (1984). Using phantom and imaginary latent variables to parameterize constraints in linear structural models. *Psychometrika*, *49*(1), 37–47. https://doi.org/10.1007/BF02294204

Rodgers, J. L. (2010). The epistemology of mathematical and statistical modeling: A quiet methodological revolution. *American Psychologist*, *65*(1), 1–12. https://doi.org/10.1037/a0018326

Rosenberg, M., Houghton, S., Hunter, S. C., Zadow, C., Shilton, T., Wood, L., & Lawrence, D. (2018). A latent growth curve model to estimate electronic screen use patterns amongst adolescents aged 10 to 17 years. *BMC Public Health*, *18*(1), 332. https://doi.org/10.1186/s12889-018-5240-0

Rovine, M. J., & Molenaar, P. C. (1999). The covariance between level and shape in the latent growth curve model with estimated basis vector coefficients. *Methods of Psychological Research Online*, *3*, 95–107.

Rovine, M. J., & Molenaar, P. C. (2005). Relating factor models for longitudinal data to quasi-simplex and NARMA models. *Multivariate Behavioral Research*, *40*(1), 83–114. https://doi.org/10.1207/s15327906mbr4001_4

Ruscio, J., & Roche, B. (2012). Determining the number of factors to retain in an exploratory factor analysis using comparison data of known factorial structure. *Psychological Assessment*, *24*(2), 282–292. https://doi.org/10.1037/a0025697

Salthe, S. (1985). *Evolving hierarchical systems: Their structure and representation*. New York, NY: Columbia University Press.

Saris, W. E., Satorra, A., & Van der Veld, W. M. (2009). Testing structural equation models or detection of misspecifications? *Structural Equation Modeling: A Multidisciplinary Journal*, *16*, 561–582. https://doi.org/10.1080/10705510903203433

Satorra, A., & Bentler, P. M. (2001). A scaled difference chi-square test statistic for moment structure analysis. *Psychometrika*, *66*(4), 507–514. https://doi.org/10.1007/BF02296192

Satorra, A., & Saris, W. E. (1985). Power of the likelihood ratio test in covariance structure analysis. *Psychometrika*, *50*(1), 83–90.

Savalei, V., & Kolenikov, S. (2008). Constrained versus unconstrained estimation in structural equation modeling. *Psychological Methods, 13*(2), 150–170. https://doi.org/10.1037/1082-989X.13.2.150

Schmitt, T. A., & Sass, D. A. (2011). Rotation criteria and hypothesis testing for exploratory factor analysis: Implications for factor pattern loadings and interfactor correlations. *Educational and Psychological Measurement, 71*(1), 95–113. https://doi.org/10.1177/0013164410387348

Schwarz, G. (1978). Estimating the dimension of a model. *The Annals of Statistics, 6*, 461–464. https://doi.org/10.1214/aos/1176344136

Shadish, W. R., Cook, T. D., & Campbell, D. T. (2002). *Experimental and quasi-experimental designs for generalized causal inference*. Boston, MA: Houghton-Mifflin.

Sher, K. J., & Rutledge, P. C. (2007). Heavy drinking across the transition to college: Predicting first-semester heavy drinking from precollege variables, *Addictive Behaviors, 32*(4), 819–835. https://doi.org/10.1016/j.addbeh.2006.06.024

Shrout, P. E., & Fleiss, J. L. (1979). Intraclass correlations: Uses in assessing rater reliability. *Psychological Bulletin, 86*(2), 420–428. https://doi.org/10.1037/0033-2909.86.2.420

Slife, B. D. (1993). *Time and psychological explanation*. Albany, NY: State University of New York Press.

Snyder, H. R., Young, J. F., & Hankin, B. L. (2017). Strong homotypic continuity in common psychopathology-, internalizing-, and externalizing-specific factors over time in adolescents. *Clinical Psychological Science, 5*(1), 98–110. https://doi.org/10.1177/2167702616651076

Sörbom, D. (1974). A general method for studying differences in factor means and factor structure between groups. *British Journal of Mathematical and Statistical Psychology, 27*(2), 229–239. https://doi.org/10.1111/j.2044-8317.1974.tb00543.x

Steegen, S., Tuerlinckx, F., Gelman, A., & Vanpaemel, W. (2016). Increasing transparency through a multiverse analysis. *Perspectives on Psychological Science, 11*(5), 702–712. https://doi.org/10.1177/1745691616658637

Steiger, J. H. (1989). *EzPATH: A supplementary module for SYSTAT and SYGRAPH*. Evanston, IL: SYSTAT.

Steiger, J. H. (1990). Structural model evaluation and modification: An interval estimation approach. *Multivariate Behavioral Research, 25*(2), 173–180. https://doi.org/10.1207/s15327906mbr2502_4

Steiger, J. H., & Lind, J. C. (May 1980). Statistically-based tests for the number of common factors. Paper presented at the *Annual Meeting of the Psychometric Society*, Iowa City, IA.

Steyer, R., Schmitt, M., & Eid, M. (1999). Latent state–trait theory and research in personality and individual differences. *European Journal of Personality, 13*(5, Spec Issue), 389–408. https://doi.org/10.1002/(SICI)1099-0984(199909/10)13:5<389::AID-PER361>3.0.CO;2-A

Thurstone, L. L. (1947). *Multiple factor analysis: A development and expansion of vectors of the mind*. Chicago, IL: University of Chicago Press.

Toga, A. W., & Mazziotta, J. C. (2002). *Brain mapping: The methods*. San Diego, CA: Academic Press.

Tucker, L. R. (1951). *A method for synthesis of factor analysis studies*. Personnel Research Section Report No. 984, Department of the Army.

Tucker, L. R. (1958). Determination of parameters of a functional relation by factor analysis. *Psychometrika, 23*(4), 19–23. https://doi.org/10.1007/BF02288975

Tucker, L. R., Koopman, R. F., & Linn, R. L. (1969). Evaluation of factor analytic research procedures by means of simulated correlation matrices. *Psychometrika, 34*(4, pt. 1), 421–459. https://doi.org/10.1007/BF02290601

Van de Schoot, R., Lugtig, P., & Hox, J. (2012). A checklist for testing measurement invariance. *European Journal of Developmental Psychology*, *9*(4), 486–492. https://doi.org/10.1080/17405629.2012.686740

Vandenberg, R. J., & Lance, C. E. (2000). A review and synthesis of the measurement invariance literature: Suggestions, practices, and recommendations for organizational research. *Organizational Research Methods*, *3*(1), 4–69. https://doi.org/10.1177/109442810031002

Velicer, W. F. (1976). Determining the number of components from the matrix of partial correlations. *Psychometrika*, *41*(3), 321–327. https://doi.org/10.1007/BF02293557

von Eye, A., Rovine, M. J., & Wood, P. K. (1990). A comparison of tests for residual analysis. In K. Berk & L. Malone (Eds.), *Computer Science and Statistics: Proceedings of the 21st Symposium on the Interface* (pp. 355–358). Alexandria, VA: The American Statistical Association.

Vrieze, S. I. (2012). Model selection and psychological theory: A discussion of the differences between the Akaike information criterion (AIC) and the Bayesian information criterion (BIC). *Psychological Methods*, *17*(2), 228–243. https://doi.org/10.1037/a0027127

Wald, A. (1950). *Statistical decision functions*. Oxford, England: Wiley.

Weisberg, S. (1985). *Applied regression analysis*. New York, NY: Wiley.

Whewell, W. (1837). *History of the inductive sciences form the earliest to the present time* (3rd ed.). New York, NY: Appleton.

Whewell, W. (1840). *Philosophy of the inductive sciences founded upon their history*. London, UK: J. W. Parker.

Widaman, K. F., & Reise, S. P. (1997). Exploring the measurement invariance of psychological instruments: Applications in the substance use domain. In K. J. Bryant & M. Windle (Eds.), *The science of prevention: Methodological advances from alcohol and substance abuse research*. (pp. 281–324). Washington, DC: American Psychological Association.

Womack, S. R., Beam, C. R., Giangrande, E. J., Turkheimer, E., & Emery, R. E. (2023). Nonlinear catch-up growth in height, weight, and head circumference from birth to adolescence: A longitudinal twin study. *Behavior Genetics*, *53*(5–6), 385–403. https://doi.org/10.1007/s10519-023-10151-0

Wood, P. K. (2023). New frontiers in prevention research models: Commentary on the special issue. *Prevention Science*, *24*(3), 517–524.

Wood, P. K., Sher, K. J., Erickson, D. J., & Debord, K. A. (1997). Predicting academic problems in college from freshman alcohol involvement. *Journal of Studies on Alcohol*, *58*(2), 200–210. https://doi.org/10.15288/jsa.1997.58.200

Wood, P. K., Steinley, D., & Jackson, K. M. (2015). Right-sizing statistical models for longitudinal data. *Psychological Methods*, *20*(4), 470.

Wood, P. K., Wiedermann, W., & Steinley, D. (2023). Equal precision measurement structural models. *Structural Equation Modeling: A Multidisciplinary Journal*, *30*(1), 32–39.

Wood, P. K., Wiedermann, W., & Steinley, D. (2025). Right-sizing growth mixture models as multi-group growth and confirmatory factor models. *Behavior Research Methods, 57*(243). https://doi.org/10.3758/s13428-025-02722-2

Wood, P. K., Wiedermann, W., & Wood, J. K. (2024). Latent growth mixture models as latent variable multigroup factor models: Comment on McNeish et al. (2023). *Psychological Methods.* Advance online publication. https://doi.org/10.1037/met0000693

Wright, S. (1928). The methods of path coefficients. *Annals of Mathematical Statistics, 5*(3), 161–215.

Yang-Wallentin, F., & Jöreskog, K. G. (2001). Robust standard errors and chi-squares for interaction models. In G. A. Marcoulides & R. E. Schumacker (Eds.), *New developments and techniques in structural equation modeling* (pp. 159–171). Mahwah, NJ: Lawrence Erlbaum Associates.

Yates, A. (1987). *Multivariate exploratory data analysis: A perspective on exploratory factor analysis.* New York, NY State University of New York Press.

Yung, Y. F., & Bentler, P. M. (1994). Bootstrap-corrected ADF test statistics in covariance structure analysis. *British Journal of Mathematical and Statistical Psychology, 47*(2), 177–196. https://doi.org/10.1111/j.2044-8317.1994.tb01039.x

Zwick, W. R., & Velicer, W. F. (1982). Factors influencing four rules for determining the number of components to retain. *Multivariate Behavioral Research, 17*(2), 253–269. https://doi.org/10.1207/s15327906mbr1702_5

Zwick, W. R., & Velicer, W. F. (1986). Comparison of five rules for determining the number of components to retain. *Psychological Bulletin, 17*(3), 253–269. https://doi.org/10.1037/0033-2909.99.3.432

# *Index*

Pages in *italics* refer to figures and pages in **bold** refer to tables.

## N



## O